Automation, Operation and Maintenance of Control and Communication Systems

Automation, Operation and Maintenance of Control and Communication Systems

Editors

Piotr Szymak
Stanisław Hożyń
Paweł Piskur

MDPI • Basel • Beijing • Wuhan • Barcelona • Belgrade • Manchester • Tokyo • Cluj • Tianjin

Editors

Piotr Szymak
Polish Naval Academy
Gdynia, Poland

Stanisław Hożyń
Polish Naval Academy
Gdynia, Poland

Paweł Piskur
Polish Naval Academy
Gdynia, Poland

Editorial Office
MDPI
St. Alban-Anlage 66
4052 Basel, Switzerland

This is a reprint of articles from the Special Issue published online in the open access journal *Electronics* (ISSN 2079-9292) (available at: https://www.mdpi.com/journal/electronics/special_issues/ASMOR_electronics).

For citation purposes, cite each article independently as indicated on the article page online and as indicated below:

LastName, A.A.; LastName, B.B.; LastName, C.C. Article Title. *Journal Name* **Year**, *Volume Number*, Page Range.

ISBN 978-3-0365-8424-9 (Hbk)
ISBN 978-3-0365-8425-6 (PDF)

Contents

About the Editors

Piotr Szymak

Piotr Szymak (Professor) has been serving in Polish Naval Academy (PNA) in the following positions: assistant, lecturer, adjunct, head of department, assistant professor, associate professor. He was a director of Institute of Electrical Engineering and Automatics in PNA in the years 2015–2019. Since 2019 he has been Polish naval Captech National Cooridnator in European Defence Agency. In 2004 he received PhD degree in Polish Naval Academy and in 2016 he obtained the academic degree of habilitated doctor in Cracow University of Technology. His research area is strictly connected with modeling and control of marine objects, especially unmanned surface and underwater vehicles, including recently biomimetic underwater vehicles.

Stanisław Hożyń

Stanisław Hożyń (Assistant Professor) is an Engineering graduate, with a major in Electrical Systems Operation, from the Polish Naval Academy. He earned his MA in Electro-Automation from the Faculty of Marine Electrical Engineering, Gdynia Maritime University, in 2010. He served on two Polish warships as a Control Engineer and an Electrical Group Commander for twelve years. Currently, he is a PhD holder from the Polish Naval Academy. His PhD focused on the control of unmanned underwater vehicles. He has been an Assistant Professor at the Department of Ship Automation since 2019. His main research interests include unmanned underwater vehicles and computer vision.

Paweł Piskur

Paweł Piskur (Head of Ship Automation Department) graduated from the Military University of Technology in Warsaw in 2004 with a master's degree in airplane studies. He subsequently worked in the Polish Army's Marine Aviation Base. In 2010 he received a PhD degree in mechanical engineering from the Koszalin University of Technology. Since 2017 he has been working at the Polish Naval Academy. His research area is strictly connected with unmanned underwater vehicles, especially biomimetic propulsion systems.

electronics

MDPI

Editorial

Automation, Operation and Maintenance of Control and Communication Systems

Piotr Szymak *, Paweł Piskur and Stanisław Hożyń

Polish Naval Academy, Federation of Military Academies, 81-127 Gdynia, Poland; p.piskur@amw.gdynia.pl (P.P.); s.hozyn@amw.gdynia.pl (S.H.)
* Correspondence: p.szymak@amw.gdynia.pl

Abstract: The special issue is devoted to selected papers, which results of the research were presented during the XVIII Conference on Automation, Operation and Maintenance of Control and Communication Systems ASMOR 2022. The conference was conducted from the 12th to the 14th of October 2022 in Władysławowo, Poland. The scope of the conference is quite wide, focusing mainly on automation and exploitation aspects of the control and communications systems. Due to the specified scope of the Electronics, only the papers selected by the ASMOR Scientific Committee and the Electronics Editors were published after an extensive reviewing and revising process.

Keywords: automation of control systems; operation and maintenance of control systems; automation of communication systems; operation and maintenance of communication systems

1. Introduction

Autonomous control systems for air, land, surface and underwater objects have become increasingly popular. The articles presented in this special issue, titled "Automation, Operation and Maintenance of Control and Communication Systems", focus on the current developments in mechanics, energy supply, automation and control problems, especially for robotics and autonomous vehicles. The topics include the latest techniques, solutions, and their applications in various civilian and military fields. The issue presents four theoretical and experimental papers on automation and robotics, autonomous vehicles, mechatronic systems, and renewable energy, as well as the modelling and simulation of dynamic objects.

This Special Issue is dedicated to control and communication systems research, which results were presented during the XVIII Conference on Automation and Exploitation of Control and Communication Systems ASMOR 2022 (Figure 1). With their provision of safety and autonomy, they have become critical in various civilian and military operations. Recently, with this given role, they have generated substantial research investigation. This has resulted in significant progress in their automation and operation of sophisticated tasks using advanced algorithms, methods and state-of-the-art devices. The Special Issue is devoted to displaying noteworthy studies concerned with recent advancements in control and communication system technological aspects.

This Special Issue includes 13 original research papers that address advancements in automation, operation, and maintenance of control and communication systems and 1 review concerning a significant topic from the underwater robotics domain, i.e., obstacle detection and avoidance.

The papers report research and real applications covering topics including, but not limited to:

- Autonomous systems;
- Mechatronic systems;
- Telecommunication systems;
- Diagnostic and measurement systems;

Citation: Szymak, P.; Piskur, P.; Hożyń, S. Automation, Operation and Maintenance of Control and Communication Systems. *Electronics* **2023**, *12*, 3119. https://doi.org/10.3390/electronics12143119

Received: 6 July 2023
Accepted: 11 July 2023
Published: 18 July 2023

- Modeling and simulation of dynamic objects

  The details of the papers are presented in the next section of the work.

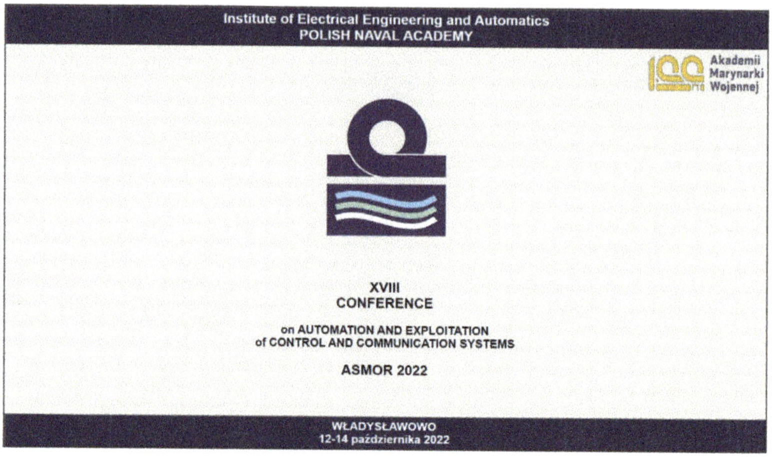

Figure 1. The details about the Automation and Exploitation of Control and Communication Systems 2022 conference (accessed on 10 July 2023): https://asmor.amw.gdynia.pl/.

2. Overview of Contributions

The first paper [1] entitled "Review of Collision Avoidance and Path Planning Algorithms Used in Autonomous Underwater Vehicles", considers the newest results of verifying collision-avoidance and path-planning algorithms in real applications together with a comparison of the difficulties encountered during simulations and their practical implementation. Analysing the last 20 years of AUV development, it can be seen that classical methods dominate experiments in a real environment. In the case of simulation studies, artificial intelligence (AI) methods are used as often as classical methods. In simulation studies, the APF approach is most often used among classical methods, whereas reinforcement learning and fuzzy logic methods are used among AI algorithms. The most used approach for real applications is reactive behaviours, and AI algorithms are rarely used in real implementations. This article provides a general summary, future works, and a discussion of the limitations that inhibit further development in this field.

The second paper [2] undertakes the problem of analyzing surfaces of non-stable, rapidly changing materials such as waxes or adhesive materials. NIR spectroscopy using a Digital Light Projection (DLP) spectrometer was used to obtain their characteristic spectra. Based on earlier experiences and the current state of the art, Artificial Neural Networks (ANNs) were used to process spectral sequences to proceed with an enormous value of spectra gathered during measurements.

The third paper [3] presents the novel concept of creating a 3D map based on the adaptive Monte-Carlo location (AMCL) and the extended Kalman filter (EKF). This approach is intended for inspection or rescue operations in a closed or isolated area with a risk to humans. The proposed solution uses particle filters and data from onboard sensors to estimate the local position of the robot. Its global position is determined through the Rao–Blackwellized technique. The developed system was implemented on a wheeled mobile robot equipped with a sensing system consisting of a laser scanner (LIDAR) and an inertial measurement unit (IMU) and was tested in the real conditions of an underground mine. One of the contributions of this work is to propose a low-complexity and low-cost solution to real-time 3D map creation.

The following paper [4] introduces a proposal for an Autonomous Navigation System for Unmanned Surface Vessels. A special emphasis is carried out on collision avoidance and manoeuvre auto-negotiation. For the purpose of manoeuvre auto-negotiation, the concept of multi-agent systems has been applied. The algorithm developed for the task of collision avoidance is briefly described and the results of the simulation tests, confirming the effectiveness of the applied method, are also given. Presented outcomes include solutions of test scenarios from the perspectives of different ships taking part in the considered situations, confirming the applicability of the collision avoidance algorithm in the process of manoeuvre auto-negotiation.

The next paper [5] presents a neuro-evolutionary algorithm called Hill Climb Assembler Encoding (HCAE), which is a light variant of Hill Climb Modular Assembler Encoding (HCMAE). While HCMAE, as the name implies, is dedicated to modular neural networks, the target application of HCAE is to evolve small/mid-scale monolithic neural networks, which, despite the great success of deep architectures, are still in use, for example, in robotic systems. The paper analyses the influence of different mechanisms incorporated into HCAE on the effectiveness of evolved neural networks and compares it with several rival algorithms. The paper entitled "A Concept of Autonomous Multi-Agent Navigation System for Unmanned Surface Vessels" introduces a proposal for an Autonomous Navigation System for Unmanned Surface Vessels. The system architecture is presented with a special emphasis on collision avoidance and manoeuvre auto-negotiation. For manoeuvre auto-negotiation, the concept of multi-agent systems has been applied. Presented outcomes include solutions of test scenarios from the perspectives of different ships taking part in the considered situations, confirming the applicability of the collision avoidance algorithm in the process of manoeuvre auto-negotiation.

The paper devoted "LTE and NB-IoT Performance Estimation Based on Indicators Measured by the Radio Module" presents research results leading to the development of a decision algorithm, called Multilink—ML, dedicated to the presented device [6]. This algorithm enables the selection between LTE and NB-IoT interfaces for packet transmission without burdening the communication system with additional transmissions.

The paper [7] entitled "Selection of Robotic Machining Parameters with Pneumatic Feed Force Progression" concerns the robotic deburring of the V2500 diffuser's sharp edges. The paper's authors propose a procedure for carrying out work allowing for the selection of suboptimal process parameters. In the analyzed case, these parameters are the speed of movement of the characteristic point of the tool (TCP) and the tool/workpiece contact force. The proposed procedure for determining the parameters of the force and speed of movement allowed for indicating a set of parameters ensuring the product's performance by the requirements defined in the documentation.

The paper [8] presents an innovative approach to measuring, comparing the results with a pattern, sending the deviations to a neural decision-making system, selecting the forces and send the results to a robot controller for adaptive machining. The presented proprietary solution includes a data acquisition system, a neural decision-making system and a robot that carries out the machining process via force control. The proposed solution was verified on aviation components. During the process parameter optimization stage for the diffuser and ADT gearbox, the points describing the change in width of the chamfer being performed and the blade thickness in the control sections were approximated.

The next paper [9] is devoted to "Development of a Dedicated Application for Robots to Communicate with a Laser Tracker". It presents the concept of operation and methods of using laser trackers in robotics. The developed solution is based on the software development kit (SDK) provided by Leica and the Python language. The structure and functioning of the developed software were described in detail. The software meets the goals set at the beginning of the design process regarding online communication with the tracker and using the universal, popular TCP/IP standard. The functioning of the developed software was shown in the paper in a few examples related to manipulating robots and mobile robots. The capabilities of the developed software were described, as well as the planned work on its development.

In the paper "TCP Parameters Monitoring of Robotic Stations", one representative process is selected, namely machining performed with various tools by ABB robots [10].

For the robotic process to be controlled, it is necessary to compare the defined path with the speed profile. Then, the speed parameters can be controlled and corrected. The proposed approach allows for improving the quality of implemented robotic processes. It presents the available IT tools for station monitoring and how to use them. The proposed solutions' advantages and limitations are shown in the examples of the implementation of robotic stations in the industry.

The paper entitled "Quaternion Attitude Control System of Highly Maneuverable Aircraft" presents an extension to research on spacecraft attitude control [11]. The article extends existing concepts and applies them to the control problem of aircraft operating in Earth's atmosphere. The controller synthesis is described using quaternion algebra. The quaternion-based attitude controller is then compared with a classical Euler-based attitude controller. The methodology for comparison and performance evaluation of both controllers is described.

In the paper "Analysis of Impulse Responses Measured in Motion in a Towing Tank", the authors present the results of the measurements realized in a towing tank where the transmitter could move with a precisely set velocity and show that the analyzed channel was non-stationary, even during the time of the transmission of a single chirp signal [12]. The article presents an evaluation method of channel stationarity at the time of the chirp transmission, which should be treated as a novelty. There is also an analysis of the impulse responses measured in motion in a towing tank.

The paper [13] is focused on "Selection of the Depth Controller for the Biomimetic Underwater Vehicle". Its main aim is to select a depth controller for innovative biomimetic underwater vehicle drives. In optimizing depth controller settings, two classical controllers were used, i.e., the proportional–integral–derivative (PID) and the sliding mode controllers (SM). The parameters of the regulators' settings were obtained as a result of optimization by three methods of the selected quality indicators in terms of the properties of the control signal. The starting point for the analysis was simulations conducted in the MATLAB environment for the three optimization methods on three types of indicators for three different desired depth values. The article describes the methods and quality indicators in detail. The paper presents the results of the fitness function obtained during the optimization. Moreover, the time courses of the vehicle position relative to the desired depth, the side fin deflection angles, the calculated parameters of the control signals, and the observations and conclusions formulated in the research were presented.

This issue ends with the paper entitled "Dynamics of Separation of Unmanned Aerial Vehicles from the Magnetic Launcher Cart during Takeoff" [14] describing the process of modelling and analyzing the dynamical properties of a launch cart of an innovative prototype launcher, which employs a passive magnetic suspension with high-temperature superconductors, developed under the GABRIEL project. The developed mathematical model of the magnetic catapult cart was employed to conduct numerical studies of the longitudinal and lateral movement of the cart, as well as the configuration of the UAV–cart system during UAV takeoff under variable atmospheric conditions. An essential aspect of the research involved experimentally determining the magnetic levitation force generated by the superconductors as a function of the gap. The results obtained demonstrate that the analyzed catapult design enables safe UAV takeoff. External factors and potential vibrations resulting from uneven mass distribution in the UAV–cart system are effectively balanced by the magnetic forces arising from the Meissner effect and the flux pinning phenomenon. Compared to commercial catapults, the primary advantage of the magnetic levitation catapult lies in its ability to provide a reduced and consistent acceleration throughout the entire takeoff process.

3. Conclusions

The editors hope all readers of the papers published in this special issue will find that the authors' detailed and careful presentation of ideas, methods and results broadens their knowledge. We want to thank all the authors of the submitted papers and the reviewers who provided constructive comments and suggestions.

Conflicts of Interest: The authors declare no conflict of interest.

References

1. Kot, R. Review of Collision Avoidance and Path Planning Algorithms Used in Autonomous Underwater Vehicles. *Electronics* **2022**, *11*, 2301. [CrossRef]
2. Gąsiorowski, M.; Szymak, P.; Patryn, A.; Naus, K. Monitoring Time-Non-Stable Surfaces Using Mobile NIR DLP Spectroscopy. *Electronics* **2022**, *11*, 1945. [CrossRef]
3. Buratowski, T.; Garus, J.; Giergiel, M.; Kudriashov, A. Real-Time 3D Mapping in Isolated Industrial Terrain with Use of Mobile Robotic Vehicle. *Electronics* **2022**, *11*, 2086. [CrossRef]
4. Lazarowska, A.; Żak, A. A Concept of Autonomous Multi-Agent Navigation System for Unmanned Surface Vessels. *Electronics* **2022**, *11*, 2853. [CrossRef]
5. Praczyk, T. Hill-Climb-Assembler Encoding: Evolution of Small/Mid-Scale Artificial Neural Networks for Classification and Control Problems. *Electronics* **2022**, *11*, 2104. [CrossRef]
6. Burczyk, R.; Czapiewska, A.; Gajewska, M.; Gajewski, S. LTE and NB-IoT Performance Estimation Based on Indicators Measured by the Radio Module. *Electronics* **2022**, *11*, 2892. [CrossRef]
7. Burghardt, A.; Muszyńska, M.; Gierlak, P.; Kurc, K.; Szybicki, D.; Ornat, A.; Uliasz, M. Selection of Robotic Machining Parameters with Pneumatic Feed Force Progression. *Electronics* **2022**, *11*, 3211. [CrossRef]
8. Kurc, K.; Burghardt, A.; Gierlak, P.; Muszyńska, M.; Szybicki, D.; Ornat, A.; Uliasz, M. Application of a 3D Scanner in Robotic Measurement of Aviation Components. *Electronics* **2022**, *11*, 3216. [CrossRef]
9. Szybicki, D.; Obal, P.; Penar, P.; Kurc, K.; Muszyńska, M.; Burghardt, A. Development of a Dedicated Application for Robots to Communicate with a Laser Tracker. *Electronics* **2022**, *11*, 3405. [CrossRef]
10. Burghardt, A.; Szybicki, D.; Gierlak, P.; Kurc, K.; Muszyńska, M.; Ornat, A.; Uliasz, M. TCP Parameters Monitoring of Robotic Stations. *Electronics* **2022**, *11*, 3415. [CrossRef]
11. Gołąbek, M.; Welcer, M.; Szczepański, C.; Krawczyk, M.; Zajdel, A.; Borodacz, K. Quaternion Attitude Control System of Highly Maneuverable Aircraft. *Electronics* **2022**, *11*, 3775. [CrossRef]
12. Czapiewska, A.; Luksza, A.; Studanski, R.; Zak, A. Analysis of Impulse Responses Measured in Motion in a Towing Tank. *Electronics* **2022**, *11*, 3819. [CrossRef]
13. Przybylski, M. Selection of the Depth Controller for the Biomimetic Underwater Vehicle. *Electronics* **2023**, *12*, 1469. [CrossRef]
14. Ładyżyńska Kozdraś, E.; Sibilska-Mroziewicz, A.; Sibilski, K.; Potoka, D.; Żyluk, A. Dynamics of Separation of Unmanned Aerial Vehicles from the Magnetic Launcher Cart during Takeoff. *Electronics* **2023**, *12*, 2883. [CrossRef]

electronics

MDPI

Article

Monitoring Time-Non-Stable Surfaces Using Mobile NIR DLP Spectroscopy

Marek Gąsiorowski [1], Piotr Szymak [2,*], Aleksy Patryn [1] and Krzysztof Naus [2]

[1] Faculty of Electronics and Computer Science, Koszalin University of Technology, 75-453 Koszalin, Poland; gasiorowski.marek@o2.pl (M.G.); aleksy.patryn@tu.koszalin.pl (A.P.)
[2] Polish Naval Academy, Federation of Military Academies, 81-127 Gdynia, Poland; k.naus@amw.gdynia.pl
* Correspondence: p.szymak@amw.gdynia.pl

Abstract: In recent years, Near Infrared (NIR) spectroscopy has increased in popularity and usage for different purposes, including the detection of particular substances, evaluation of food quality, etc. Usually, mobile handheld NIR spectroscopy devices are used on the surfaces of different materials, very often organic ones. The features of these materials change as they age, leading to changes in their spectra. The ageing process often occurs only slowly, i.e., corresponding reflection spectra can be analyzed each hour or at an even longer interval. This paper undertakes the problem of analyzing surfaces of non-stable, rapidly changing materials such as waxes or adhesive materials. To obtain their characteristic spectra, NIR spectroscopy using a Digital Light Projection (DLP) spectrometer was used. Based on earlier experiences and the current state of the art, Artificial Neural Networks (ANNs) were used to process spectral sequences to proceed with an enormous value of spectra gathered during measurements.

Keywords: NIR DLP spectroscopy; reflectance time-non-stable spectra; artificial neural network

Citation: Gąsiorowski, M.; Szymak, P.; Patryn, A.; Naus, K. Monitoring Time-Non-Stable Surfaces Using Mobile NIR DLP Spectroscopy. *Electronics* **2022**, *11*, 1945. https://doi.org/10.3390/electronics11131945

Academic Editor: Hamid Reza Karimi

Received: 25 May 2022
Accepted: 18 June 2022
Published: 22 June 2022

1. Introduction

Nowadays, Near Infrared (NIR) spectroscopy has become very popular. Modern computers with more and more computing power and appropriate data analysis software make it possible to obtain a powerful tool that enables efficient processing of thousands of data points to obtain useful information about a tested sample. NIR spectroscopy has many significant advantages: the non-destructive nature of the test, no need to prepare a sample, simplicity and speed of measurements, and low cost. Some disadvantages bring limitations: the useful information is not directly available, and the sample may be described by thousands of variables. It often turns out that the spectra obtained are not repeatable [1].

Today, there are several examples of using NIR spectroscopy being used for various purposes, including the evaluation of food quality. One of them is to use a portable NIR spectrometer working in the range of 1396–2396 [nm] to collect the spectra of breast milk samples for quality evaluation [2], which is an essential matter for newborn children. The authors use different chemometrics to calculate and then develop 18 calibration models with and without using derivatives and the standard normal variate. Once the calibration models were developed, the best treatments were selected according to the correlation coefficients and prediction errors. The other example of using NIR to estimate food quality is included in [3]. The authors examined Visible (VS) and NIR spectroscopy usage to monitor grape composition within a vineyard to facilitate the decision-making process with regards to grape quality sorting and harvest scheduling. Measurements of grape clusters were acquired in the field using a VS/NIR spectrometer, operating in the 570–990 [nm] spectral range, from a motorised platform moving at 5 km/h. To analyse the obtained spectra, they used classical methods, e.g., a correlation function. In [4], the authors examined a novel prototype NIR instrument designed to measure dry matter content in single potatoes. The instrument is based on interaction measurements to measure deeper into the potatoes. It measures rapidly, up to 50 sizes per second. The device also enables several distances to be recorded for each measurement. The instrument was calibrated based on three different potato varieties, and the calibration measurements were done in a process plant, making the calibration model suitable for production line use.

The other type of NIR spectroscopy usage is connected with the need to detect undesired substances. One of the examples is non-destructive detection of tomato pesticide residues using VS/NIR spectroscopy and prediction models such as ANNs [5]. The authors used VS/NIR spectral data from 180 samples of non-pesticide tomatoes (used as a control treatment) and samples impregnated with a pesticide with a concentration of 2/1000 [L], recorded by a spectroradiometer working in the range 350–1100 [nm], to train and then verify ANNs. The other example is included in [6]. The authors used NIR spectroscopy and characteristic variables selection methods to develop a quick way of determining cellulose, hemicellulose, and lignin contents in Sargassum horneri, i.e., the species of brown macroalgae that is common along the coast of Japan and Korea. Calibration models for cellulose, hemicellulose, and lignin in Sargassum horneri were established using partial least square regression methods with full variables. The last example showed the detection of measuring vitamin C and ellagic acid in wild-harvested Kakadu plum fruit samples [7]. The results of this study demonstrated the ability to predict vitamin C and ellagic acid in whole and pureed Kakadu plum fruit samples using a handheld NIR spectrophotometer. In the next paper [8], the NIR spectroscopy method was developed to analyze the oil and moisture contents of the plant Camellia gauchowensis Chang and C. semiserrata Chi seeds kernels. The authors used principal component analysis (PCA) and partial least squares (PLS) regression methods for calibration and validation. Finally, they obtained correlation coefficients of 0.98 and 0.95 for oil, and 0.92 and 0.89 for moisture, respectively, for calibration and validation.

More often, NIR spectroscopy is used with other methods to obtain the expected solution. One example is presented in [9]. The authors used NIR spectroscopy coupled with chemometric tools and obtained a fast and low-cost alternative solution for evaluating wood properties and quality categories. The obtained results of research showed that NIR spectroscopic data combined with powerful multivariate statistic tools and artificial intelligence solutions provided a fast and reliable tool, helpful in the decision-making process. The other opportunities are connected with NIR-absorbing organic semiconductors, especially for organic photovoltaic cells (OPV) [10]. OPV has increased its popularity in the field of renewable energy due to its lightweight, flexibility, and relatively low cost. To find new OPV materials, experiments with different types of NIR materials as active layers were conducted. Good results have also been achieved using NIR spectroscopy in conjunction with machine learning methods [11]. In this study, the authors used VIS/NIR spectroscopy for the effective discrimination of genetically modified (GM) and non-GM Brassica napus, B. rapa, and F1 hybrids (B. rapa X GM B. napus). As a classification method, the convolutional ANNs were used with success. More references to the advances in NIR spectroscopy and related computational methods can be found in [12,13].

Progress in the development of spectroscopy has led to the creation of a class of miniaturized spectrometers, including the VS and NIR [14,15]. The miniaturization of devices and systems is related to the tendency to perform non-laboratory measurements, including the adaptation of sizes to the in-line version [16,17]. An essential advantage of using smaller instruments is the potential possibility of implementing distributed measurement schemes and approaches to "remote" monitoring of environmental and "field" measurements.

As can be seen over a dozen different examples, the use of NIR spectroscopy, often supported by other classical methods and artificial intelligence methods, has been analyzed. This proves the great potential of NIR spectroscopy supported by other methods for data analysis. The factor connecting the above examples of NIR spectroscopy applications are the relatively slow processes of change taking place in the tested materials and substances, usually demanding monitoring every hour or at even longer intervals. Therefore, in this article the problem of using NIR spectroscopy supported by ANNs for the analysis of rapidly changing processes of the transition from liquid to solid is discussed. The authors of this article did not find any other article dealing with this subject using the commonly used NIR spectroscopy. Waxes and adhesives commonly used in everyday life were selected as test objects. As shown in [18], features of the materials change in different environment conditions, e.g., ambient temperature. In this paper, the method of analyzing optical reflection spectra of objects whose properties change quickly over time using the Digital Light Projection (DLP) measurement technique was proposed. The surfaces of several different materials were selected as test objects, the optical properties of which can vary even within a few seconds and which can be first approximated as prototypes of materials used in electron technology. At the current stage, these test objects were: quick-drying glue, two-component epoxy glue, and natural beeswax. Spectroscopic measurements were carried out using the DLP NIRSCAN Nano EVM Spectrometer by Texas Instruments. To process the obtained spectra, ANNs were used. The earlier results of the research with the ANNs are included in [19,20].

The test measurements were carried out to estimate the possibilities and effectiveness of further mobile spectroscopic measurements of the surface of materials, emphasising mobility and the possibility of carrying out measurements in various conditions, especially apart from stationary laboratories presented.

A method of measuring reflectance with the spectrometer window positioned practically in direct contact with the tested surface was chosen. In-line measurement results were saved in .dat or .xlms files for processing with Matlab. The recording time of one reflectance spectrum in the 900–1700 nm range was 2.67 s, while the spectrum was recorded six times and averaged.

The kit version of DLP NIRSCAN NANO and the primary measurement schemes were used in the same way as included in [21,22]. The modernisation of the device allowed it to work more efficiently after using its own housing design. The test object in the configuration is typically placed on the windows at the top of the monochromator.

In the next section, the applied measurement method and stand were described. Then, the results of measurements were presented. At the end, the discussion on received results and final conclusions were included.

2. Methods

The spectra were recorded using DLP NIRSCAN NANO, i.e., a small-size Texas Instrument spectrometer operating in the light wavelength range from 900 to 1700 nm with optical resolution equal to 10 nm [23]. The measurements were carried out several times at a given measuring point, depending on the selected option. The dimensions of the device, equal to 58 × 62 × 36 mm allowed us to perform mobile studies of reflectance spectra. The device can work in the reflective mode. In Figure 1, the general view and data flow in the measurement stand are visualised. The hardware part of the stand consists of the DLP NIR EVM spectrometer connected with a PC using USB. The spectrometer produces light of a specific wavelength illuminating the examined sample and then measures the intensity of light that passes through the sample. The software part of the stand includes the DLP NIRscan Nano GUI (used to control the device, simple visualisation of measurements, collecting and saving of data in the appropriate format) and the Matlab program (used for data processing employing ANNs). The spectra obtained from the first software can be displayed and saved in two formats: CSV and DAT. Matlab accepts both file formats.

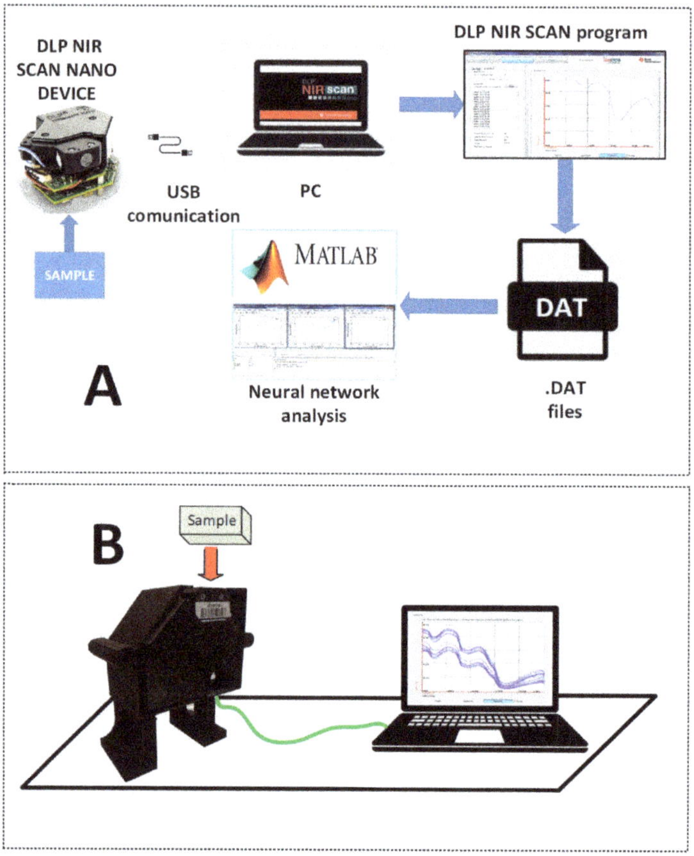

Figure 1. The measurement stand based on the DLP NIR EVM spectrometer: (**A**)—a scheme of the data flow, (**B**)—a general view of the stand.

3. Results

The research results were divided into two groups, depending on the type of examined materials:

1. Wax materials,
2. Adhesive materials.

Both types of materials changing from liquid to solid (solidification time) at different rates. Therefore, the research was conducted with different time scales to observe the essential parts of the processes.

The measurements results of both material groups are presented in the two following subsections. In the next subsection, the results of research using Artificial Neural Networks (ANNs) are included.

3.1. Wax Materials

Paraffin and beeswax were selected as representatives of the first group of materials. These materials are characterised by a quick transition from liquid to solid phase of approx. 30 [s]. Therefore, the time interval between consecutive measurements equal to 2.67 [s] was chosen. In Figure 2, the obtained results of seizing reflectance R for paraffin and natural beeswax for subsequent time steps and different wavelengths of light are illustrated. The reflectance is given without a physical unit. If R achieves 1, it means that all the light was reflected. At first glance, the transition of the paraffin is similar to beeswax. However, as it turned out, the spectra of these two substances differed significantly in reflectance levels during the change of the state of aggregation. Still, the characteristics are very similar. Both materials have characteristic peaks around 1200 [nm] and 1400 [nm]. Some differences between the materials may be because due to their different colours, i.e., the tested beeswax is yellow while the paraffin is white. The total observation time ranged from 0 to 28.6 s, during which time the materials changed from liquid to solid. The value t_1 is equal to 0 [s], and t_{11} is equal to 28.6 [s].

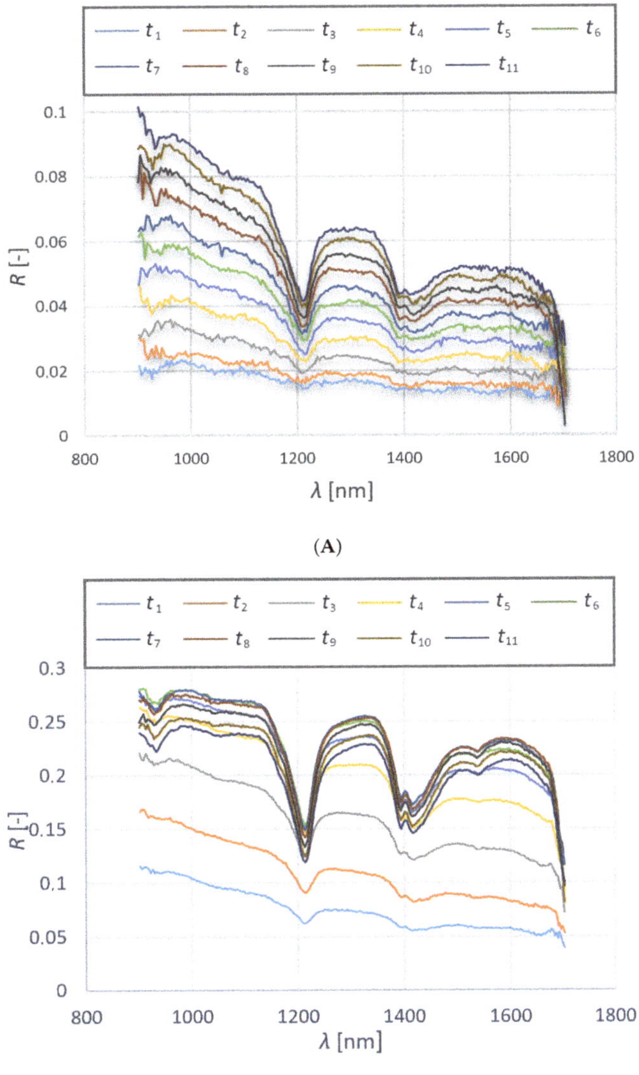

Figure 2. Reflectance spectra of two wax materials: (**A**)—paraffin, (**B**)—beeswax.

3.2. *Adhesive Materials*

The next group of tested materials were adhesives. General-purpose adhesives available on the commercial market and used in households. The two following kinds of such adhesives were selected and then investigated using DLP NIR spectrometry:

1. Cyanoacrylate glue, so called "superglue",
2. Two-component epoxy glue.

Due to the longer time of transition from liquid to solid phase of the second group of materials, a longer time interval between successive measurements was selected, i.e., 60 [s]. All the measurements took 10 [min]. In Figure 3, the results of changes in reflection spectra over time were illustrated, i.e., over the course of the solidification process. The changes during the superglue test turned out to be not very expressive. Even after hardening, the glue was characterised by high transparency, and no apparent changes over time could be observed. Significant differences were observed only at the edge of the measurement ranges. A considerable fluctuation can be seen in the long-wave region. It may be due to the operating range of the device. A more extensive measuring range would allow a more precise material analysis.

Considering the reflectance spectrum of epoxy glue (Figure 3), similar to wax materials, characteristic peaks around 1200 [nm] and 1400 [nm] can be observed. After a detailed comparison, it should be stated that the characteristics peaks are more distant from each other than for the wax materials, i.e., the first peak can be seen below a wavelength of 1200 [nm] and the second at wavelengths higher than 1400 [nm].

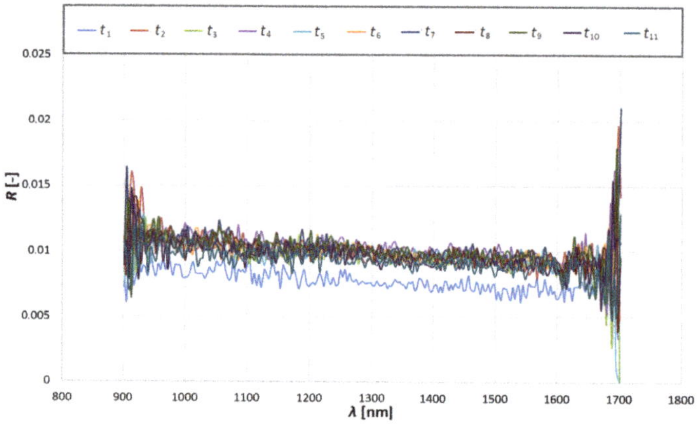

(**A**)

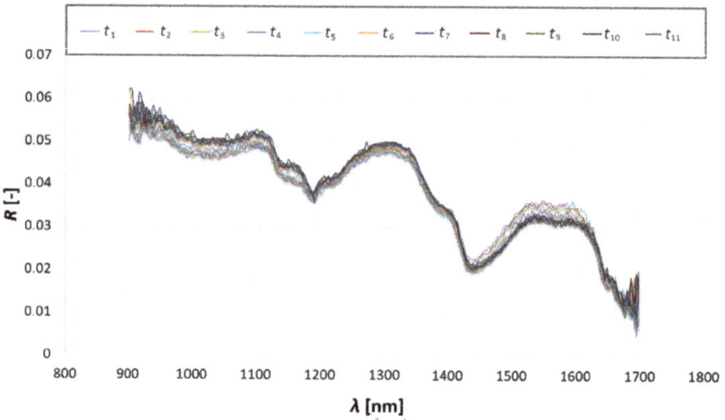

(**B**)

Figure 3. Reflectance spectra of two adhesive materials: (**A**)—superglue, (**B**)—epoxy glue.

Moreover, it is tough to differentiate the lines corresponding to the successive time steps. The lines are overlapped, and their order changes for various wavelength ranges, e.g., the range from 900 [nm] to 1430 [nm] and the range from 1430 [nm] to 1700 [nm]. It is possible that analysis of the adhesive materials requires a spectrometer offering a broader wavelength.

The obtained in the following time steps reflectance spectra of selected waxes and adhesives showed that the spectra course and their change over time are unique for each material. It can be used for determining relationships between specific spectra and the progress of the solidification process of the particular material. However, the measurement of adhesives needs to use a spectrometer offering a wider wavelength than used in research. It results from the observations that it is tough to differentiate the lines corresponding to the successive time steps. Moreover, the lines are overlapped and their order changes for various wavelength ranges.

3.3. Different Variants of ANNs

The obtained results (reflection spectra of individual materials) were uploaded to the Matlab program (Figure 4) to train the neural networks based on the obtained series of measurements, which can be used to identify the solidification stage (stage) of the material observed. However, in the first place, it was necessary to check the efficiency of the available learning methods and the correct minimum number of hidden neurons n_h to minimise the Mean Square Error (MSE). This is essential because the level of this error tells us about the degree of network training (the ability to recognise objects with a relatively low time error).

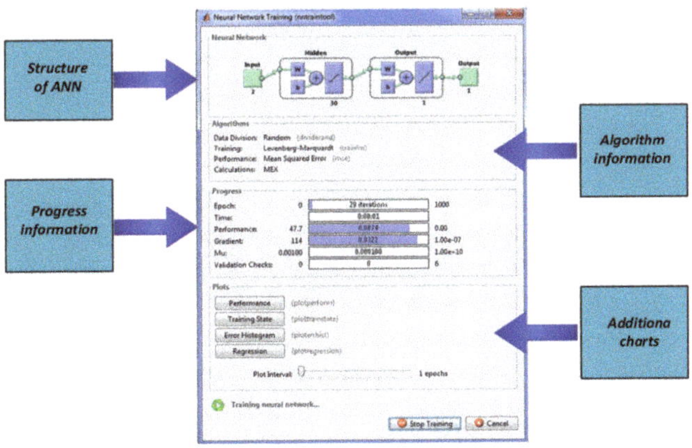

Figure 4. Matlab toolbox for ANN training and verification.

Different training methods for ANNs available in Matlab were also compared to find the method which reduces the MSE error to a minimum with the lowest possible number of hidden neurons. For example, the analysis for a network containing 200 neurons using a typical PC (4 GB operating memory, 3.30 GHz processor) takes about 2 h. During the tests, eleven methods of learning the network were tested based on up to 80 measurement spectra for each of the selected test objects.

The general structure of ANNs is illustrated in Figure 5. ANNs consists of three inputs, a hidden layer including the specific number of neurons, and one output layer. As can be seen, the number of neurons in the hidden layer was changed from 10 to 200 at step 10 of the research.

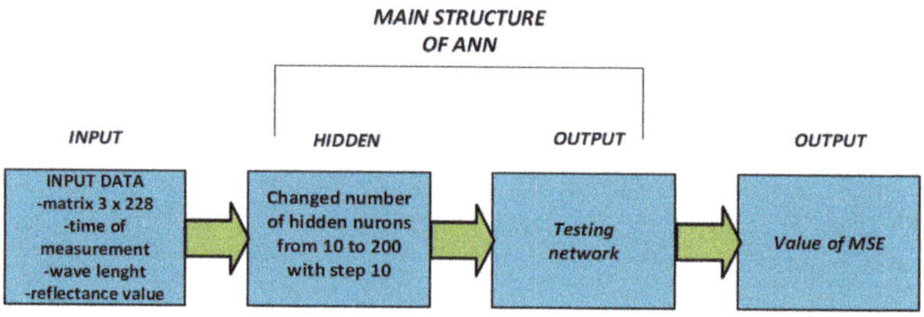

Figure 5. General structure of ANNs.

4. Discussion

In Figure 6, the ANNs verification results have been presented. The ANNs were trained by different learning methods on two measurement series for beeswax and paraffin materials. For the appropriate methods, the network in the first case achieves a satisfactory level of MSE with about 40 hidden neurons, which directly translates into a short analysis time. However, when analysing paraffin, the course of the MSE change is quite different, despite their being similar materials. A satisfactorily low error rate is obtained with about 20 hidden neurons. The mathematical description of the learning methods, using abbreviated names from Matlab, is included in [24]. In Table 1, the abbreviated names are presented with the corresponding full name of the methods.

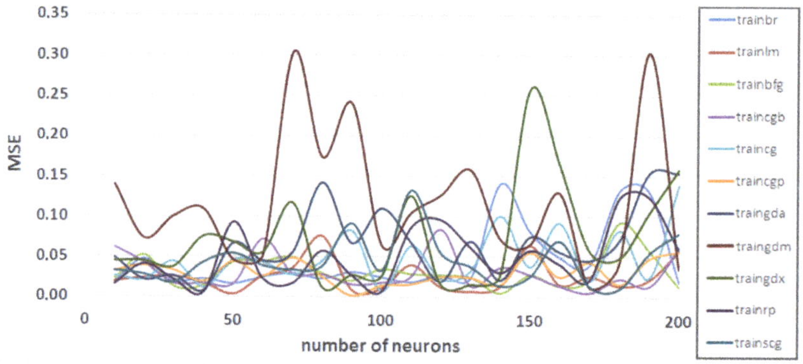

(**A**)

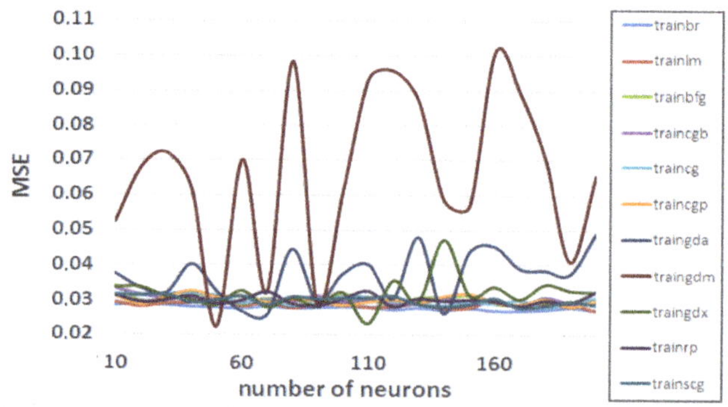

(**B**)

Figure 6. MSE for ANN's verification for different learning methods and number on neurons in the hidden layer for wax materials: (**A**)—beeswax, (**B**)—paraffin.

Table 1. The abbreviated and full names of the learning methods for ANNs.

No of Method	Abbreviated Name	Full Name
1	trainbr	Bayesian regularization backpropagation
2	trainlm	Levenberg–Marquardt backpropagation
3	trainbfg	BFGS quasi-Newton backpropagation
4	traincgb	Conjugate gradient backpropagation with Powell–Beale restarts
5	traincgf	Conjugate gradient backpropagation with Fletcher–Reeves updates
6	traincgp	Conjugate gradient backpropagation with Polak–Ribiére updates
7	traingda	Gradient descent with adaptive learning rate backpropagation
8	traingdm	Gradient descent with momentum backpropagation
9	traingdx	Gradient descent with momentum and adaptive learning rate backpropagation
10	trainrp	Resilient backpropagation
11	trainscg	Scaled conjugate gradient backpropagation

The trangdm method (Gradient descent with momentum backpropagation) achieved a lower error rate with about 50 hidden neurons (Table 2). Nevertheless, the course of MSE changes as a function of the number of neurons is chaotic. The network learning method that achieves a significantly lower MSE rate for the value of around 110 hidden neurons is traingdx (gradient descent with momentum and adaptive learning rate backpropagation) method. An essential piece of information resulting from comparing these two allegedly similar materials is that, as can be observed, increasing the number of hidden neurons does not significantly improve the MSE value. It only causes a significant extension of the analysis time. In addition, it turns out that the bottom learning method optimal for a given material (measurement series of reflection spectra) will not be suitable for another measurement series of a different material, even for materials supposedly similar to each other.

Table 2. The obtained minimum MSE and corresponding number of neurons n_h in the hidden layer for the selected learning methods and the wax materials.

No of Method	Beeswax: MSE	Beeswax: n_h	Paraffin: MSE	Paraffin: n_h
1	0.016	50	0.027	120
2	0.003	50	0.027	200
3	0.005	140	0.028	80
4	0.006	170	0.029	60
5	0.007	100	0.028	180
6	0.001	90	0.028	100
7	0.007	40	0.026	70
8	0.012	170	0.022	50
9	0.012	80	0.023	110
10	0.006	40	0.028	90
11	0.010	180	0.028	140

The idea for the further development of the research is presented in Figure 7. The use of the statistical Principal Component Analysis (PCA) method is also considered, which would allow for a reduction in the statistical data set and speed up the analysis time. The idea behind the algorithm is to monitor and save the most significant variances in the data set. Elements with low or zero variance of the set are ignored [25]. However, as already mentioned, for a given measurement series of a specific material, it will be necessary to select a particular method and determine the optimal number of hidden neurons to shorten the analysis time as much as possible.

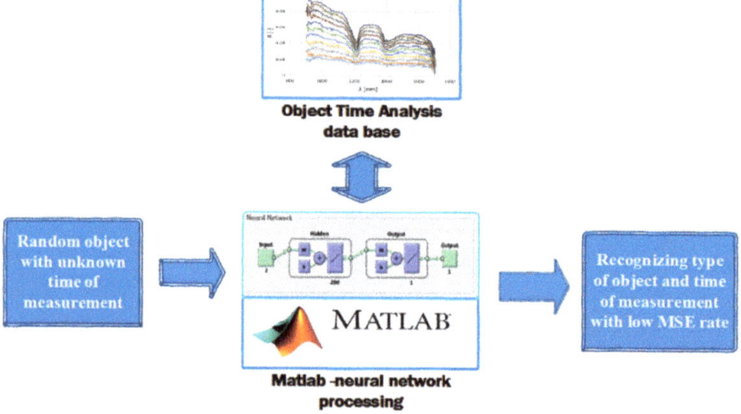

Figure 7. Conception of future development of software in Matlab.

5. Conclusions

Measurements of optical properties of materials whose aggregate state is unstable can be effectively carried out using the DLP method using DPL NIR Scan equipment. To increase the certainty of analysis of rapidly changing spectra, the method of spectra analysis using neural networks was tested. Four materials were tested in the study: epoxy, cyanoacrylate glue, paraffin, and beeswax. The first two materials are adhesives, while the second two are waxes. Based on the results of the research, the following conclusions can be stated:

1. Samples from the adhesive materials showed almost no changes in the optical spectrum in the range from 900 to 1700 nm during the solidification process. Therefore, their further analysis using neural networks is pointless. The lack of visible changes will not allow determining of the characteristic features for a given moment of the solidification process. A spectrometer with a wider wavelength range would possibly give better results.
2. The results of measurement of waxes showed significant changes in optical properties in the measurement range mentioned above. Along with the solidification process, the reflectance level changes slightly and the characteristic shape changes.
3. ANN analysis showed MSE values for individual learning methods with a relatively small number of 40–50 neurons at satisfactory levels. It is possible to describe sets of spectra using ANNs with appropriate learning methods, reducing the number of hidden neurons to a minimum. This is vital because it significantly affects the analysis time.

Thanks to the DLP NIRSCAN Nano device, it is possible to monitor the reflection spectra and thus the changes taking place in various types of materials. Analysing the spectra makes it possible to observe the course and dynamics of changes, which may be helpful information during technological processes where the given materials are used. The analysis of the spectra of objects with variable optical parameters using ANNs tells us that the described learning methods implemented in Matlab can learn entire series of measurements without significant problems and then recognise specific measures while maintaining a relatively small error (the described learning methods used about 80% of the measurement series for learning and about 20% of the series were used for network verification and MSE determination). During the analysis, it turned out that there is no universal method of training the networks, even for similar materials. The method and the optimal number of neurons should be matched to a given material, which in the future may allow for the recognition of materials or determining their condition. The next planned stage of the study is the modification of the program. After reading the given measuring series of the spectra of a given object, it will be possible to determine the time and stage of setting of a random measured adhesive sample. Thanks to this solution, it will be possible to determine the time (solidification stage) with a small error in any technological processes where wax is used. Such information can be beneficial in solidifying glues, resins, and other materials.

Author Contributions: Conceptualization, M.G. and A.P.; methodology, M.G. and A.P.; software, P.S. and K.N.; validation, M.G. and P.S.; formal analysis, A.P.; investigation, M.G.; resources, M.G.; data curation, M.G.; writing—original draft preparation, M.G. and P.S.; writing—review and editing, P.S.; visualization, M.G. and P.S.; supervision, A.P.; project administration, A.P.; funding acquisition, P.S. and K.N. All authors have read and agreed to the published version of the manuscript.

Funding: This research received no external funding.

Institutional Review Board Statement: Not applicable.

Informed Consent Statement: Not applicable.

Data Availability Statement: Not applicable.

Acknowledgments: Authors thank Leszek Bychto for useful discussion and consultations.

Conflicts of Interest: The authors declare no conflict of interest.

Abbreviations

The following abbreviations are used in this manuscript:

ANN	Artificial Neural Network
CSV	Comma-Separated Values
DAT	generic DATa file
DLP	Digital Light Projection
DMD	Digital Micromirror Device
EVM	Evaluation Module
MSE	Mean Square Error
NIR	Near Infrared Spectroscopy
PC	Personal Computer
USB	Universal Serial Bus
VS	Visible

References

1. Wrosz, P. Application of near infrared and classical infrared spectroscopy in qualitative and quantitative analysis of raw materials in the pharmaceutical and cosmetic industries. Ph.D. Thesis, Gdansk University of Technology, Faculty of Chemistry, Gdansk, Poland, 2012.
2. Melendreras, C.; Forcada, S.; Fernández-Sánchez, M.L.; Fernández-Colomer, B.; Costa-Fernández, J.M.; López, A.; Ferrero, F.; Soldado, A. Near-Infrared Sensors for Onsite and Noninvasive Quantification of Macronutrients in Breast Milk. *Sensors* **2022**, *22*, 1311. [CrossRef] [PubMed]
3. Fernández-Novales, J.; Tardáguila, J.; Gutiérrez, S.; Diago, M.P. NIR Spectroscopy as a Reliable Monitoring Tool for Grape Composition within the Vineyard. *Molecules* **2019**, *24*, 2795. [CrossRef] [PubMed]
4. Wold, J.P.; O'Farrell, M.; Andersen, P.V.; Tschudi, J. Optimization of Instrument Design for In-Line Monitoring of Dry Matter Content in Single Potatoes by NIR Interaction Spectroscopy. *Foods* **2021**, *10*, 828. [CrossRef] [PubMed]
5. Nazarloo, A.S.; Sharabiani, V.R.; Gilandeh, Y.A.; Taghinezhad, E.; Szymanek, M. Evaluation of Different Models for Non-Destructive Detection of Tomato Pesticide Residues Based on Near-Infrared Spectroscopy. *Sensors* **2021**, *21*, 3032. [CrossRef] [PubMed]
6. Ai, N.; Jiang, Y.; Omar, S.; Wang, J.; Xia, L.; Ren, J. Rapid Measurement of Cellulose, Hemicellulose, and Lignin Content in Sargassum horneri by Near-Infrared Spectroscopy and Characteristic Variables Selection Methods. *Molecules* **2022**, *27*, 335. [CrossRef] [PubMed]

7. Bobasa, E.; Phan, A.D.T.; Netzel, M.; Smyth, H.E.; Sultanbawa, Y.; Cozzolino, D. The Use of a Micro Near Infrared Portable Instrument to Predict Bioactive Compounds in a Wild Harvested Fruit—Kakadu Plum (Terminalia ferdinandiana). *Sensors* **2021**, *21*, 1413. [CrossRef] [PubMed]
8. Zhang, Y.; Zhang, L.; Wang, J.; Tang, X.; Wu, H.; Wang, M.; Zeng, W.; Mo, Q.; Li, Y.; Li, J.; et al. Rapid Determination of the Oil and Moisture Contents in Camellia gauchowensis Chang and Camellia semiserrata Chi Seeds Kernels by Near-infrared Reflectance Spectroscopy. *Molecules* **2018**, *23*, 2332. [CrossRef] [PubMed]
9. Costa, L.R.; Ramalho, F.M.G.; Hein, P.R.G. 16–Evaluating biofibers' properties and products by NIR spectroscopy. In *Advances in Bio-Based Fiber*; Rangappa, S.M., Puttegowda, M., Parameswaranpillai, J., Siengchin, S., Gorbatyuk, S., Eds.; The Textile Institute Book Series; Woodhead Publishing: Cambridge, UK, 2022; pp. 367–392. [CrossRef]
10. Meng, D.; Zheng, R.; Zhao, Y.; Zhang, E.; Dou, L.; Yang, Y. Near-Infrared Materials: The Turning Point of Organic Photovoltaics. *Adv. Mater.* **2022**, *34*, 2107330. [CrossRef] [PubMed]
11. Sohn, S.I.; Pandian, S.; Zaukuu, J.L.Z.; Oh, Y.J.; Park, S.Y.; Na, C.S.; Shin, E.K.; Kang, H.J.; Ryu, T.H.; Cho, W.S.; et al. Discrimination of Transgenic Canola (*Brassicanapus* L.) and their Hybrids with B. rapa using Vis-NIR Spectroscopy and Machine Learning Methods. *Int. J. Mol. Sci.* **2022**, *23*, 220. [CrossRef]
12. Beć, K.B.; Huck, C.W. Advances in Near-Infrared Spectroscopy and Related Computational Methods. *Molecules* **2019**, *24*, 4370. [CrossRef] [PubMed]
13. Ozaki, Y.; Huck, C.W.; Beć, K.B. Chapter 2—Near-IR Spectroscopy and Its Applications. In *Molecular and Laser Spectroscopy*; Gupta, V., Ed.; Elsevier: Amsterdam, The Netherlands, 2018; pp. 11–38. [CrossRef]
14. Pruett, E. Latest developments in Texas Instruments DLP near-infrared spectrometers enable the next generation of embedded compact, portable systems. In Proceedings of the The International Society for Optical Engineering, SPIE, Baltimore, MD, USA, 3 June 2015; Volume 9482. [CrossRef]
15. Coates, J. A review of new small-scale technologies for near infrared measurements. *Am. Pharm. Rev.* **2014**, *17*. Available online: https://www.americanpharmaceuticalreview.com/Featured-Articles/163573-A-Review-of-New-Small-Scale-Technologies-for-Near-Infrared-Measurements/ (accessed on 5 May 2022).
16. Laske, S. In-line quality control of polymer nanocomposites using near-infrared spectroscopy. In Proceedings of the NanoConf 2009, Roznov pod Radhostem, Czech Republic, 20–22 October 2009.
17. Barbas, J.; Machado, A.; Covas, J. In-line near-infrared spectrometry for the characterization of dispersion in polimer-clay nanocomposites. *Polim. Test.* **2012**, *31*, 527–536. [CrossRef]
18. Imran, M.; Sarwar, S.; Abdullah, M.; Khan, I. An analysis of the semi-analytic solutions of a viscous fluid with old and new definitions of fractional derivatives. *Chin. J. Phys.* **2018**, *56*, 1853–1871. [CrossRef]
19. Szymak, P.; Gasiorowski, M. Using Pretrained AlexNet Deep Learning Neural Network for Recognition of Underwater Objects. *Naše More* **2020**, *67*, 9–13. [CrossRef]
20. Szymak, P.; Piskur, P.; Naus, K. The Effectiveness of Using a Pretrained Deep Learning Neural Networks for Object Classification in Underwater Video. *Remote Sens.* **2020**, *12*, 3020. [CrossRef]
21. Gąsiorowski, M.; Patryn, A.; Bychto, L. Possibilities and area of application of the small size DLP NIRSCAN NANO spectrometer for instant spectral measurements. *Sci. J. Koszal. Univ. Tehchnol.* **2020**, *15*, 58–68.
22. Gąsiorowski, M.; Szymak, P.; Bychto, L.; Patryn, A. Application of Artificial Neural Networks in Analysis of Time-Variable Optical Reflectance Spectra in Digital Light Projection Spectroscopy. *Coatings* **2022**, *12*, 37. [CrossRef]
23. Texas Instruments. *DLP NIRscan Nano EVM User's Guide*; Texas Instruments: Dallas, TX, USA.
24. Demund, H.; Beale, M. Neural Network Toolbox for Use with Matlab—User's Guide. MathWorks. 2001. Available online: https://www.mathworks.com/support/search.html?q=Neural%20Network%20Toolbox%20for%20use%20with%20Matlab&page=1 (accessed on 5 May 2022).
25. Ventura, D. Manifold Learning Examples—PCA, LLE and ISOMAP. Neural Network and Machine Learning Laboratory. 2008. Available online: https://www.mathworks.com/support/search.html?q=%20Neural%20Network%20and%20Machine%20Learning%20Laboratory&page=1 (accessed on 5 May 2022).

electronics

MDPI

Article

Real-Time 3D Mapping in Isolated Industrial Terrain with Use of Mobile Robotic Vehicle

Tomasz Buratowski [1], Jerzy Garus [2], Mariusz Giergiel [1,*] and Andrii Kudriashov [1]

1 Faculty of Mechanical Engineering and Robotics, AGH University of Science and Technology, 30-059 Krakow, Poland; tburatow@agh.edu.pl (T.B.); akudr@agh.edu.pl (A.K.)
2 Faculty of Mechanical and Electrical Engineering, Polish Naval Academy, 81-103 Gdynia, Poland; j.garus@amw.gdynia.pl
* Correspondence: giergiel@agh.edu.pl

Abstract: Simultaneous localization and mapping (SLAM) is a dual process responsible for the ability of a robotic vehicle to build a map of its surroundings and estimate its position on that map. This paper presents the novel concept of creating a 3D map based on the adaptive Monte-Carlo location (AMCL) and the extended Kalman filter (EKF). This approach is intended for inspection or rescue operations in a closed or isolated area where there is a risk to humans. The proposed solution uses particle filters together with data from on-board sensors to estimate the local position of the robot. Its global position is determined through the Rao–Blackwellized technique. The developed system was implemented on a wheeled mobile robot equipped with a sensing system consisting of a laser scanner (LIDAR) and an inertial measurement unit (IMU), and was tested in the real conditions of an underground mine. One of the contributions of this work is to propose a low-complexity and low-cost solution to real-time 3D-map creation. The conducted experimental trials confirmed that the performance of the three-dimensional mapping was characterized by high accuracy and usefulness for recognition and inspection tasks in an unknown industrial environment.

Keywords: SLAM; 3D mapping; mobile robot; underground inspection

Citation: Buratowski, T.; Garus, J.; Giergiel, M.; Kudriashov, A. Real-Time 3D Mapping in Isolated Industrial Terrain with Use of Mobile Robotic Vehicle. *Electronics* **2022**, *11*, 2086. https://doi.org/10.3390/electronics11132086

Academic Editor: Hamid Reza Karimi

Received: 5 June 2022
Accepted: 29 June 2022
Published: 3 July 2022

1. Introduction

Mobile robotics have seen an increase in interest in the last few decades, especially due to their ability to be deployed in a hazardous environment without endangering humans. Currently, it is common to use mobile robotic vehicles to accomplish missions such as environmental recognition, the inspection of urbanized and industrial terrains, and search and rescue operations. In the military field, they are employed in tasks such as reconnaissance missions, surveillance, intelligence gathering, hazardous-site exploration, and more. The robot can explore an area of interest, teleoperated from a remote area or autonomously. Knowing the obstacles in the vicinity of the robot is essential for the successful completion of a mission in a working area, avoiding any collisions that could make the robot unusable.

If the map is given a priori, the robot can self-localize by matching some features of the environment observed at the given moment to features of the same type existing in the known map. A feature suitable for self-localization is a static, salient object (or part of an object) in the environment, which can be described with respect to some other co-ordinate frame. On the other hand, the predefined map can be perceptually incompatible, i.e., it may not properly reflect all features in the environment perceived with the given sensing modality. Therefore, the robot should be able to build its own model of the environment.

In recent decades, many designs for mobile robotic vehicles created to explore a partially or fully unknown environment have been developed and demonstrated in both academia and industry [1–7]. The simultaneous localization and mapping (SLAM), currently one of the main research topics in robotics, appears to be the attractive solution to this problem [8–16]. The objective of SLAM is to concurrently build a map of investigated environment and allow the moving vehicle to localize itself within this environment using its on-board sensors.

At present, the most widely used SLAM frameworks are based on laser sensors, because these kinds of sensors can detect the distances of the obstacles around the robot with high accuracy. However, 2D-mapping methods based on laser scanners still have some problems. The obstacles can only be detected when they are in the same height with the sensors. Therefore, 3D-mapping methods based on multiple-laser sensors or 3D-distance sensors have been developed because these methods can reflect the shapes of an area and objects in the environment and allow the robot to plan its path with collision avoidance.

This paper focuses on a map creation for the recognition and inspection purposes of an isolated industrial environment carried out by a mobile wheeled robotic vehicle. The map is generated based on data sets obtained from a laser scanner (LIDAR) and an inertial measurement unit (IMU) installed on its board. We propose a robust real-time mapping concept based on probabilistic SLAM formulation, including both extended Kalman filters and particle filters, to build a 3D map of the surrounding area. The effectiveness of the developed solution has been confirmed by mapping experiments in the underground environment of the mine for different operational conditions.

The remainder of this paper is organized as follows. In Section 2, a brief description of the SLAM problem is provided. Details of the implementation of the proposed approach are presented in Section 3. Section 4 provides the results of the experimental studies carried out in an experimental underground mine. Finally, the conclusions and future works are summarized in Section 5.

2. SLAM Mapping Overview

Mapping approaches are generally divided into a 2D and 3D representation of the environment. Each type can be used for different tasks, e.g., navigation or workspace-state representation purposes. A planar map is good enough for navigation tasks. However, from the inspection point of view, a three-dimensional picture of the surroundings is definitely more useful, but it impacts on the calculation loading of the SLAM process and requires more advanced sensory systems.

There are many ways to describe the SLAM problem represented in the literature [17–24]. Currently, the most common way is to define SLAM as a probability density function, which can be described by the following generic form [25–27]:

$$p(X_{0:t}, m | Z_{1:t}, U_{1:t}) \tag{1}$$

where:

$X_{0:t} = \{X_0, X_1, \ldots, X_t\}$—a sequence of the mobile vehicle's poses (passed path) in a sampling time interval <0, *t*>;
$U_{1:t} = \{U_1 U_2 \ldots U_t\}$—a sequence of control signals;
$Z_{1:t} = \{Z_1 Z_2 \ldots Z_t\}$—a sequence of relative observations.

The observations $Z_{1:t}$ are made by sensors, which measure the distance to the closest objects in all directions from the poses $X_{1:t}$ at the same time instant.

Such a form of expression (1), where wanted values are reconstructed for all previous states, is called "Full SLAM". The opposite approach, where only a recent position is estimated, called "Online SLAM", can be calculated by recursive integration [23]. It is possible to apply Bayes and Markov rules to (1) and define the probability density function as a recursive process of predictions and corrections of the robot's localization in the map m, which depends on motion (kinematic constrains and controls) and observation models [19]:

$$p(X_{0:t}, m | Z_{1:t}, U_{1:t}) = \alpha \cdot p(Z_t | X_{0:t}, m, Z_{1:t-1}, U_{1:t} \cdot p(X_{0:t}, m | Z_{1:t-1}, U_{1:t}) \tag{2}$$

where α is a normalization constant coefficient.

The recursive SLAM definition (2) is very convenient, since estimators such as the extended Kalman filter (EKF) [12,28] and particle filters [29–31] began to be applied to solve the estimation problem. The Kalman filter has the ability to use data from different sources and then efficiently fuse them into precise estimated values. It is one of its more valuable features. On the other hand, the biggest disadvantage of the EKF is its dependency on previous states that might cause huge average error decreases in the case of an unexpected measurement disaster. The use of EKF for the SLAM problem was first proposed in the work in reference [32], after which, many researchers further improved this method [12,33,34].

The particle filter [18,35,36] is a recursive filter using a Monte-Carlo algorithm to estimate the pose and movement path of the moving object. They are good and fast in global localization but are limited in their use of data from only one domain. Their wide usage for SLAM came from the brilliant idea of the Rao–Blackwellization method, which was introduced into regular map-building algorithms [37]. Although the particle filter algorithm can be used as an effective method to solve the SLAM problem [9,13,18,21,38], the main problem is that a large number of samples are required to approximate the posterior probability density of the system. The more complex the environment that the robot meets, the more samples that are needed to describe the posterior probability, and the higher the complexity of the algorithm.

Regarding the map construction of SLAM, there are three commonly used maps [39], i.e., feature-based maps, topological maps, and grid maps. The last one, also called an occupancy map, is the most common way for robots to describe the model of the environment. It divides the workspace into a series of grids, where each cell in the grid corresponds to binary random variables that describe the occupancy probability [8,20,40–42]. Hence, the map m is given by a product occupancy probability of each cell mi, which is associated with a certain point in three-dimensional Cartesian space [23,34]:

$$p(m) = \prod_i p(m_{x_i,y_i,z_i})$$

From the SLAM perspective, Equation (3) can be formulated as a probabilistic problem of the recursive pose estimation from given observations. There are many different SLAM techniques based on occupancy grids for map building in 2D or 3D space [10,22]. One of possible representations of the 3D map as an occupancy grid is octrees, which instead of the 2D quadtrees representation presented directly in references [8,42], uses an octree hierarchical data structure, where each volume-named node has eight child connections with inner nodes [25]. Currently, in the scientific literature, a rich number of occupancy grids-based SLAM techniques can be found [18,20,26,43]. Nevertheless, in any case, the best SLAM algorithm for a particular environment depends on hardware restrictions, the size of the map to be built, and the optimization criterion of the processing time.

The algorithms depend heavily on the sensors with which the robot perceives its surrounding. The selection and installation of sensors determine the specific form of observation results, and also affect the difficulty of SLAM problems. In case of mobile robots performing in indoor environments, exploration cameras or LIDARs are mostly used. Some approaches use a fusion of laser and visual data [34]. As mentioned before, LIDARs provide data that are more accurate, robust, less noisy, and sensitive to changes in the lighting. For these reasons, it is a very reliable data acquisition mechanism for industrial site mapping [44]. However, due to the limited number of dimensions, the laser data are not sufficient to estimate the robot's pose on uneven ground with six degrees of freedom (DOF) [45]. Therefore, most algorithms need auxiliary data from other sensors, such as an inertial measurement unit (IMU) or odometer. In recent years, along with the rapid development of artificial intelligence (AI), learning methods are also used to solve the SLAM problem [46–48].

3. Hybrid Mapping Concept for Inspection Purposes

The map-building approach applied in this paper has its basis in a robot's pose estimation, which is presented in references [42,49], called the AMCL-EKF, which connects two separate paradigms—Kalman and particle filters. The pose is estimated by the extended Kalman filter by using measurements data from LIDAR and IMU sensors and adaptive Monte-Carlo localization (AMCL).

In the case of a planar motion, the local pose is described by a state vector $\boldsymbol{X}^{(2)}$ consisting of three components:

$$\boldsymbol{X}^{(2)} = \begin{bmatrix} x & y & \psi \end{bmatrix}^T \tag{3}$$

where x, y are Cartesian positions and ψ is a yaw angle.

However, from the inspection point of view, tridimensional maps showing environment features in their natural scale are more preferred. Due to the fact that 3D mapping requires information about the global location of the robotic vehicle with six degrees of freedom (DOF), it is necessary to recreate missing coordinates. It may be performed by taking the needed coordinates from the local 3D pose that the EKF performs [49,50] (see Figure 1). Then, the state vector $\boldsymbol{X}^{(3)}$ can be written in the following form:

$$\boldsymbol{X}^{(3)} = \left\{ \boldsymbol{X}^{(2)} \quad z \quad \varphi \quad \theta \right\}^T \tag{4}$$

where z is the Cartesian position and φ, θ are roll and pitch angles. The global localization can be solved by a Rao–Blackwellized 2D SLAM algorithm [18].

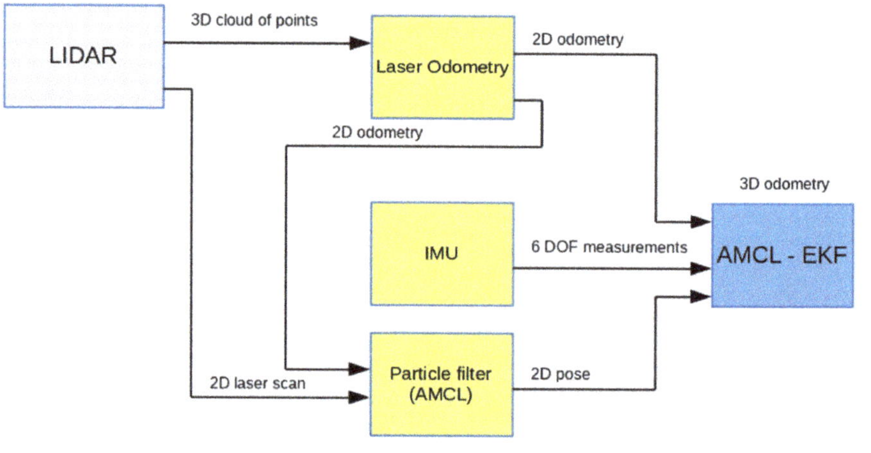

Figure 1. Schematic diagram of local 3D pose estimation by AMCL-EKF algorithm.

As far as $\boldsymbol{X}^{(3)}$ is delivered, it allows for the creation of a three-dimensional map. In this way, the 2D SLAM algorithm is used continuously to efficiently build the 2D map, and at the same time the 3D SLAM algorithm (the delivered precise $\boldsymbol{X}^{(3)}$ and recent cloud of points data allows to create the 3D map (see Figure 2)). This task can be performed both online as a separate task or offline postprocess based on previously collected sensor data.

The efficiency and accuracy of the developed method, called "hybrid 3D SLAM", for locating the robot in unknown terrains, have been confirmed in several simulation experiments presented in our previous works [19,50,51], where the 3D-mapping algorithm was implemented in accordance with the Robot Operation System framework, and was verified with using the V-REP simulation environment. The carried-out investigations affirmed the legitimacy of the proposed approach to 3D-map building for inspection tasks.

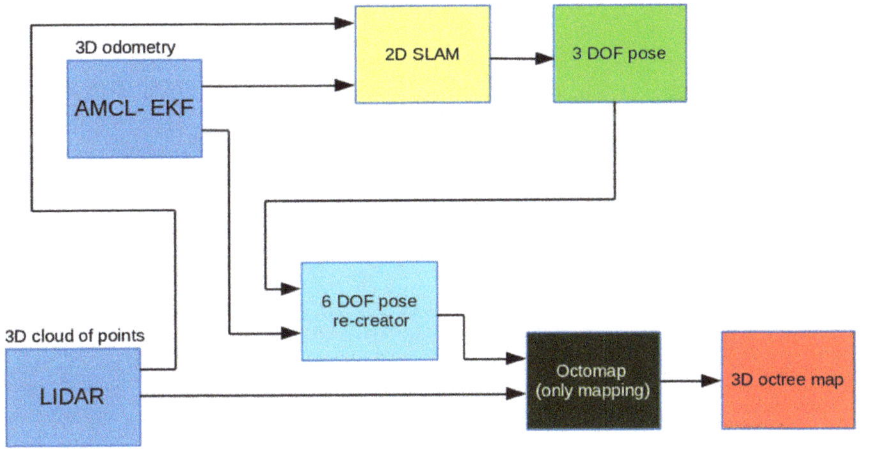

Figure 2. Block diagram of hybrid 3D-map-building bases on 2D SLAM and reduced Octomap algorithms.

4. Environment Description and Experimental Results

4.1. Robotic Vehicle and Environment

The SLAM strategy presented in the previous section was implemented in the Dr Robot Jaguar 4×4 mobile vehicle constructed by the Canadian company Dr Robot Inc. (see Figure 3). The robot is equipped with the LIDAR (Velodyne VPL-16) and the IMU (Pololu MinIMU-9) sensors.

Figure 3. Mobile robotic vehicle used for tests.

The robot is a four-wheeled robotic platform designed for tough indoor and outdoor operation, and has the following operational parameters: mass 20 kg, max speed 11 km/h, and a ground clearance 88 mm. The Velodyne VLP-16 LIDAR uses an array of 16 infra-red lasers paired to detect and measure distances to objects. The array provides 300,000 data points per second in real-time. The IMU sensor Pololu MinIMU-9 is a compact board that combines a 3-axis gyroscope, a 3-axis accelerometer, and a 3-axis magnetometer for measuring robot data parameters.

Investigations were conducted in a real industrial environment—the experimental underground mine being a part of the AGH University of Science and Technology infrastructure. It was assumed that the robotic vehicle had to inspect the site in two operational cases, i.e., before and after the mining disaster. This task required the robot to move on both smooth and rough surfaces that were sometimes blocked by degraded infrastructure. Figure 4 shows two fragments of real operational scenarios.

(a) (b)

Figure 4. Underground inspection scenarios: (**a**) smooth terrain before accident; (**b**) rough terrain after accident.

4.2. Experiments

Carried out trials were divided into two stages, i.e., creating the map of the mine area for normal working conditions and after shutdown due to an accident. In accordance with the methodology described in Section 3, the 2D maps were first built using the AMCL-EKF technique, and then the 3D maps were estimated by applying the Rao–Blackwellized SLAM approach.

The results of the map creation for the terrain before accident are presented in Figure 5. Both 2D and 3D maps precisely recreate the real state of the mine's corridors with all observable main features. The 2D map (Figure 5a) clearly shows free spaces and therefore can be fully used for navigation tasks, whereas the 3D map (Figure 5b) contains all features of the passed corridors and therefore is much more useful for a precise and safe inspection of the underground environment. It is especially important when some parts of the terrain have become inaccessible.

(a)

Figure 5. *Cont.*

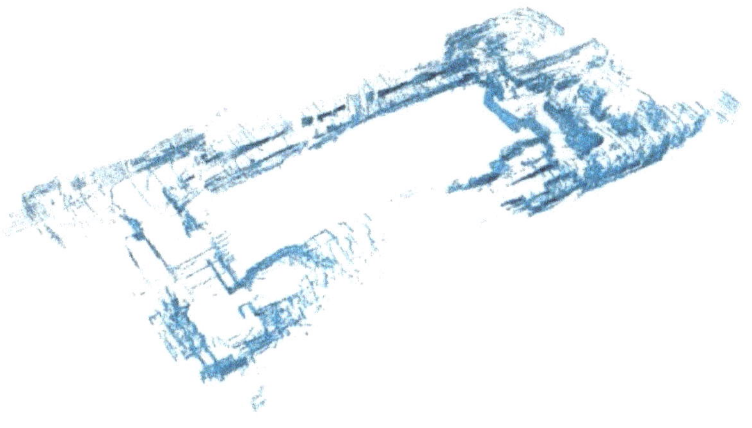

(b)

Figure 5. Maps obtained before accident: (**a**) created with the use of 2D SLAM technique; (**b**) created with the use of proposed hybrid 3D SLAM approach.

Such a situation is presented in Figure 6, where a photographic image of the corridor and destroyed infrastructure is shown. Due to the disaster, the main road was partially blocked, and the terrain became rough.

Figure 6. Photographic image of the mine corridor after accident.

The results of the map creation are depicted in Figure 7. The depicted maps show the examined area from the same perspective in which it was captured by the camera. The comparison of the scenes from the photo and the generated maps shows a high compliance of the presented mine's environment. The general differences between 2D and 3D maps, shown in Figure 7a,b, respectively, are similar to those described for the previous case. It can be seen that the 3D map is accurate enough to assess the real state of the environment.

The performed investigations showed a good compliance of the obtained 2D map with the geodetic map of the laboratory ground mine. The final product, which is a 3D-space map, properly reflects the current of the working environment.

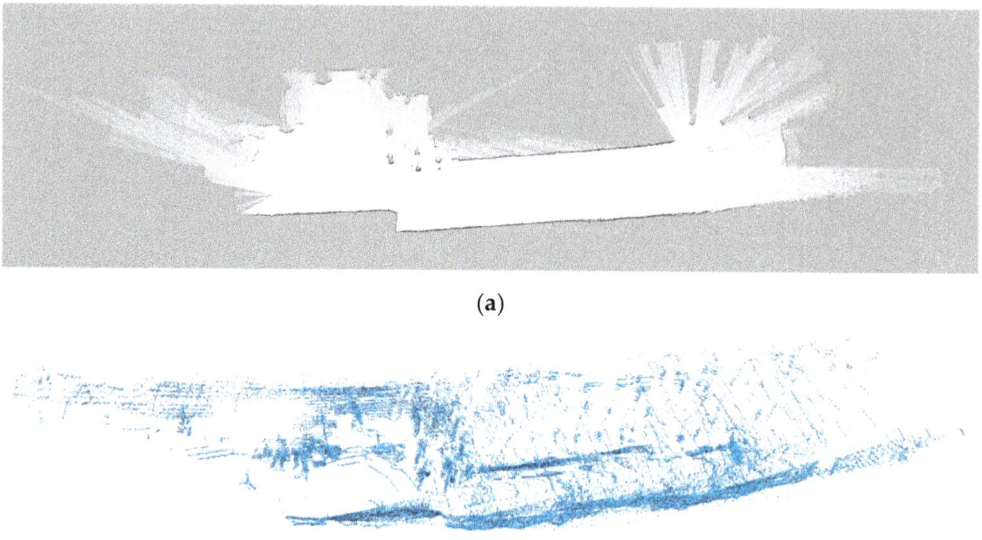

(**a**)

(**b**)

Figure 7. Maps obtained during underground inspection in the mine terrain after accident: (**a**) created with the use of 2D SLAM technique; (**b**) created with the use of proposed hybrid 3D SLAM approach.

However, its quality and practical usability is limited by specific problems related to the limitations connected with real working conditions. Although the AMCL provides a global location based on reference points, the quality of the pose estimation process strongly depends on errors concerned with the robot's slippage or skidding. A problem with LIDAR can be long spaces with a limited number of potential reference points, such as a tunnel with single surrounding features.

5. Conclusions

The method for the concurrent construction of a 3D map of the industrial terrain realized with use of a mobile wheeled robot is presented and discussed in the paper. Since the three-dimensional map shows the complete picture of the explored terrain, it is more convenient for use in environmental inspection tasks.

The main contribution of this work is to develop a real-time method for creating a 3D map in both structured and unstructured industrial environments. We propose to solve this problem by using LIDAR to preprocess the scanning and IMU to generate odometries.

After mapping the environment, we created a grid map based on the 2D SLAM, the AMCL, and the EKF. The robot's global position was determined by means of the Rao–Blackwellized technique. This approach significantly reduced computational costs in comparison with the Full 3D SLAM.

The efficiency, accuracy, and robustness of the developed hybrid 3D SLAM approach for 3D-map creation was validated by a number of real experiments conducted in the different scenarios of the underground mine. All obtained results confirmed the correctness of the developed solution for the creation of the 3D map in the indoor environment. The findings also demonstrated that the worked-out algorithm can be run in conditions similar to real ones by using a mobile robotic vehicle equipped with modest hardware.

Further works will be devoted to examining the applicability of the developed solution to create a 3D map in other types of real environments, such as underwater or aerial.

Author Contributions: Conceptualization, T.B., J.G., M.G. and A.K.; methodology, M.G.; software, A.K.; validation, T.B. and A.K.; formal analysis, T.B., J.G., M.G. and A.K.; investigation, T.B., M.G. and A.K.; resources, M.G. and A.K.; data curation T.B. and A.K.; writing—original draft preparation, J.G. and M.G.; writing—review and editing, T.B., J.G., M.G. and A.K.; visualization, A.K.; supervision, J.G. and M.G.; project administration, M.G. All authors have read and agreed to the published version of the manuscript.

Funding: This research received no external funding.

Data Availability Statement: All necessary data are given in this article.

Acknowledgments: The authors express their gratitude to the AGH UST Center of Energy for enabling the use of scientific and laboratory infrastructure for experiments.

Conflicts of Interest: The authors declare no conflict of interest.

References

1. Murphy, R.R.; Kravitz, J.; Stover, S.L.; Shoureshi, R. Mobile robot in mine rescue and recovery. *IEEE Robot. Autom. Mag.* **2009**, *16*, 91–103. [CrossRef]
2. Michael, N.; Shen, S.; Mohta, K.; Mulgaonkar, Y.; Kumar, V.; Nagatani, K.; Okada, Y.; Kiribayashi, S.; Otake, K.; Yoshida, K.; et al. Collaborative mapping of an earthquake-damaged building via ground and aerial robots. *J. Field Robot.* **2012**, *29*, 832–841. [CrossRef]
3. Kruijff, G.J.; Kruijff-Korbayová, I.; Keshavdas, S.; Larochelle, B.; Janíček, M.; Colas, F.; Liu, M.; Pomerleau, F.; Siegwart, R.; Neerincx, M.A.; et al. Designing, developing, and deploying systems to support human-robot teams in disaster response. *Adv. Robot.* **2014**, *28*, 1547–1570. [CrossRef]
4. Pradhan, D.; Sen, J.; Hui, N.B. Design and development of an automated all-terrain wheeled robot. *Adv. Robot. Res.* **2014**, *1*, 21–39. [CrossRef]
5. Pierzchała, M.; Giguère, P.; Astrup, A. Mapping forests using an unmanned ground vehicle with 3D LIDAR and graph-SLAM. *Comput. Electron. Agric.* **2018**, *145*, 217–225. [CrossRef]
6. Saponara, S. Sensing and connection systems for assisted and autonomous driving and unmanned vehicles. *Sensors* **2018**, *18*, 1999. [CrossRef]
7. Zhang, K.; Yang, Y.; Fu, M.; Wang, M. Traversability assessment and trajectory planning of unmanned ground vehicles with suspension systems on rough terrain. *Sensors* **2019**, *19*, 4372. [CrossRef]
8. Elfes, A. Using occupancy grids for mobile robot perception and navigation. *Computer* **1989**, *6*, 46–57. [CrossRef]
9. Mihálik, M.; Malobický, B.; Peniak, P.; Vestenický, P. The new method of active SLAM for mapping using LiDAR. *Electronics* **2022**, *11*, 1082. [CrossRef]
10. Fabrizi, E.; Saffiotti, A. Extracting topology-based maps from gridmaps. In Proceedings of the IEEE International Conference on Robotics and Automation ICRA'2000, San Francisco, CA, USA, 24–28 April 2000; pp. 2972–2978.
11. Fox, D.; Burgard, W.; Thrun, V. Markov localization for mobile robots in dynamic environments. *J. Artif. Intell. Res.* **1999**, *11*, 391–427. [CrossRef]
12. Schmidt, A. The EKF-based visual SLAM system with relative map orientation measurements. In Proceedings of the 2014 International Conference on Computer Vision and Graphics, Warsaw, Poland, 15 September 2014; pp. 182–194.
13. Montemerlo, M.; Thrun, S. A scalable method for the simultaneous localization and mapping problem in robotics. In *FastSLAM 2.0*; Springer: Berlin/Heidelberg, Germany, 2007; pp. 63–90.
14. Thrun, S.; Burgard, W.; Fox, D.; Dellaert, F. Robust Monte Carlo localization for mobile robots. *Artif. Intell.* **2001**, *1–2*, 99–141. [CrossRef]
15. Thrun, S.; Montemerlo, M. The graph SLAM algorithm with applications to large-scale mapping of urban structures. *Int. J. Robot. Res.* **2006**, *5–6*, 403–429. [CrossRef]
16. Ismail, H.; Roy, R.; Sheu, L.J.; Chieng, W.H.; Tang, L.C. Exploration-based SLAM (e-SLAM) for the indoor mobile robot using Lidar. *Sensors* **2022**, *2*, 1689. [CrossRef] [PubMed]
17. Bresson, G.; Alsayed, Z.; Yu, L.; Glaser, S. Simultaneous localization and mapping: A survey of current trends in autonomous driving. *IEEE Trans. Intell. Veh.* **2017**, *3*, 194–220. [CrossRef]
18. Grisetti, G.; Stachniss, C.; Burgard, W. Improved techniques for grid mapping with Rao-Blackwellized particle filters. *IEEE Trans. Robot.* **2007**, *1*, 34–46. [CrossRef]
19. Kudriashov, A.; Buratowski, T.; Giergiel, M. Multi-level exploration and 3D mapping with octrees and differential drive robots. Warsaw University of Technology Press. *Electronics* **2018**, *192*, 491–500.
20. Milstein, A. Occupancy grid maps for localization and mapping. In *Motion Planning*; Jing, X.J., Ed.; InTech: London, UK, 2008; pp. 382–408.
21. Montemerlo, M.; Thrun, S.; Koller, D.; Wegbreit, B. FastSLAM: A factored solution to the simultaneous localization and mapping problem. In Proceedings of the AAAI National Conference on Artificial Intelligence, Edmonton, Canada, 28 July–1 August 2002; pp. 318–329.

22. Nuss, D.; Reuter, S.; Thom, M.; Yuan, T.; Krehl, G.; Maile, M.; Gern, A.; Dietmayer, K. A random finite set approach for dynamic occupancy grid maps with real-time application. *arXiv* **2016**, arXiv:1605.02406. [CrossRef]
23. Chang, B.; Kaneko, M.; Lim, H. Robust 2D mapping integrating with 3d information for the autonomous mobile robot under dynamic environment. *Electronics* **2019**, *8*, 1503.
24. Wiliams, S.B.; Dissanayake, V.; Durant-Whyte, H.F. An efficient approach to the simultaneous localisation and mapping problem. In Proceedings of the IEEE International Conference on Robotics and Automation, Washington, DC, USA, 11–15 May 2002; pp. 4006–4411.
25. Hornung, A.; Wurm, K.M.; Bennewitz, M.; Stachniss, C.; Burgard, W. Octomap: An efficient probabilistic 3D mapping framework based on octrees. *Auton. Robot.* **2013**, *3*, 189–206. [CrossRef]
26. Peasley, B.; Birchfield, S.; Cunningham, A.; Dellaert, F. Accurate on-line 3D occupancy grids using manhattan world constraints. In Proceedings of the IEEE/RSJ International Conference on Intelligent Robots and Systems IROS'2012, Vilamoura, Portugal, 7–12 October 2012; pp. 5283–5290.
27. Skrzypczynski, P. Simultaneous localization and mapping: A feature-based probabilistic approach. *Int. J. Appl. Math. Comp. Sci.* **2009**, *4*, 575–588. [CrossRef]
28. Grewal, M.S. Kalman filtering. In *International Encyclopaedia of Statistical Science*; Lovric, M., Ed.; Springer: Berlin, Germany, 2011; pp. 705–708.
29. Fayyad, J.; Jaradat, M.A.; Gruyer, D.; Najjaran, H. Deep learning sensor fusion for autonomous vehicle perception and localization: A review. *Sensors* **2020**, *15*, 4220. [CrossRef] [PubMed]
30. Fox, D.; Burgard, W.; Thrun, V.; Dellaert, F. Particle Filters for Mobile Robot Localization. In *Sequential Monte Carlo Methods in Practice*; Doucet, A., Freitas, N., Gordon, N., Eds.; Springer: Berlin, Germany, 2001; pp. 401–428.
31. Smith, S.R.; Self, M.; Cheesman, P. Estimating uncertain spatial relationships in robotics. In Proceedings of the Second Conference on Uncertainty in Artificial Intelligence, Philadelphia, PA, USA, 8 August 1990.
32. Julier, S.J.; Uhlmann, J.K. Unscented filtering and nonlinear estimation. *Proc. IEEE* **2004**, *92*, 401–422. [CrossRef]
33. Zhang, Y.; Wang, L.F.; Yu, J.H. Depth-image based 3D map reconstruction of indoor environment for mobile robots. *J. Comput. Appl.* **2014**, *34*, 3438–3445.
34. Grisetti, G.; Stachniss, C.; Burgard, W. Improving grid-based SLAM with Rao-Blackwellized particle filters by adaptive proposals and selective resampling. In Proceedings of the IEEE International Conference on Robotics and Automation, Barcelona, Spain, 18 April 2005.
35. Grisetti, G.; Tipaldi, G.D.; Stachniss, C.; Burgard, W.; Nardi, D. Fast and accurate SLAM with Rao-Blackwellized particle filters. *Robot. Auton. Syst.* **2007**, *55*, 30–38. [CrossRef]
36. Godsill, S. Particle filtering: The first 25 years and beyond. In Proceedings of the IEEE International Conference on Acoustics, Speech and Signal Processing (ICASSP), Brighton, UK, 12–17 May 2019; pp. 7760–7764.
37. Doucet, A.; Freitas, N.; Murphy, K.; Russell, S. Rao-Blackwellised particle filtering for dynamic Bayesian networks. In Proceedings of the 16th Conference on Uncertainty in Artificial Intelligence, Stanford, CA, USA, 30 June–3 July 2000; pp. 176–183.
38. Yu, S.; Fu, C.; Gostar, A.K.; Hu, M. A review on map-merging methods for typical map types in multiple-ground-robot SLAM solutions. *Sensors* **2020**, *20*, 6988. [CrossRef]
39. Li, H.; Tsukada, M.; Nashashibi, F.; Parent, M. Multivehicle cooperative local mapping: A methodology based on occupancy grid map merging. *IEEE Trans. Intell. Transp. Syst.* **2014**, *15*, 2089–2100. [CrossRef]
40. Birk, A.; Carpin, S. Merging occupancy grid maps from multiple robots. *Proc. IEEE* **2006**, *94*, 1384–1397. [CrossRef]
41. Sünderhauf, N. Robust Optimization for Simultaneous Localization and Mapping. Ph.D. Thesis, Chemnitz University of Technology, Chemnitz, Germany, 2012.
42. Sobczak, L.; Filus, K.; Domanski, A.; Domanska, J. LiDAR point cloud generation for SLAM algorithm evaluation. *Sensors* **2021**, *21*, 3313. [CrossRef]
43. Ren, Z.; Wang, L.; Bi, L. Robust GICP-based 3D LiDAR SLAM for underground mining environment. *Sensors* **2019**, *19*, 2915. [CrossRef]
44. Li, M.; Zhu, H.; You, S.; Wang, L.; Tang, C. Efficient laser-based 3D SLAM for coal mine rescue robots. *IEEE Access* **2018**, *7*, 14124–14138. [CrossRef]
45. Parisotto, E.; Chaplot, D.S.; Zhang, J.; Salakhutdinov, R. Global pose estimation with an attention-based recurrent network. In Proceedings of the IEEE/CVF Conference on Computer Vision and Pattern Recognition Workshops (CVPRW), Salt Lake City, UT, USA, 18 June 2018.
46. Wang, S.; Clark, R.; Wen, H.; Trigoni, N. Towards end-to-end visual odometry with deep recurrent convolutional neural networks. In Proceedings of the IEEE International Conference on Robotics and Automation (ICRA), Singapore, 29 May 2017.
47. Bruno, H.M.S.; Colombini, E.L. LIFT-SLAM: A deep-learning feature-based monocular visual SLAM method. *Neurocomputing* **2021**, *455*, 97–110. [CrossRef]
48. Moravec, H.P. Sensor fusion in certainty grids for mobile robots. *AI Mag.* **1988**, *9*, 61–71.
49. Kudriashov, A.; Buratowski, T.; Giergiel, M. Hybrid AMCL-EKF filtering for SLAM-based pose estimation in rough terrain. In Proceedings of the 15th IFToMM World Congress on Advances in Mechanism and Machine Science, Krakow, Poland, 30 June–4 July 2019; pp. 2819–2828.

50. Kudriashov, A.; Buratowski, T.; Garus, J.; Giergiel, M. 3D environment exploration with SLAM for autonomous mobile robot control. *WSEAS Trans. Control Sys.* **2021**, *16*, 451–456. [CrossRef]
51. Kudriashov, A.; Buratowski, T.; Giergiel, M. Robot's pose estimation in environment exploration process with SLAM and laser techniques. *Nauk. Notatki* **2017**, *58*, 204–212.

electronics

Article

Hill-Climb-Assembler Encoding: Evolution of Small/Mid-Scale Artificial Neural Networks for Classification and Control Problems

Tomasz Praczyk

Department of Computer Science, Polish Naval Academy, Smidowicza 69, 81-127 Gdynia, Poland; t.praczyk@amw.gdynia.pl

Abstract: The paper presents a neuro-evolutionary algorithm called Hill Climb Assembler Encoding (HCAE) which is a light variant of Hill Climb Modular Assembler Encoding (HCMAE). While HCMAE, as the name implies, is dedicated to modular neural networks, the target application of HCAE is to evolve small/mid-scale monolithic neural networks which, in spite of the great success of deep architectures, are still in use, for example, in robotic systems. The paper analyses the influence of different mechanisms incorporated into HCAE on the effectiveness of evolved neural networks and compares it with a number of rival algorithms. In order to verify the ability of HCAE to evolve effective small/mid-scale neural networks, both feed forward and recurrent, it was tested on fourteen identification problems including the two-spiral problem, which is a well-known binary classification benchmark, and on two control problems, i.e., the inverted-pendulum problem, which is a classical control benchmark, and the trajectory-following problem, which is a real problem in underwater robotics. Four other neuro-evolutionary algorithms, four particle swarm optimization methods, differential evolution, and a well-known back-propagation algorithm, were applied as a point of reference for HCAE. The experiments reported in the paper revealed that the evolutionary approach applied in the proposed algorithm makes it a more effective tool for solving the test problems than all the rivals.

Keywords: evolutionary neural networks; hill climb; control; classification

Citation: Praczyk, T. Hill-Climb-Assembler Encoding: Evolution of Small/Mid-Scale Artificial Neural Networks for Classification and Control Problems. *Electronics* **2022**, *11*, 2104. https://doi.org/10.3390/electronics11132104

Academic Editor: Boris Andrievsky

Received: 9 June 2022
Accepted: 2 July 2022
Published: 5 July 2022

1. Introduction

Deep neural networks (DNN) are computing systems that have now become practically irreplaceable in many areas. It is currently difficult to imagine image/video processing without convolutional neural networks (CNN), and the analysis of time series data without gated recurrent units (GRU) and long short-term memories (LSTM). Thanks to the DNNs, it has become possible to solve many problems that were previously either unsolvable or were solvable, however, at an unsatisfactory level.

However, DNNs are not a magic formula for all problems. There are tasks that do not require large-scale neural architectures and that can be effectively performed by a class of simpler traditional neural networks, having up to five hundred neurons [1] which, in the context of this paper, are called small/mid-scale networks. Low/high-level robot control [2–7], prediction of complex object behaviour [8–10], and robot navigation [11–13] are examples of tasks that are well-suited for this class of networks.

In order to train ANNs, gradient decent methods are usually applied. Back propagation (BP), with many different variants, is the most popular algorithm that is used for that purpose. However, the drawback of BP is its susceptibility to get stuck in the local optima of an objective function that is optimized during ANN training. In effect, to find a satisfactory neural solution, the algorithm has to be run many times from different starting points which, for complex objective functions with many local optimums, significantly prolongs the training process, and, what is more, it does not give a guarantee of ultimate success.

An alternative solution to the one outlined above is to apply global optimization methods instead of local ones such as BP. Neuro-evolution (NE) seems to be the most popular global optimization approach which, together with BP, is used to train ANNs. However, the problem of NE is the exponentially increasing search space with respect to the number of neurons in the ANN. Even searching for small/mid-scale ANNs involves very large search spaces, including all possible connection weights between the neurons, and the parameters of each neuron: the bias and the transfer function. In turn, complex optimization domains usually need to perform a large number of fitness function evaluations which, in the case of robotic systems, means either huge costs (evaluation on real robots) or a time-consuming process of training the network (evaluation on a complex robot simulation model).

In order to cope with the above problem, the NE methods try to somehow reduce search space, for example, through indirect network representations or some restrictions imposed on possible network architectures. However, all the state-of-the-art NE methods, not only those meant for small/mid-scale networks, are based on the solution, adapted from natural systems, of a one-to-one relation between the network and its genome, i.e., all the information needed to create the network is encoded in a single network representation. As a consequence, simple networks need simple representations, whereas complex networks require complex representations. Meanwhile, the goal of the current paper is to prove that, at least for small/mid-scale networks, an approach that forms neural networks from pieces that evolve in successive evolutionary runs, each of which is responsible for a single piece of the network, not for the entire network, is more efficient than the traditional model of the NE with a one-to-one relation between the network and its genome.

In the paper, an NE method is presented, called Hill Climb Assembler Encoding (HCAE), which is a light version of Hill Climb Modular Assembler Encoding (HCMAE) [14] and whose target application, unlike its modular counterpart, is the evolution of small/mid-scale neural networks. The main assumption of both HCAE and HCMAE is the gradual growth in the networks and the use of the evolutionary algorithm to generate successive network increments encoded in the form of simple programs that do not grow over time. Instead of more and more complex network representations which are difficult to process effectively, it builds the networks from small pieces that evolve in many consecutive evolutionary runs. The task of each run is to make an improving step in a neural network space. The first run performs in the same way as other traditional NE algorithms. The most effective network from this run is the first step of HCAE. The following runs create the next steps through expanding the network with new neurons and connections or modifying connections that already exist. In contrast to other NE algorithms, HCAE does not process encoded networks but encoded pieces of the networks that are short and simple in effective processing.

HCAE is an improvement of Assembler Encoding (AE) [15], and Assembler Encoding with Evolvable Operations (AEEO) [6,10,12,16], i.e., variations of genetic programming (GP) that represent networks in the form of simple programs similar to assembler programs. The major difference between HCAE and its predecessors is the applied model of the evolution. Both AE and AEEO represent the traditional model with the central role of cooperative co-evolutionary GA (CCEGA) [17,18], whereas HCAE is in fact a hill climber, whose each step is made by means of CCEGA.

In order to determine how different solutions applied in HCAE affect the results of the algorithm compared with the original solutions used in AE/AEEO, four different variants of HCAE were examined during the tests. The first two variants implemented the new model of the evolution and original mechanisms applied in AE/AEEO, whereas the next two variants implemented improved HCAE mechanisms and the new model of the evolution as well.

After the tests with AE/AEEO, the most effective HCAE variant from the previous experiments was also compared with eight other algorithms. Neuro-Evolution of Augmenting Topologies (NEAT) [19] and Cooperative Synapse Neuro-Evolution (CoSyNE) [20] were the first HCAE competitors from outside the AE family, and the reason why they were chosen is that they are regarded as state-of-the-art NE methods with proven effectiveness. Classical Particle Swarm Optimization (PSO) [21–24], Levy Flight Particle Swarm Optimization (LFPSO) [25], modified LFPSO [26] called PSOLF, and Mantegna Levy Flight, Particle Swarm Optimization, and Neighbourhood Search (LPSONS) [27] were PSO-based methods applied as the next points of reference for HCAE. The Differential Evolution (DE) [28–30] and classic BP algorithm were the last HCAE rivals.

The comparison between HCAE, NEAT, CoSyNE, AE, and AEEO was made on challenging variants of classical identification and control problems, i.e., on the two-spiral problem (TSP) [31] and inverted-pendulum problem (IPP). In the experiments, the difficulty of both problems was intentionally increased compared with their original variants. TSP is a binary classification problem, and to solve it, feed-forward ANNs (FFANN) are sufficient, whereas the IPP is a control problem that needs recurrent ANNs (RANN). The average complexity of both problems was a deciding factor for why they were selected as a testbed for HCAE and its rivals.

The comparison of HCAE, LFPSO, PSOLF, LPSONS, and BP was made based on the results of experiments reported in [27]. To this end, thirteen datasets from the UCI machine-learning repository were applied, each of which defines a real classification problem.

In turn, HCAE, DE, and PSO were compared on the trajectory-following problem (TFP). In this case, the task of FFANNs and RANNs was to control an autonomous underwater vehicle (AUV) along a desired trajectory defined spatially and temporally. The TFP is a real problem in underwater robotics, if the coordinated operation of a team of AUVs in time and space is required.

The contributions of the paper are as follows:

- The scheme of network evolution applied in HCAE, in which the genome represents a piece of the network, is compared with the traditional model in which the genome represents the entire network.
- HCAE is qualitatively evaluated on a number benchmark-classification and control problems.
- HCAE is compared with ten other algorithms: AE, AEEO, NEAT, CoSyNE, PSO, LFPSO, LFPSO, LPSONS, DE, BP.

The rest of the paper is organized as follows: section two presents related work, section three details the HCAE, section four reports experiments, section five shows direction of further research, and the final section concludes the paper.

2. Related Work

As already mentioned, the search space of the NE algorithms increases exponentially with every new neuron in the network. For the direct NE methods, which directly and explicitly specify every neuron and connection in the genome, at some point, the search space becomes too large to provide a reasonable chance to design an effective network at all. In order to cope with this problem, in NEAT [19], which seems to be, currently, the most popular direct algorithm, the genome contains only gens that correspond to existing connections in the network—the connections that do not exist are not represented in the genome. This way, sparsely connected networks can be encoded in the form of short chromosomes, which simplifies the task of NEAT. However, the problem returns for fully or densely connected networks.

To some extent, the above problem can be solved by applying indirect encoding of the networks, in which the genome includes the recipe of how to form the network. The recipe can take different forms, for example, the form of a production rule or a program. By repeatedly using the same rule or a piece of program code in different areas of the evolved neural network, it is possible to generate large-scale networks. In other words, the indirect encoding leads to the compression of the phenotype to a smaller genotype, providing a smaller search space. However, such an approach is only beneficial if the problem solved is modular and has regularities. Otherwise, the indirect representations are just as complex as the direct representations.

In [32,33], a developmental approach is presented that is based on production rules that decide how neuron synapses grow and how mother neurons divide. Initially, the mother neurons are placed in 2D space, and then the repeated process of synapse growth and neuron division starts which, after a specified number of iterations, defines the final architecture of the network.

A similar solution is proposed in [34–41]. In this case, a grammar is defined including terminals with the network parameters, non-terminals, and the production rules that map non-terminals into terminals or other non-terminals. The networks are represented in the form of encoded starting rules which are run for a number of iterations generating, ultimately, a complete specification of the network.

In order to reduce the dimensionality of fixed topology network representations, the algorithm proposed by [42] does not work in the weight domain but in the frequency domain. At the phenotypic level, each network is represented as a weight matrix, which, at the genotypic level, is encoded as a matrix of Fourier-transform coefficients, however, without the high-frequency part, which is ignored. To obtain the network, the inverse Fourier transform is applied.

An indirect counterpart of NEAT is HyperNEAT, that is, Hypercube NeuroEvolution of Augmented Topologies [43]. In the HyperNEAT, networks are produced by other networks called compositional pattern producing networks (CPPNs), which evolve according to NEAT. To generate a network—theoretically, of any size—its neurons are first placed in n-dimensional space, and, then, weights of inter-neuron connections are determined by a single CPPN. To fix a weight between a pair of neurons, the CPPN is supplied with the coordinates of the neurons. An output signal of the CPPN indicates the weight of the connection. Motivated by this approach, a number of variants were proposed to evolve even larger-scale networks [44–47]. Moreover, algorithms were also proposed that replace NEAT as the CPPN constructor with genetic programming [48].

The NE methods based on genetic programming (GP) are a separate class of indirect solutions that represent neural networks in the form of programs. By repeatedly executing the same code in different regions of the network or by cloning neurons/connections, simple programs can represent even large-scale neural networks. In the classical GP, the programs take the form of trees whose nodes include instructions that operate on neurons/connections and control transfer instructions to direct the flow of execution, e.g., sequential/parallel division of a neuron, modification of a weight/bias, jump, and loop [49–52].

Gene expression programming (GEP) is a variation of GP in which chromosomes take the form of fixed-length linear strings that, like their tree-structured counterparts, include network-building instructions. The application of GEP to evolve neural networks is presented, among other things, in [53,54].

Assembler encoding (AE) [15] is the next representative of GP. In AE, a network is represented in the form of a linearly organized Assembler Encoding Program (AEP) whose structure is similar to the structure of a simple assembler program. The AEP is composed of two parts, i.e., a part including operations and a part including data. The task of each AEP is to create a Network Definition Matrix (NDM) which includes all the information necessary to produce the network. Unlike instructions of tree-based GP and GEP which work, rather, on the level of individual neurons/connections, each operation of AE can modify even the whole NDM, and, consequently, the network. This way, even a few operations working together within one AEP can form very complex neural structures.

Assembler Encoding with Evolvable Operations (AEEO) [55] and its modular/ensemble variants [16,56] are the successors of AE. Instead of applying hand-made algorithmic operations with evolved parameters, AEEO, like HyperNEAT, uses operations which take the form of simple directly encoded networks, called ANN operations. Just like in the classic variant of the AE, the task of ANN operations, which are run one after the other, is to change the content of the NDM.

Another solution to cope with the problem of large search spaces is the gradual growth in representations and the networks, which is applied, for example, in NEAT [19], AE [15], and AEEO [55]. In this case, the evolution usually starts with light representations and simple networks and increases their complexity in the course of time. The idea is to evolve the networks little by little, and to focus, at each stage of evolution, mainly on genotype/phenotype increments. Initially, small networks and their representations grow if the evolution cannot make progress for a time. Beneficial sub-structures in both genotypes and phenotypes are preserved and gradually expanded with new sub-structures. However, along with the increase in the size of the representations, the complexity of the task of the evolutionary algorithm automatically grows as well. Often, in the late phases of the evolutionary process, the change in the size of chromosomes does not entail any progress in network effectiveness.

There are also NE algorithms that restrict the search space by applying constraints on the network architectures and the evolutionary process. The constraints are usually based on domain knowledge about the problem to be solved. Their task is to focus the search only on allowed network architectures, making the search space for the evolutionary algorithm smaller and its task simpler. In [1,57], constraint functions are proposed which directly manipulate neural networks and implement structure, functional, and evolutionary constraints whose task is, for example, to evolve symmetries and repetitive structures, to insert predefined functional units to evolved networks, or to restrict the range of evolutionary operators.

Surrogate model-based optimization [58,59] is a solution that does not simplify the task of NE algorithm but alleviates the burden associated with the evaluation of neural solutions scattered over large areas of high-dimensional network space. This is of particular importance in the case of robotic systems whose evaluation requires expansive simulations or real-world experiments. By using surrogate models that mimic the behaviour of the simulation model as closely as possible, being at the same time and computationally cheaper to evaluate, NE algorithms can explore their search spaces more intensely, and thus increase the chance of finding optimal networks.

3. Hill Climb Assembler Encoding

As already mentioned, HCAE is a light variant of HCMAE and it originates from both AE and AEEO. All the algorithms are based on three key components, i.e., a network definition matrix (NDM), which represents the neural networks, assembler encoding program (AEP), which operates on NDM, and evolutionary algorithm, whose task is to produce optimal AEPs, NDMs, and, consequently, the networks. All the three components are described below.

3.1. Network Definition Matrix

To represent a neural network, HCAE, like its predecessors, uses a matrix called network definition matrix (NDM). The matrix includes all the parameters of the network, including the weights of inter-neuron connections, bias, etc. The matrix which contains non-zero elements above and below the diagonal encodes a recurrent neural network (RANN), whereas the matrix with only the content above the diagonal represents a feed-forward network (FFANN) [14] .

3.2. Assembler Encoding Program

In all the AE family, filling up the matrix, and, consequently, constructing an ANN is the task of an assembler encoding program (AEP) which, like an assembler program, consists of a list of operations and a sequence of data. Each operation implements a fixed algorithm and its role is to modify a piece of NDM. The operations are run one after another and their working areas can overlap, which means that modifications made by one operation can be overwritten by other operations which are placed further in the program. AEPs can be homogeneous or heterogeneous in terms of applied operations. In the first case, all operations in AEP are of the same type and they implement the same algorithm whereas, in the second case, AEPs can include operations with different algorithms. The first solution is applied in HCAE and AEEO, whereas the second one in AE [14] .

The way each operation works depends, on the one hand, on its algorithm and, on the other hand, on its parameters. Each operation can be fed with its "private" parameters, linked exclusively to it, or with a list of shared parameters concentrated in the data sequence. Parametrization allows operations with the same algorithm to work in a different manner, for example, to work in different fragments of NDM.

HCAE uses two types of operations, say, `Oper1` and `Oper2`. `Oper1` is an adaptation of a solution applied in AEEO. It is of a global range, which means that it can modify any element of NDM, and it uses a small feed-forward neural network, say, ANN operation, in the decision-making process. The task of ANN operation is to decide which NDM items are to be updated and how they are to be updated (see Figure 1). The architecture of each ANN operation is determined by parameters of `Oper1`, whereas inputs to the ANN operation are taken from the data sequence of AEP. A pseudo-code of `Oper1` is given in Algorithm 1.

Each ANN operation has two inputs and five outputs. The inputs indicate individual items of NDM. In AEEO, ANN operations are fed with coordinates of items to be modified, that is, with numbers of columns and rows, for example, in order to modify item $[i, j]$, an ANN operation is supplied with i and j. In HCAE, a different approach is used, namely, instead of i, j, ANN operations are fed with data items which correspond to i and j, that is, with `row` $[i]$ and `column` $[j]$ (lines (10) and (11) in Algorithm 1). Vectors `row` and `column` are filled with appropriate data items (lines (2) and (5) in Algorithm 1).

The outputs of ANN operation decide whether to modify a given item of NDM or to leave it intact—outputs no. 1 and no. 2 (line (13) in Algorithm 1), and then, whether to reset the item or to assign it a new value—outputs no. 3 and no. 4 (line (14) in Algorithm 1), the new value is taken from the fifth output of the ANN operation (line (15) in Algorithm 1). Parameter M is a scaling parameter.

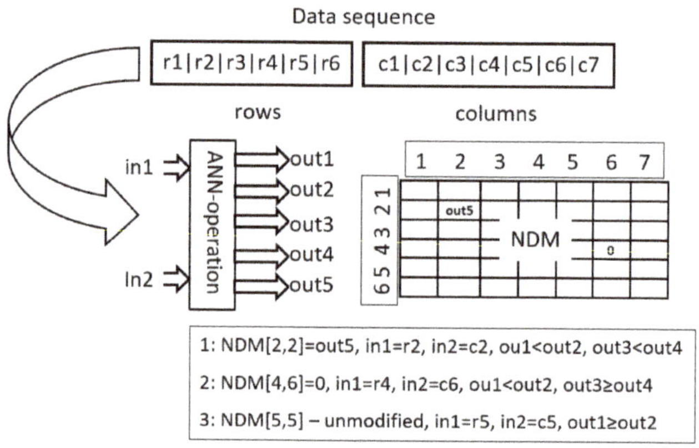

Figure 1. Applying ANN operations in `Oper1`: ANN operation is run for each item in NDM, one after another, and it can change the value of each item. The figure shows applying the network to determine the value of three items: NDM[2,2], NDM[4,6] and NDM[5,5]. In the first case, the item is modified to the value $out5$, which is the response of the network to the input $r2, c2$. $r2, c2$ are data items that correspond to the second row and column, that is, to the location of the modified item. The value $out5$ is inserted into NDM because $out1 < out2$ and $out3 < out4$. In the second case, the item receives the value 0 because $out1 < out2$ and $out3 \geq out4$. In addition, in the third case, the item is left unchanged because $out1 \geq out2$.

Algorithm 1 Pseudo-code of `Oper1`.

Input: operation parameters (p), data sequence (d), NDM
Output: NDM

```
for i∈<0..NDM.numberOfRows) do
    row[i] ← d[i mod d.length];
end for
for i∈<0..NDM.numberOfColumns) do
    column[i] ← d[(i+NDM.numberOfRows) mod d.length];
end for
ANN-oper ← getANN(p);
for i∈<0..NDM.numberOfColumns) do
    for j∈<0..NDM.numberOfRows) do
        ANN-oper.setIn(1,row[j]);
        ANN-oper.setIn(2,column[i]);
        ANN-oper.run();
        if ANN-oper.getOut(1) <ANN-oper.getOut(2) then
            if ANN-oper.getOut(3) < ANN-oper.getOut(4) then
                NDM[j,i] ← M*ANN-oper.getOut(5);
            else
                NDM[j,i] ← 0;
            end if
        end if
    end for
end for
Return NDM
```

Like resultant ANNs, ANN operations are also represented in the form of NDMs, say, NDM operations. To generate an NDM operation, and consequently, an ANN operation, `getANN(p)` is applied whose pseudo-code is depicted in Algorithm 2. It fills all matrix items with subsequent parameters of `Oper1` divided by a scaling coefficient, N. If the number of parameters is too small to fill the entire matrix, they are used many times.

Algorithm 2 Pseudo-code of `getANN`.

Input: operation parameters (p)
Output: ANN-operation

```
NDM-operation ← 0;
noOfItem ← 0;
for i∈<0..NDM-operation.numberOfColumns) do
    for j<i //feed-forward ANN do
        NDM-operation[j,i] ← p[noOfItem mod p.length]/N;
        noOfItem++;
    end for
end for
Return ANN-operation encoded in NDM-operation.
```

Unlike `Oper1`, `Oper2` works locally in NDM, and is an adaptation of a solution applied in AE. Pseudo-code of `Oper2` is given in Algorithm 3 and 4. It does not use ANN-operations; instead, it directly fills NDM with values from the data sequence of AEP: where NDM is updated, and which and how many data items are used, are determined by operation parameters. The first parameter indicates the direction according to which NDM is modified, that is, whether it is changed along columns or rows (lines (4) and (10) in Algorithm 3). The second parameter determines the size of holes between NDM updates, that is, the number of zeros that separate consecutive updates (line (3) in Algorithm 4). The next two parameters point out the location in NDM where the operation starts to work, i.e., they indicate the starting row and column (line (1) in Algorithm 4). The fifth parameter determines the size of the altered NDM area, in other words, it indicates how many NDM items are updated (line (1) in Algorithm 4). Additionally, the last, sixth parameter points out location in the sequence of data from where the operation starts to take data items and put them into the NDM (line (2) in Algorithm 3) [14] .

Algorithm 3 Pseudo-code of `Oper2` [14].

Input: operation parameters (p), data sequence (d), NDM
Output: NDM

```
1: filled ← 0;
2: where ← p[6];
3: holes ← 0;
4: if p[1] mod 2 = 0 then
5:     for k∈<0..NDM.numberOfColumns) do
6:         for j∈<0..NDM.numberOfRows) do
7:             NDM[j,k] ← fill(k,j,param,data,filled,where,holes);
8:         end for
9:     end for
10: else
11:     for k∈<0..NDM.numberOfRows) do
12:         for j∈<0..NDM.numberOfColumns) do
13:             NDM[k,j] ← fill(j,k,p,d,filled,where,holes);
14:         end for
15:     end for
16: end if
17: Return NDM.
```

Algorithm 4 Pseudo-code of `fill()` [14].

Input: number of column (c), number of row (r), operation parameters (p), data sequence (d), number of updated items (f), starting position in data (w), number of holes (h)
Output: new value for NDM item

```
1: if f < p[5] and c ≥ p[4] and r ≥ p[3] then
2:     f++;
3:     if h = p[2] then
4:         h ← 0;
5:         w++;
6:         Return d[w mod d.length];
7:     else
8:         h++;
9:         Return 0.
10:    end if
11: end if
```

3.3. Evolutionary Algorithm

The common characteristic of all AE-based algorithms is the use of cooperative co-evolutionary GA (CCEGA) [17,18] to evolve AEPs, that is, to determine the number of operations (AE,AEEO), the type of each operation (AE), the parameters of the operations (all algorithms), the length of the data sequence (AE,AEEO), and its content (all algorithms). As already mentioned, the implementations of operations are predefined. According to CCEGA, each evolved component of AEP evolves in a separate population, that is, an AEP with n operations and the sequence of data evolves in $n+1$ populations (see Figure 2) [14] .

To construct a complete AEP, NDM, and finally, a network, the operations and the data are combined together according to the procedure applied in CCEGA. An individual (for example, an operation) from an evaluated population is linked to the best leader individuals from the remaining populations that evolved in all previous CCEGA iterations. Each population maintains the leader individuals, which are applied as building blocks of all AEPs constructed during the evolutionary process. In order to evaluate newborn individuals, they are combined with the leader individuals from the remaining populations [14].

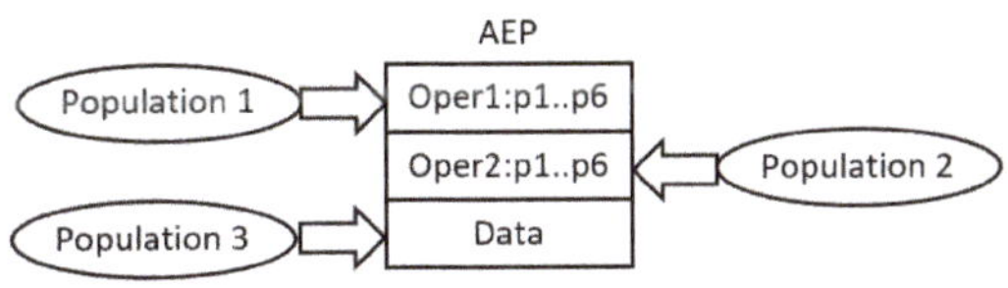

Figure 2. Evolution of AEPs according to CCEGA

Even though all the AE family applies CCEGA to evolve neural networks, HCAE does it in a different way from the remaining AE algorithms. In AE/AEEO, the networks evolve in one potentially infinite loop of CCEGA. Throughout the evolution, AEPs can grow or shrink, that is, they adjust their complexity to the task by changing in size. Each growth or shrinkage entails a change in the number of populations in which AEPs evolve. Unfortunately, such an approach appeared to be ineffective for a greater number of operations/populations. Usually, an increase in the number of operations/populations to three or more does not improve results, which is due to the difficulties in the coordination of a greater number of operations.

In contrast to AE/AEEO, HCAE is a hill climber whose each step is made by CCEGA (see Algorithm 5). A starting point of the algorithm is a blank network represented by a blank NDM (line (1)). Then, the network as well as NDM are improved in subsequent evolutionary runs of CCEGA (line (5)). Each next run works on the best network/NDM found so far by all earlier runs (each AEP works on its own copy of NDM), is interrupted after a specified number of iterations without progress (*MAX_ITER_NO_PROG*), and delegates outside, to the HCAE main loop, the best network/NDM that evolved within the run (tempNDM). If this network/NDM is better than those generated by earlier CCEGA runs, a next HCAE step is made—each subsequent network/NDM has to be better than its predecessor (line (7) [14]).

Algorithm 5 Evolution in HCAE [14].

Input: CCEGA parameters, for example crossover probability
Output: Neural network

```
1:  NDM ← 0;
2:  numberOfIter ← 0;
3:  fitness ← evaluation of NDM;
4:  while numberOfIter < maxEval and fitness < acceptedFitness do
5:      tempNDM ← CCEGA.run(NDM,MAX_ITER_NO_PROG);
6:      if tempNDM.fitness > fitness then
7:          NDM ← tempNDM;
8:          fitness ← tempNDM.fitness;
9:      end if
10:     numberOfIter ← numberOfIter + 1;
11: end while
12: Return Neural network decoded from NDM.
```

In order to avoid AE/AEEO problems with the effective processing of complex AEPs, HCAE uses constant-length programs of a small size. They include, at most, two operations and the sequence of data; the number of operations does not change over time. Such a construction of AEPs affects the structure of CCEGA. In this case, AEPs evolve in two or, at most, three populations; the number of populations is invariable. One population includes sequences of data, i.e., chromosomes data, whereas the remaining populations contain encoded operations, i.e., chromosomes operations. The operations are encoded as integer strings, whereas the data as real-valued vectors, which is a next difference between

HCAE and AE/AEEO that apply binary encoding. Both chromosomes operations and chromosomes data are of constant length.

In HCAE, like in AE/AEEO, the evolution in all the populations takes place according to simple canonical genetic algorithm with a tournament selection. The chromosomes undergo two classical genetic operators, i.e., one-point crossover and mutation. The crossover is performed with a constant probability P_c, whereas the mutation is adjusted to the current state of the evolutionary process. Its probability (P_m^d—probability of mutation in data sequences; P_m^o—probability of mutation in operations) grows once there is no progress for a time and it decreases once progress is noticed [14] .

The chromosomes data and chromosomes operations are mutated differently, and they are performed according to Equations (1) and (2) [14].

$$d_{new} = \begin{cases} d + \text{randU}(-a, a) \text{ if randU}(0,1) \leq P_m^d \\ d \text{ otherwise} \end{cases} \tag{1}$$

$$o_{new} = \begin{cases} o + \text{randI}(-b, b) \text{ if randU}(0,1) \leq P_m^o \text{ and} \\ \quad \text{randU}(0,1) \geq P_m^{o,zero} \\ 0 \text{ if randU}(0,1) \leq P_m^o \text{ and randU}(0,1) \leq P_m^{o,zero} \\ o \text{ otherwise} \end{cases} \tag{2}$$

where d—is a gene in a chromosome-data; o—is a gene in a chromosome-operation; randU($-a, a$)—is a uniformly distributed random real value from the range $< -a, a >$; randI($-b, b$)—is a uniformly distributed random integer value from the range $< -b, b >$; $P_m^{o,zero}$—is a probability of a mutated gene to be zero.

3.4. Complexity Analysis

Although algorithms no. 1, 2 and 3 present the traditional iterative implementation style, which is due to the ease of analysis of such algorithms, the actual HCAE implementation is parallel. This means that the algorithm can be divided into three parallel blocks executed one after the other, namely: the genetic algorithm (CCEGA + CGA), the AEP program and the evaluation of neural networks. The complexity of the algorithm can therefore be defined as O(O(CCEGA + CGA) + O(AEP) + O(Fitness)). The parallel implementation of the genetic algorithm requires, in principle, three steps, i.e., selection of parent individuals, crossover and mutation, which means that we obtain O(3). In addition, it also requires $l(n + n1n2)$ processors or processor cores, where l is the number of chromosomes in a single CCEGA population, $n1$ is the number of AEP operations, and n and $n2$ are the number of genes in chromosome data and chromosomes operations, respectively. The AEP program is executed in $n1$ steps (O($n1$)) and requires a maximum of Z processors/cores, where Z is the number of cells of the NDM matrix. The last block of the algorithm is the evaluation of neural networks, the computational complexity of which depends on the problem being solved. Ignoring the network evaluation, it can be concluded that the algorithm complexity is O(3 + $n1$) and requires $max(l(n + n1n2), Z)$ processors/cores.

4. Experiments

As already mentioned in the introduction, HCAE was tested on fourteen classification problems and one control problem against eight rival methods. First, HCAE(AE) and HCAE(AEEO) algorithms were tested on the two-spiral problem (TSP) and the inverted-pendulum problems (IPP) against original AE and AEEO. HCAE(AE) and HCAE(AEEO) are algorithms that evolve networks according to the HCAE algorithm depicted in Algorithm 5; however, at the same time, they apply AE and AEEO operations, binary encoding, variable-size chromosomes, and CCEGA mechanisms of adapting size of AEPs to a problem—AEPs can change their size throughout the evolution. The goal of this phase of the experiments was to examine the effectiveness of the new model of evolution applied in HCAE.

In the second stage of the experiments, HCAE(AE) and HCAE(AEEO) competed with HCAE(`Oper1`) and HCAE(`Oper2`) with the goal of testing the new operations applied in HCAE. Again, TSP and IPP constituted a testbed for the compared algorithms.

The third phase of the experiments was devoted to the comparison between HCAE, NEAT (C++ implementation of NEAT was taken from [60]), and CoSyNE (C++ implementation of CoSyNE was taken from [61]). Only the most effective variant of HCAE took part in the tests. As before, the task of the selected algorithms was to evolve neural solutions to TSP and IPP problems.

Finally, HCAE was also compared with four PSO-based algorithms considered in [27] and with backpropagation (BP). This time, the algorithms were put to the test on thirteen classification problems defined in the UCI machine-learning repository. The objective of both of the last phases of the experiments was to present HCAE against other algorithms in the field of NE.

4.1. Two-Spiral Problem

TSP, which is a well-known benchmark for binary classification, was selected as a starting problem for HCAE and its rivals. Even though it is not a new problem, its average complexity corresponds to the complexity of problems that can be effectively solved with small/medium-size neural networks and learning algorithms dedicated to such networks.

In TSP, the task is to split into (x, y) data points that form intertwined spirals which cannot be linearly separated into two classes. The learning set consists of 194 points, 97 for each spiral. To generate the first spiral, the following formulas can be used: $(x, y) = (r\cos(\phi), r\sin(\phi))$, $r = \frac{6.5(104-i)}{104}$, $\phi = \frac{i}{16}\pi$, $i = 0.96$. The second spiral can be obtained by the negation of first spiral coordinates, that is, by: $(-x, -y)$.

To solve the above-mentioned problem, feed-forward neural networks with two inputs, two outputs, and maximally thirty-six hidden neurons were applied. The inputs were fed with the (x, y) coordinates of data points, whereas outputs were responsible for identification, one output for one class. If the output signal of the first output neuron was greater than the output signal of the second output neuron, an input data point was assigned to the first class, otherwise, to the second class. All neurons in the networks used a hyperbolic tangent activation function.

When evolving the networks, the algorithms were allowed to make, maximally, 3,000,000 evaluations. In order to evaluate each evolved neural network, the following fitness function was applied:

$$F(ANN) = S + \frac{1}{1 + E(i)} \tag{3}$$

where S—is the number of correct classifications up to the first wrong decision; in the case of a wrong decision, the evaluation process was immediately interrupted; $E(i)$—is an error in i-th learning iteration in which the first wrong decision was made; in the learning process two classes were presented alternately to the network, which means that points from the first class even had indexes in the learning set ($i \bmod 2 = 0$), whereas points from the second class had odd indexes ($i \bmod 2 = 1$), o_1 and o_2 are outputs of the neural network; the formal definition of $E(i)$ is given below:

$$E(i) = \begin{cases} 100 \text{ if } o_1 = o_2 \\ abs(1 - o_1) + abs(-1 - o_2) \text{ if } i \bmod 2 = 0 \\ \quad \text{and } o_1 \neq o_2 \\ abs(1 - o_2) + abs(-1 - o_1) \text{ otherwise} \end{cases} \tag{4}$$

The fitness function (3) introduces additional difficulties compared with the original problem defined in [31]. Originally, in order to evaluate the network, classification results on all training points are used. This corresponds to the situation in which the learning algorithm has complete knowledge of the effectiveness of the network in the whole range of the input space (considered in the learning set). Meanwhile, according to (3), the learning algorithm is forced to rely on partial knowledge. If the network fails at some point, the evaluation process is interrupted with the consequence that the algorithm has only insight into fragmentary capabilities of the evaluated network and its evaluation can be misleading. For example, a network successful in the first n points will be more fit than a network successful in m points in total; $m > n$, however, in only k consecutive points from the top of the learning point list, $k < n$.

Error (4) has three options. The first option $E = 100$ drastically reduces the network fitness if $o_1 = o_2$. In the absence of this option, function (3) would have a deep local maximum for $o_1 = o_2$, at which the algorithm would often get stuck. The other two options are just the sum of the errors on each network output. For the first class, the ideal situation is $o_1 = 1$, $o_2 = -1$, while for the second class, $o_1 = -1$, and $o_2 = 1$.

4.2. Inverted-Pendulum Problem

In order to assess how HCAE copes with evolving recurrent neural networks (RANN) dedicated for control problems of average complexity, and to compare its performance in this regard with the performance of other algorithms, experiments in the inverted-pendulum problem (IPP) were carried out. Even though the original variant of IPP was defined quite a long time ago [20], the modified version applied in the experiments is enough to achieve the goal of the research mentioned above.

In this case, the networks deal with a wheeled cart moving on a finite length track and with two poles installed on the cart; one pole is shorter and lighter, whereas the second one is longer and heavier. The task of the networks is to indefinitely balance the poles and to keep the cart within track boundaries. To accomplish the task, the cart has to be pushed left or right with a fixed force. The decision about the direction and the strength of each move is made based on the information about the state of the cart-and-pole system. The complete state vector includes the following parameters: the position of the cart (x), the velocity of the cart ($\dot{x}$), angles of both poles (θ_1, θ_2), and angular velocities of both poles ($\dot{\theta}_1$, $\dot{\theta}_2$). To model behaviour of the cart-and-pole system, the following equations are used [20]:

$$\ddot{x} = \frac{F - \mu_c(\dot{x}) + \sum_{i=1}^{2} \widetilde{F_i}}{M + \sum_{i=1}^{2} \widetilde{m_i}} \tag{5}$$

$$\ddot{\theta}_i = -\frac{3}{4l_i}\left(\ddot{x}\cos\theta_i + g\sin\theta_i + \frac{\mu_{pi}\dot{\theta}_i}{m_i l_i}\right) \tag{6}$$

where $\ddot{x}$—acceleration of cart; $\ddot{\theta}_i$—acceleration of ith pole; F—force put to cart; M—mass of cart; m_i—mass of ith pole; l_i—half length of ith pole; μ_c—coefficient of friction of cart on track; μ_{pi}—coefficient of friction of ith pole's hinge; g—gravity.

$$\widetilde{F_i} = m_i l_i \theta_i^2 \sin\theta_i + \frac{3}{4} m_i \cos\theta_i \left(\frac{\mu_{pi}\dot{\theta}_i}{m_i l_i} + g\sin\theta_i\right) \tag{7}$$

$$\widetilde{m_i} = m_i\left(1 - \frac{3}{4}\cos^2\theta_i\right) \tag{8}$$

The state of the cart-and-pole system in subsequent points in time of simulation, i.e., every t seconds, where t is a step size, is determined by means of Euler's method:

$$x = x + t\dot{x} \tag{9}$$

$$\dot{x} = \dot{x} + t\ddot{x} \tag{10}$$

$$\theta = \theta + t\dot{\theta} \tag{11}$$

$$\dot{\theta} = \dot{\theta} + t\ddot{\theta} \tag{12}$$

To control the cart-and-pole system, the networks were fed with the state parameters scaled to the range $< -1, 1 >$. The force F used to control the system was calculated as a product of network output (one output neuron) and value 10. To make the task of the networks more difficult, they had access to only two parameters, i.e., x, and θ_1 (two input neurons). This means that they had no information about the shorter pole, and they did not know the direction of movement and velocity of the cart and the longer pole. To effectively control the cart-and-pole system, all missing information had to be reconstructed by the networks in the subsequent steps of the simulation.

The above-mentioned setting was not the only impediment the networks had to face. In contrast to the original problem, each network was evaluated four times, each time for other starting conditions, and each single evaluation was interrupted once the cart went off the path or any pole fell down, that is, exceeded a failure angle. In order for the task of the networks to be even harder in relation to the original problem [20], the failure angels for both poles were decreased to 8 degrees for the longer pole, and 20 degrees for the shorter one. The other experiment settings remained unchanged in relation to [20] and are given in Table 1.

The networks applied in the experiments had the following architecture: two inputs, one output, and, maximally, forty hidden units. When evolving the networks, the algorithms were allowed to make, maximally, 30,000 evaluations. In order to evaluate each evolved neural network, the following fitness function was applied:

$$F_{IP}(MANN) = \sum_{i=1}^{4} W_i \tag{13}$$

where W_i is the duration of ith simulation that was measured as the number of steps the cart-and-pole system remained controlled. The maximum duration of each single simulation was equal to 10,000 control steps, which means that the maximum fitness that could be achieved was 40,000. The task of function (13) is simply to obtain a network capable of holding the pendulum up for as long as possible in each of the four simulations.

Table 1. Parameter setting for inverted-pendulum problem.

Parameter	Description	Value
	length of path	4.8 m
l	length of pole	l_1 = 0.5 m l_2 = 0.05 m
m	mass of pole	m_1 = 0.1 kg m_2 = 0.01 kg
θ	angle of pole	$\theta_1 =< -8, 8 >$ deg, $\theta_2 =< -20, 20 >$ deg
	initial position of poles	$(\theta_1, \theta_2) =< (4,0), (-4,0), (0,4), (0,-4) >$deg
x	position of cart	$<-2.4, 2.4>$ m
	initial position of cart	0 m
M	mass of cart	1 kg
F	force applied to cart	$< -10, 10 >$N but no less than $\pm 1/256 \times 10$ N
g	gravity	9.8 m/s^2
μ_c	coefficient of friction of cart on track	0.0005
μ_p	coefficient of friction of ith pole's hinge	0.000002
t	step size	0.02 s

4.3. *Trajectory-Following Problem*

In this case, the task of the neural networks was to guide the AUV along a desired trajectory defined spatially and temporally. Given that the AUV trajectory is determined by a set of way points in 2D space (it is assumed that the AUV moves on the horizontal plane without depth control) and each straight segment between two way points has a march velocity (V_M) assigned, at each point in time, it is possible to determine a desirable position of the AUV and two errors in the position, i.e., the position error calculated along the trajectory—E_1, and the position error perpendicular to the trajectory—E_2. E_1 indicates the timing error, that is, whether the AUV is late or ahead of time, whereas E_2 corresponds to the distance to the right trajectory—see Figure 3a.

The neural network controls the vehicle by determining its heading H and speed V in such a way as to minimize the errors E_1 and E_2. The new parameters of the AUV movement are defined every 0.1 s. The heading of the vehicle is determined from point D (see Figure 3a), where the vehicle is currently located, to point C, which is the local destination of the AUV. To find point C, the neural network determines the distance d from point B towards the next way point.

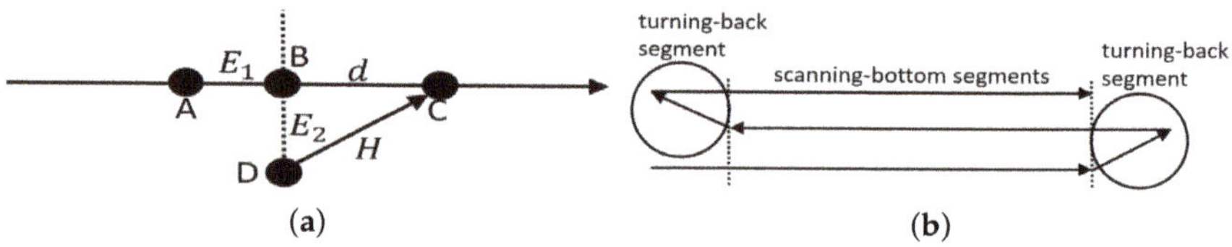

Figure 3. (**a**) A—desired position of the AUV, B—true position of the AUV projected on the trajectory, C—local destination point of the AUV, D—true position of the AUV, H—heading to point C, d—distance from point B to point C, E_1, E_2—errors, (**b**) trajectory of the vehicle.

Speed V is the sum of the marching speed V_M associated with each straight section of the vehicle route and speed V_1 determined by the neural network.

Both FFANN and RANN networks were used during the research. In both cases, the networks were supplied with E_1 and E_2 errors. The E_1 error was positive if the vehicle was late and negative—if it was in a hurry. The E_2 error was always positive. FFANN networks had four inputs, two inputs corresponding to errors from the current time moment and two inputs corresponding to errors from the previous time moment. The RANN networks had only two inputs corresponding to the errors that occurred at the current time.

Apart from two output neurons (d and V_1) and two or four input neurons, both types of networks had 10 hidden neurons. All the neurons had the same activation function—the hyperbolic tangent function.

The tests were carried out in simulation conditions using a kinematic vehicle model implemented in the MOOS-IvP application called uSimMarine [62]. This model has a number of parameters determining the speed of the vehicle, its manoeuvrability or inertia. The description of the parameters with their values used during the simulation can be found at the end of the paper, in Appendix B.

During the simulation, the behaviour of the vehicle was also influenced by the low-level control system, whose task was to convert high-level decisions made by the neural network into low-level decisions for the propellers (engine) and rudder of the vehicle. The low-level system consisted of four PID controllers, each with three parameters: P, I, and D, the values of which are given in Appendix B.

The task of the AUV was to follow a lawn-mower trajectory consisting of eight way points and seven straight segments, i.e., three 100-m long scanning-bottom segments and four turning-back segments—see Figure 3b. The march speed V_M along the scanning-bottom segments amounted to 1 m/s, whereas, along the turning–back segments, it amounted to 0.5 m/s.

For the conditions of the experiments to be maximally similar to the real ones, the AUV was subject to a sea current which had random duration, direction and strength. It appeared regularly every 6 s at most, and its maximum duration amounted to 2 s. The direction of the current amounted to $45° \pm 10°$, whereas its strength amounted, maximally, to 0.8 m/s.

Each of the evolutionarily formed networks was evaluated k times, each time with a different influence of the sea current; k_{max} = 40. The evaluation of the network was always interrupted ($k < k_{max}$) if the error E_1 or E_2 exceeded the threshold value equal to 3 m and 2 m, respectively. The following fitness function was used to evaluate the networks:

$$F_{TFP}(ANN) = \frac{1}{1 + E_1^{max} + 5E_2^{max}} \tag{14}$$

where E_1^{max} and E_2^{max} are the maximum errors obtained during the entire network-evaluation process, while the AUV was moving along the scanning-bottom segments of the trajectory. Short turning-back segments were not included in the evaluation. In order to obtain the precise trajectory of the vehicle along the entire scanning-bottom segments, function (14) was oriented to minimize maximum errors. Error E_2^{max} has a greater influence on function (14), which means that more emphasis is on minimizing the spatial deviation from the trajectory than the temporal deviation.

4.4. Results in the First Phase of Experiments

As already mentioned, the objective of this phase was to evaluate the effectiveness of the new model of evolution applied in HCAE. To this end, original AE and AEEO were compared with HCAE(AE) and HCAE(AEEO) which are variants of AE/AEEO with the HCAE model of evolution.

At the beginning, AE and AEEO were tuned to both testing problems (TSP and IPP), mainly by adjusting the probability of mutation. Other parameters were either manually tailored to the tasks, for example: size of NDM-operation = 10 × 11 − 10 neurons in total, one extra column for biases, crossover probability $P_c = 0.5$, size of each population = 100, size of tournament = 3, length of chromosome-data = 80, initial number of operations = 1, length of chromosome-operations = 12 or 6, number of CCEGA iterations without progress followed by a change of AEP size = 100,000 (TSP) or 2500 (IPP), or they were the same as those applied in [15,16].

In order to compare the algorithms, each of them were run thirty times. Then, thirty of the most effective networks from all the runs, one network per run, were applied to compare the algorithms. Four criteria were used to measure the performance of each algorithm, i.e., the average, maximum, and minimum fitness, and the total number of successes in the learning process (TSP—all learning data points classified correctly, IPP—the poles successfully balanced in all attempts). The final results of the tests are summarized in Tables 2 and 3.

Table 2. Results of the first phase of experiments—TSP.

Method	Mean	Std	Maximum	Minimum	Successes
AEEO	115.3	32.7	194	65.4	1
HCAE(AEEO)	124.4	28.5	194	62.1	1
AE	105.7	18.7	145.6	62.3	0
HCAE(AE)	155.3	32.2	194	66.5	5

Table 3. Results of the first phase of experiments—IPP.

Method	Mean	Std	Maximum	Minimum	Successes
AEEO	10,289	1128	12,530	8445	0
HCAE(AEEO)	12,611	1851	13,481	9560	0
AE	9620	1292	10,571	7402	0
HCAE(AE)	36,522	7845	40,000	1202	24

Both tables clearly show the positive effect of HCAE model of evolution on the performance of AE/AEEO. The greatest increase in efficiency is noticeable when comparing results of HCAE(AE) and AE. A smaller improvement is observed in the case of HCAE(AEEO) and AEEO.

The result proves, at least with regard to evolutionary neural networks and the AE family, that the model in which the evolution is conducted in many separate evolutionary runs and each earlier run produces input data for its successors, when the solutions grow gradually in many subsequent increments, and when the complexity of the genome is replaced with a sequence of simple genomes, outperforms a traditional model in which the complexity of the problem corresponds to the complexity of the genome and the whole problem has to be solved within one evolutionary super run.

Two main factors seem to decide the success of the HCAE model: firstly, the strategy of gradual growth, and secondly, the simplicity of representation. However, as the above results show, the success depends on the interference between subsequent evolutionary runs and between operations inside AEP—the greater the interference, the smaller the success. To put it simply, if operations within one AEP, or AEPs executed one after another, strongly interfere with each other, it is difficult to achieve the effect of gradual growth.

One objective of the HCAE model is to simplify AEPs as much as possible with the task to simplify cooperation between operations derived from different populations. It is simply much easier to match two or three components of a solution together than to do it with, say, five or ten components. In other words, the interference between two or three operations is smaller than the interference between five or ten operations. To make matters worse, the interference can also apply to operations from different evolutionary runs.

However, the interference between operations depends not only on the number of operations in AEP but also on the range of operations, which is different in HCAE(AE) and HCAE(AEEO). In the first case, the operations work locally in a selected region of NDM, whereas in the second case, they are allowed to update each element of NDM. In consequence, operations applied in HCAE(AE) rarely interfere with each other, whereas those used in HCAE(AEEO) do it much more often. This, in turn, means that the strategy "little by little" implemented in HCAE has a greater chance for success in combination with AE than AEEO.

4.5. Results in the Second Phase of Experiments

In this phase, the objective was to evaluate the effectiveness of the remaining solutions applied in HCAE compared with their originals from AE and AEEO. The results of this phase are given in Tables 4 and 5.

Table 4. Results of the second phase of experiments—TSP.

Method	Mean	Std	Maximum	Minimum	Successes
HCAE(AEEO)	124.4	28.5	194	62.1	1
HCAE(`Oper1`)	128.3	26.6	194	68.2	3
HCAE(AE)	155.3	32.2	194	66.5	5
HCAE(`Oper2`)	169.2	18.4	194	125.8	10

Table 5. Results of the second phase of experiments—IPP.

Method	Mean	Std	Maximum	Minimum	Successes
HCAE(AEEO)	12,611	1851	13,481	9560	0
HCAE(`Oper1`)	12,023	6224	40,000	8966	1
HCAE(AE)	36,522	7845	40,000	1202	24
HCAE(`Oper2`)	40,000	0	40,000	40,000	30

As it turned out, new solutions implemented in HCAE(`Oper2`) appeared to have an advantageous influence on the algorithm performance compared with HCAE(AE). In turn, for those applied in HCAE(`Oper1`), if they brought any positive effect it is visible only in TSP. In IPP, the results of HCAE(`Oper1`) and HCAE(AEEO) are roughly at the same level.

The new solutions implemented in HCAE can be divided into two categories, i.e., the ones that are common for both HCAE variants and the ones that are variant-specific. Real and integer encodings, instead of binary encoding, and fixed-length chromosomes belong to the first category, whereas new operations belong to the second category. The difference in results of HCAE(`Oper1`) against HCAE(AEEO) and HCAE(`Oper2`) against HCAE(AE) indicates that solutions from the first category do not constitute a valuable replacement for original solutions. The same applies to `Oper1`, the added value of this operation, if it exists, seems to be rather symbolic. A different situation is with `Oper2`, in this case, an increase in efficiency is clearly noticeable.

There are only two differences between Oper2 and operations used in AE which, as it turned out, had an influence on the results. Firstly, AEPs in AE can contain three different types of operations: modification of a single row in NDM, modification of a single column in NDM, and modification of a rectangular patch in NDM, whereas in HCAE(Oper2) there is no choice; each AEP consists of operations of only one type. This way, evolution in HCAE(AE) has a slightly harder task than in HCAE(Oper2). It has to match operations to the task, and to adjust parameters to selected operations. In HCAE(Oper2), in turn, operations do not differ in implementation, so the only difficulty is to appropriately set the parameters.

However, it seems that, in this case, it is more significant how individual operations modify NDM, and, consequently, the network. In HCAE(AE), all the three operations have no possibility of deactivating connections by setting their weights to zero, with the effect that all networks produced by HCAE(AE) are densely connected. They simply copy connection weights from the data sequence to NDM; zero weight could only appear in the network if it also appeared in the data sequence. To overcome this problem, each Oper2 periodically fills NDM with zeros; they appear in the matrix alternately with connection weights from the data sequence.

To sum up, the experiments in this phase showed that two factors are crucial for HCAE success: firstly, a local range of the operations, and secondly, their ability to establish sparsely connected sub-networks.

4.6. Results in the Third Phase of Experiments

The objective of this phase was to compare HCAE with other algorithms from the field of NE. Two algorithms were selected as a point of reference for the proposed algorithm, i.e., NEAT and CoSyNE. Both rivals represent the same class of algorithms as HCAE, that is, the algorithms meant for constructing small/medium-size neural networks.

Before the tests, NEAT and CoSyNE, like all the previous algorithms, were also tuned to testing problems. Again, TSP and IPP constituted a testbed for the compared algorithms. A detailed parameter setting after the tuning process is given in the structures `mutation_rate_container` and `cosyneArgs` in Appendices A and B.

In contrast to the AE family, the evolution in NEAT took place in a single population with 300 individuals. In HCAE, AE and AEEO, the evolution is conducted in many populations, at least in two, and each population has 100 individuals, which means 200 or 300 network evaluations in each evolutionary iteration. Meanwhile, NEAT works in a single population, and the number of individuals in the population corresponds to the number of evaluations and networks generated in a single evolutionary iteration. In order to equalize the chances of NEAT, and HCAE, the number of individuals in a single NEAT population was the same as the maximum number of individuals in all HCAE populations. Moreover, in order for NEAT to evolve networks of the same maximum size as the networks produced by HCAE (TSP: 2 input, 2 output, and a maximum of 36 hidden neurons, IPP: 2 input, 1 output, and a maximum of 40 hidden neurons), it was necessary to slightly modify the C++ code of NEAT.

The conditions of the tests for CoSyNE were the same as for the remaining algorithms, meaning the number of hidden units was set to 36 (TSP) or 40 (IPP), whereas the number of network evaluations in one evolutionary generation was equal to 300.

The results of the experiments in this phase, after thirty runs of each algorithm, are given in Table 6 (The implementation of CoSyNE provides only one type of feed-forward network, namely, multilayered perceptron with one hidden layer; the table includes results for this type of network) and Table 7 (The implementation of CoSyNE used in the tests offers five different types of recurrent neural network: SimpleRecurrentNetwork, SecondOrderRecurrentNetwork, LinearRecurrentNetwork, FullyRecurrentNetwork, and FullyRecurrentNetwork2; the table includes results of SecondOrderRecurrentNetwork which appeared to be the most effective).

Table 6. Results of experiments on two-spiral problem: average, maximum, and minimum fitness, and the number of runs ended with success.

Method	Mean	Std	Maximum	Minimum	Successes
HCAE(`Oper2`)	169.2	18.4	194	125.8	10
NEAT	122.3	5.6	132.5	112.5	0
CoSyNE	106.5	6.7	112.1	96.2	0

Table 7. Results of experiments on inverted-pendulum problem: average, maximum, and minimum fitness, and the number of runs ended with success, networks with 40 hidden units.

Method	Mean	Std	Maximum	Minimum	Successes
HCAE(`Oper2`)	40,000	0	40,000	40,000	30
NEAT	9461	954	10,368	7511	0
CoSyNE	10,563	8.2	10,568	10,545	0

As Table 6 shows, the only algorithm that coped with the TSP is HCAE(`Oper2`) (example spirals generated by HCAE(`Oper2`) networks are depicted in Figure 4). In this case, ten runs ended with fully successful networks, and the average result amounts to 169. NEAT and CoSyNE, like AE and AEEO, appeared to be insufficiently effective to evolve even a single fully successful neural network. All NEAT and CoSyNE runs got stuck in more or less the same region of fitness function, that is, for NEAT, between 112 and 132 (example spirals generated by NEAT networks are depicted in Figure 5), whereas for CoSyNE, between 96 and 106.

Figure 4. Spirals generated by ten fully successful neural networks evolved by HCAE(`Oper2`).

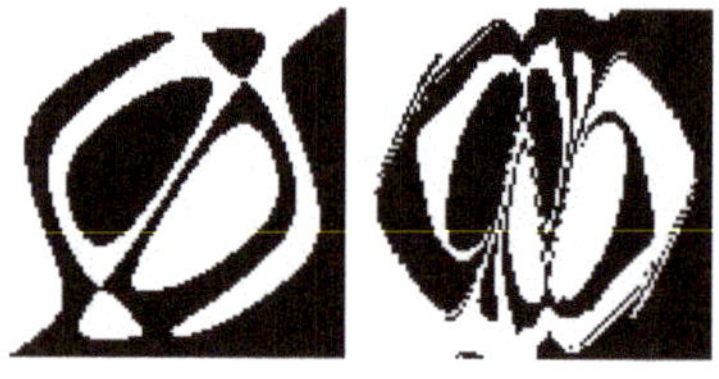

Figure 5. Example spirals generated by NEAT networks.

In IPP, HCAE(`Oper2`) again outperformed other methods. Generally, it had no problems with evolving fully successful networks that were able to balance the cart-and-pole-system for the maximal number of iterations in all four attempts. To evolve a fully successful network, it needed only 1386 network evaluations on average and 310 minimum.

Meanwhile, the rivals of HCAE, like in the previous case, could not successfully cope with the problem. The best result in both cases corresponds to the beginning of the second out of four simulations, that is, to the area of fitness function equal to 10,000. There were very few runs, in this case, that exceeded 11,000. As before, this result is very similar to those obtained by AE and AEEO.

Generally, two factors can be responsible for the difference in performance between HCAE and its rivals, i.e., the evolutionary algorithm and the encoding method. However, the most likely "suspect", in this case, seems to be the evolutionary algorithm. As already mentioned, regardless of the testing problem, the results of NEAT and CoSyNE are very similar to those obtained by AE and AEEO. Meanwhile, the only thing that all the algorithms have in common is the traditional evolutionary model. This suggests that, like in the case of AE and AEEO, the decisive factor that prevented NEAT and CoSyNE from obtaining better results is the model of the evolution applied in both algorithms.

In turn, the influence of the encoding method is visible in the networks' architecture: most networks evolved by NEAT and CoSyNE were densely connected with a maximum number of neurons. In the early stages of the evolution, NEAT produces "light" networks from "light" genetic representations. However, the representations and networks quickly become more and more complex and their effective processing more and more difficult. Exactly the same problem is observed in AE and AEEO, where the growth in genotypic complexity decreases the effectiveness of the evolution. In turn, CoSyNE, from the very beginning, evolves fully connected neurons of a fixed architecture and does not have any effective mechanism for producing sparse networks.

Meanwhile, the networks produced by HCAE(`Oper2`) often included fewer neurons than the assumed maximum size, and were very often sparse; NDMs of these networks contained many spacious holes filled with zeros.

4.7. Tests on Datasets from UCI Machine Repository

The most effective variant of HCAE was also compared with LFPSO, PSOLF, LPSONS, and BP on thirteen classification benchmarks from the UCI machine repository. To this end, HCAE was put to the test in the same conditions as those used in the experiments reported in [27]. For each classification benchmark, the following setting was applied: 30 runs of the algorithm, neural networks with five hidden units, termination of the algorithm after 13,000 network evaluations, mean square error (MSE) as a minimized objective function, training set—80% and testing set—20% of data instances (in [27]: 70%—testing set, 10% validation set, and 20% testing set).

Since the above conditions of the experiments assume a quick effect (only 13,000 network evaluations) with the use of small neural networks (only five hidden units), HCAE had to be adjusted to these conditions through appropriate parameter setting. First, the size of each NDM corresponded to networks with five hidden units. Second, it was assumed that, in order to evolve effective networks with a small number of neurons, the networks cannot be both small and sparse. In consequence, `Oper2` was modified so as not to generate holes between consecutive values in NDMs. Third, in order to quickly evolve effective networks, the stagnation of the evolutionary process in CCEGA should be detected very quickly; the first symptoms of the stagnation should interrupt one evolutionary run and start a next run. To achieve such an effect, the interruption of each single CCEGA run took place after five iterations without progress.

Tables 8 and 9 show the comparison of all the five methods in terms of the training and testing accuracy that is defined in [27] as a rate of correctly identified data instances to all instances derived either from the training or the testing set, respectively. In both tables, the best results for each benchmark are in bold.

The tables show that, in spite of unfavorable conditions for HCAE which is rather dedicated to problems which require greater networks and more effort to train them, it outperforms all rival methods in terms of both the training and testing accuracy. The number of wins (the best results in bold) for each algorithm and for each benchmark is as follows: training phase (BP = 0, PSOLF = 2, LFPSO = 3, LPSONS = 11, HCAE = 12), testing phase (BP = 6, PSOLF = 1, LFPSO = 2, LPSONS = 7, HCAE = 13), total (BP = 6, PSOLF = 3, LFPSO = 5, LPSONS = 18, HCAE = 25).

Table 8. Classification training accuracy (results of BP, PSOLF, LFPSO, and LPSONS taken from [27]).

Dataset	Param.	BP	PSOLF	LFPSO	LPSONS	HCAE
Breast Cancer	Best	0.9760	0.9790	0.9806	0.9811	**0.9821**
	Mean	0.9721	0.9721	0.9753	0.9759	**0.9786**
	Std	0.0041	0.0043	0.0046	0.0041	0.0042
Liver	Best	0.7837	0.7824	0.7740	0.7866	**0.7898**
	Mean	0.7393	0.7297	0.7535	0.7652	**0.7816**
	Std	0.0193	0.0248	0.0183	0.0177	0.0101
Wine	Best	0.9908	**1.0000**	**1.0000**	**1.0000**	**1.0000**
	Mean	0.9872	0.9832	0.9968	**1.0000**	0.9941
	Std	0.0047	0.0096	0.0056	0.0000	0.0127
Australian	Best	0.8929	0.8902	0.8944	**0.9586**	0.8804
	Mean	0.8791	0.8712	**0.8819**	0.8818	0.8736
	Std	0.0110	0.0090	0.0091	0.0279	0.0039
Hepatitis	Best	0.9853	**1.0000**	**1.0000**	**1.0000**	0.9200
	Mean	0.9721	0.9619	0.9768	**0.9885**	0.8888
	Std	0.0162	0.0305	0.0147	0.0128	0.0222
Heart	Best	**0.9392**	0.8956	0.9086	0.9183	0.9023
	Mean	**0.9116**	0.8719	0.8764	0.8894	0.8824
	Std	0.0175	0.0113	0.0170	0.0147	0.0135
Blood	Best	0.7934	0.7881	0.7977	0.8015	**0.8030**
	Mean	0.7512	0.7674	0.7807	0.7904	**0.7949**
	Std	0.0182	0.0134	0.0145	0.0077	0.0059
Iris	Best	0.9921	0.9809	0.9904	**1.0000**	0.9833
	Mean	0.9635	0.9519	0.9647	**0.9752**	0.9691
	Std	0.0323	0.0495	0.0175	0.0163	0.0209
Credit	Best	0.9046	0.8905	0.8993	**0.9059**	0.9005
	Mean	0.8825	0.8822	0.8840	0.8877	**0.8921**
	Std	0.0178	0.0065	0.0084	0.0114	0.0060
Seeds	Best	0.9775	0.9640	0.9640	**0.9784**	0.9761
	Mean	0.9587	0.9489	0.9566	**0.9647**	0.9623
	Std	0.0123	0.0115	0.0106	0.0110	0.0158
Haberman	Best	0.7810	0.7803	0.7850	0.7803	**0.7836**
	Mean	0.7635	0.7616	0.7612	0.7654	**0.7752**
	Std	0.0215	0.0136	0.0186	0.0134	0.0061
Balance	Best	0.8921	0.8949	0.9640	0.8995	**0.9023**
	Mean	0.8705	0.8741	0.8849	0.8853	**0.8909**
	Std	0.0189	0.0097	0.0104	0.0090	0.0111
Diabetes	Best	0.7163	0.7099	0.7229	**0.7345**	0.7296
	Mean	0.7082	0.7093	**0.7191**	0.7109	0.7079
	Std	0.0147	0.0093	0.0084	0.0109	0.0233

However, if the best accuracy is considered separately from the mean accuracy, the results are slightly different: training phase—best (BP = 0, PSOLF = 1, LFPSO = 2, LPSONS = 7, HCAE = 6), testing phase—best (BP = 4, PSOLF = 1, LFPSO = 2, LPSONS = 5, HCAE = 4), total—best (BP = 4, PSOLF = 2, LFPSO = 4, LPSONS = 12, HCAE = 10), training phase—mean (BP = 0, PSOLF = 0, LFPSO = 2, LPSONS = 4, HCAE = 6), testing phase—mean (BP = 2, PSOLF = 0, LFPSO = 0, LPSONS = 2, HCAE = 9), total—mean (BP = 2, PSOLF = 0, LFPSO = 2, LPSONS = 6, HCAE = 15). They show the high stability of HCAE which, regardless of the problem, the number of problem parameters, the number of data instances, and the starting point of the evolution, is able to quickly produce effective neural networks in almost every run. However, the above-mentioned result may also suggest some problems of HCAE with exploitation and finding globally optimal solutions. While the average effectiveness of HCAE is very high compared with other algorithms, its best networks were often inferior to the best networks of the rivals, especially to the networks of LPSONS.

Table 9. Classification testing accuracy (results of BP, PSOLF, LFPSO, and LPSONS taken from [27], *—a table in [27] includes a value 0.7315 instead of 0.6315, however, due to the fact that 0.7315 is greater than 0.68 which is the best value for Diabetes dataset, the value 0.7315 was considered to be a typo and it was replaced with 0.6315.

Dataset	Param.	BP	PSOLF	LFPSO	LPSONS	HCAE
Breast Cancer	Best	0.9748	0.9756	0.9707	**0.9756**	0.9640
	Mean	0.9529	0.9590	0.9590	**0.9648**	0.9630
	Std	0.0920	0.0116	0.0076	0.0072	0.0083
Liver	Best	**0.8427**	0.7323	0.7058	0.7353	0.7971
	Mean	0.6558	0.6469	0.6539	0.6638	**0.7893**
	Std	0.0495	0.0448	0.0647	0.0450	0.0182
Wine	Best	0.9259	0.9811	0.9811	**1.0000**	0.9714
	Mean	0.8667	0.9377	0.9413	0.9509	**0.9714**
	Std	0.0531	0.0427	0.0346	0.0328	0.0101
Australian	Best	0.8985	0.8985	0.8888	0.8985	**0.9130**
	Mean	0.8629	0.8604	0.8623	0.8635	**0.9045**
	Std	0.0325	0.0263	0.0184	0.0223	0.0081
Hepatitis	Best	**0.9655**	0.9117	0.8529	0.9118	0.9000
	Mean	**0.9211**	0.7913	0.7911	0.7941	0.8722
	Std	0.0462	0.0801	0.0426	0.0399	0.0194
Heart	Best	**0.9870**	0.8314	0.8764	0.8539	0.9502
	Mean	0.8293	0.8024	0.8258	0.8112	**0.8311**
	Std	0.0265	0.0183	0.0348	0.0264	0.0322
Blood	Best	0.8154	0.7991	**0.8214**	0.8125	0.8053
	Mean	0.7739	0.7713	0.7746	0.7799	**0.7908**
	Std	0.0425	0.0248	0.0339	0.0244	0.0122
Iris	Best	0.9565	0.9777	**1.0000**	**1.0000**	**1.0000**
	Mean	0.9230	0.9155	0.9711	0.9755	**0.9944**
	Std	0.0553	0.0701	0.0298	0.0286	0.0299
Credit	Best	**0.9529**	0.8775	0.8775	0.8826	0.8905
	Mean	**0.9329**	0.8418	0.8250	0.8290	0.8725
	Std	0.220	0.0259	0.0587	0.0621	0.0079
Seeds	Best	0.9808	**1.0000**	0.9833	**1.0000**	0.9761
	Mean	0.9215	0.9416	0.9250	0.9250	**0.9730**
	Std	0.0393	0.0345	0.0326	0.0479	0.0080
Haberman	Best	0.7872	0.7826	0.7826	**0.8043**	0.7540
	Mean	0.7416	0.7347	0.7360	**0.7500**	0.7453
	Std	0.0361	0.0370	0.0455	0.0251	0.0081
Balance	Best	0.8635	0.9090	0.9037	0.9144	**0.9186**
	Mean	0.8239	0.8883	0.8716	0.8883	**0.9086**
	Std	0.0424	0.0173	0.0232	0.0200	0.0111
Diabetes	Best	0.6800	0.7043	0.6795	0.7092	**0.7391**
	Mean	0.6315 *	0.6392	0.6078	0.6383	**0.7159**
	Std	0.0369	0.0233	0.0402	0.0221	0.0213

4.8. Results of Experiments on the Trajectory-Following Problem

The aim of this phase of research was to verify the ability of HCAE to solve real problems requiring neural networks of small to medium size. Since one of the target applications of HCAE are control problems, it was decided that the real problem in this phase of the research would be to control a complex non-linear object, namely, an underwater vehicle. The rivals of the proposed algorithm in this case were DE and PSO, whose parameter setting is in Appendix A, at the end of the paper. The task of each of the algorithms was the construction of the FFANN and RANN networks for the TFP problem described in Section 4.3. Each of the algorithms was run 30 times and, during each run, it was possible to evaluate a maximum of 3,000,000 neural networks. The test results in the form of E_1^{max} and E_2^{max} errors are presented in Table 10, and an example of the AUV trajectory is shown in Figure 6.

Table 10. Results of experiments on TFP.

	HCAE(FF)	HCAE(R)	DE(FF)	DE(R)	PSO(FF)	PSO(R)
Mean E_1^{max}	0.59	0.58	0.94	0.73	1.12	1.05
Max $E_1 max$	0.64	0.62	0.96	0.81	1.32	1.33
Std $E_1 max$	0.04	0.03	0.01	0.05	0.05	0.04
Mean E_2^{max}	0.44	0.45	0.59	0.72	0.82	0.79
Max E_2^{max}	0.52	0.5	0.65	0.75	0.95	0.99
Sts E_2^{max}	0.03	0.02	0.05	0.06	0.03	0.05

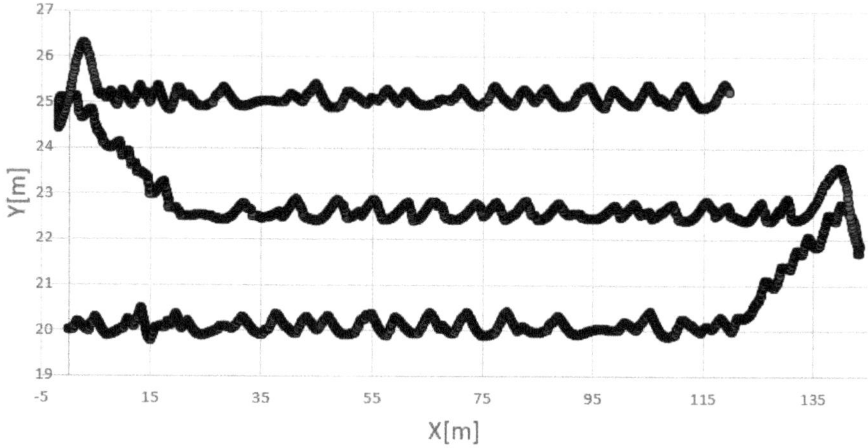

Figure 6. Example AUV trajectory.

As can be seen, also in this case, the HCAE outperforms its rivals. The average and maximum errors E_1^{max} and E_2^{max} generated for all the 30 runs of the algorithm are definitely smaller than the errors obtained for the other algorithms, which proves that HCAE is able to very effectively solve not only artificial problems but also real ones. Moreover, the obtained result confirms the high efficiency of the new model of the evolution of neural networks proposed in the paper. The algorithms used in DE and PSO along with the direct network coding method used in them turned out to be less effective.

5. Directions of Further Research

The current variant of HCAE is a classic hill climber which assumes that only better NDMs are accepted in the following algorithm iterations. In a future variant, a solution from simulated annealing can be borrowed which allows the acceptance of non-improving steps. In HCAE, the acceptance of worse matrices would depend on the phase of the algorithm. In early phases, they would be accepted with a greater probability, whereas in later phases, they would be very unlikely.

Instead of developing a single leader network, the algorithm, at each evolutionary step, can also work on a number of networks simultaneously. The increase in the number of evolved networks should have a positive effect on the chance of finding the global optimum. In the extreme case, a solution can also be imagined in which the number of programs constructed in a single evolutionary iteration is equal to the number of leader networks, and each program develops its own network.

6. Conclusions

The paper presented a novel neuro-evolutionary algorithm called Hill Climb Assembler Encoding. The algorithm encodes a neural network in the form of a matrix which is filled with an evolutionary formed program called assembler encoding program. Like every assembler program, the AEP is composed of two parts, i.e., a part with implementation and a part with a data sequence. The implementation part includes operations whose code is fixed—it does not evolve. Each operation is supplied with a vector of parameters and a sequence of data. Both data and the parameters are shaped in the evolutionary way. Their evolution is conducted according to the CCEGA algorithm, one population includes data sequences, whereas vectors of parameters are processed in two other populations.

The HCAE has two variants which differ in applied operations. One variant uses so-called ANN operations, which are small neural networks whose task is to form the matrix representing a resultant neural network. The ANN operations are supplied with sequences of data. Another variant of HCAE does not use ANN operations; instead, it simply copies data directly to the matrix. Where data is copied and how much data is copied is determined by the operation.

HCAE was tested on fourteen classification problems, including the two-spiral problem, which is a well-known benchmark for binary classification and is regarded as extremely challenging, and on the inverted-pendulum problem which is a well-known control benchmark. In the tests, apart from HCAE, eight other algorithms were also used for comparison purposes, i.e., two predecessors of HCAE: essembler encoding, and assembler encoding with evolvable operations, and six other methods: Neuro-Evolution of Augmenting Topologies which is a well-known state-of-the-art NE method, Cooperative Synapse Neuro-Evolution, Levy Flight Particle Swarm Optimization (LFPSO), modified LFPSO, Mantegna Levy Flight, Particle Swarm Optimization, and Neighbourhood Search, and the classic Back-Propagation algorithm.

The experiments revealed that HCAE outperforms all rival methods. The main factor that decided the high effectiveness of HCAE compared with other algorithms is the gradual growth in the networks and neuro-evolution conducted in a reduced search space. HCAE constructs the networks incrementally, little by little, by gradually adding new neurons and connections to the best network found so far. Each increment is a product of other evolutionary runs, each of which processes very simple HCAE programs.

In contrast, in the remaining algorithms, we deal with a one-to-one relation between genotype and phenotype, resulting in each algorithm being responsible for finding a complete network in the genotype space. If the space is large and the problem is abundant with local minima, finding the optimum is not a trivial task. This applies particularly to direct methods such as NEAT or CoSyNE, which do not scale well with larger networks.

The HCAE variant, which constructs neural networks by locally expanding them with concentrations of neurons and connections, appeared to be the most effective. Another characteristic of this variant is extreme simplicity compared with other HCAE variants and other algorithms. To form a neural network, it simply copies data directly to the matrix representation of the network.

The global nature of ANN operations seems to be the main reason for the problems of the HCAE variant based on these type of operations with generating optimal networks. HCAE assumes a slow incremental growth in the networks in many successive evolutionary runs, whereas, ANN operations disprove this assumption with serious changes introduced to the networks, which can be very destructive for the connection schemes found earlier.

Funding: The paper is supported by Polish Ministry of Defence within the framework of the program entitled "Research Grant".

Institutional Review Board Statement: Not applicable.

Data Availability Statement: Not applicable.

Conflicts of Interest: The authors declare no conflict of interest.

Abbreviations

The following abbreviations are used in this manuscript:

HCAE	Hill climb assembler encoding
HCMAE	Hill climb assembler encoding
ANN	Artificial neural network
FFANN	Feed-forward ANN
RANN	Recurrent ANN
AE	Assembler encoding
AEEO	Assembler encoding with evolvable operations
NDM	Network definition matrix
AEP	Assembler encoding program
TSP	Two-spiral problem
IPP	Inverted-pendulum problem

Appendix A. Structures Including Parameters of NEAT

```
    typedef struct {
double connection_mutate_chance = 0.6;
double perturb_chance = 0.90;
double crossover_chance = 0.7;
double link_mutation_chance = 2.0;
double node_mutation_chance = 0.50;
double bias_mutation_chance = 0.40;
double step_size = 0.1;
double disable_mutation_chance = 0.4;
double enable_mutation_chance = 0.2;
} mutation_rate_container;

typedef struct {
unsigned int population = 300;
double delta_disjoint = 2.0;
double delta_weights = 0.4;
double delta_threshold = 1.3;
unsigned int stale_species = 15;
} speciating_parameter_container;
```

Appendix B. Structures Including Parameters of CoSyNE

```
    typedef struct {
int numpops = 36;
int popsize = 300;
int init_popsize = 0;
int numevals = 1;
int stagnation = 20;
int nettype = 0; - feed-forward ANN
bool netlevel = false;
int netlevelsize = 20;
bool grownets = true;
double mutation = 0.2;
bool perc = false;
int loci = 1;
double neutral = 0.1;
bool proportional = false;
double shift = 0.1;
double wtrange = 10.0;
int crossover_type = 1;
```

```
} cosyneArgs;
```

DE:

```
size of population = 300,
differential weight: F = 0.01,
mutation probability: Cr = 0.3,
length of chromosomes in FFANNs = 120 genes,
length of chromosomes in RANNs = 196 genes
```

PSO:

```
PSO: size of population = 300,
inertia rate: θ = 0.1,
learning rates: α = 2, β = 2,
length of chromosomes in FFANNs = 120 genes,
length of chromosomes in RANNs = 196 genes
```

References

1. Rempis, C.W. Evolving Complex Neuro-Controllers with Interactively Constrained Neuro-Evolution. Ph.D. Thesis, Osnabruck University, Osnabruck, Germany, 2012.
2. Hwangbo, J.; Sa, I.; Siegwart, R.; Hutter, M. Control of a quadrotor with reinforcement learning. *IEEE Robot. Autom. Lett.* **2017**, *2*, 2096–2103. [CrossRef]
3. Koch, W.; Mancuso, R.; West, R.; Bestavros, A. Reinforcement learning for UAV attitude control. *arXiv* **2018**, arXiv:1804.04154.
4. Pi, C.-H.; Hu, K.-C.; Cheng, S.; Wu, I.-C. Low-level autonomous control and tracking of quadrotor using reinforcement learning. *Control. Eng. Pract.* **2020**, *95*, 104222. [CrossRef]
5. Praczyk, T. Neural anti-collision system for Autonomous Surface Vehicle. *Neurocomputing* **2015**, *149*, 559–572. [CrossRef]
6. Praczyk, T. Neural collision avoidance system for biomimetic autonomous underwater vehicle. *Soft Comput.* **2020**, *24*, 1315–1333. [CrossRef]
7. Quoy, M.; Moga, S.; Gaussier, P. Dynamical Neural Networks for Planning and Low-Level Robot Control. *IEEE Trans. Syst. Man Cybern. Part A Syst. Hum.* **2003**, *33*, 523–532. [CrossRef]
8. Cerkala, J.; Jadlovska, A. Application of neural models as controllers in mobile robot velocity control loop. *J. Electr. Eng.* **2017**, *68*, 39–46. [CrossRef]
9. Ciupan, E.; Lungu, F.; Ciupan, C. ANN Method for Control of Robots to Avoid Obstacles. *Int. J. Comput. Commun. Control.* **2014**, *9*, 539–554. [CrossRef]
10. Praczyk, T. Using evolutionary neural networks to predict spatial orientation of a ship. *Neurocomputing* **2015**, *166*, 229–243. [CrossRef]
11. Curro, J.A., II. Navigation with Artificial Neural Networks. Ph.D. Thesis, Air Force Institute of Technology, Dayton, OH, USA, 2018. Available online: https://scholar.afit.edu/etd/1948 (accessed on 13 September 2018).
12. Praczyk, T. Using Neuro-Evolutionary Techniques to Tune Odometric Navigational System of Small Biomimetic Autonomous Underwater Vehicle—Preliminary Report. *J. Intell. Robot. Syst.* **2020**, *12020*. [CrossRef]
13. Song, L.; Duan, Z.; He, B.; Li, Z. Application of Federal Kalman Filter with Neural Networks in the Velocity and Attitude Matching of Transfer Alignment. *Neural Netw. Complex Syst. Theory Appl.* **2018**, *18*, 3039061. [CrossRef]
14. Praczyk, T. Hill Climb Modular Assembler Encoding: Evolving Modular Neural Networks of fixed modular architecture. *Knowl.-Based Syst.* **2022**, *232*. [CrossRef]
15. Praczyk, T. Solving the pole balancing problem by means of Assembler Encoding. *J. Intell. Fuzzy Syst.* **2014**, *26*, 857–868. [CrossRef]
16. Praczyk, T. Cooperative co-evolutionary neural networks. *J. Intell. Fuzzy Syst.* **2016**, *30*, 2843–2858. [CrossRef]
17. Potter, M. The Design and Analysis of a Computational Model of Cooperative Coevolution. Ph.D. Thesis, George Mason University, Fairfax, VA, USA, 1997.
18. Potter, M.A.; Jong, K.A.D. Cooperative coevolution: An architecture for evolving coadapted subcomponents. *Evol. Comput.* **2000**, *8*, 1–29. [CrossRef] [PubMed]
19. Stanley, K.O.; Miikkulainen, R. Evolving neural networks through augmenting topologies. *Evol. Comput.* **2002**, *10*, 99–127. [CrossRef]
20. Gomez, F.; Schmidhuber, J.; Miikkulainen, R. Accelerated Neural Evolution through Cooperatively Coevolved Synapses. *J. Mach. Learn. Res.* **2008**, *9*, 937–965.

21. Chen, J.F.; Do, Q.H.; Hsieh, H.N. Training Artificial Neural Networks by a Hybrid PSO-CS Algorithm. *Algorithms* **2015**, *8*, 292–308. [CrossRef]
22. Olivas, F.; Valdez, F.; Castillo, O.; Melin, P. Dynamic parameter adaptation in particle swarm optimization using interval type-2 fuzzy logic. *Soft Comput.* **2016**, *20*, 1057–1070. [CrossRef]
23. Sengupta, S.; Basak, S.; Peters, R.A., II. Particle Swarm Optimization: A survey of historical and recent developments with hybridization perspectives. *Mach. Learn. Knowl. Extr.* **2018**, *1*, 10. [CrossRef]
24. Wang, D.; Tan, D.; Liu, L. Particle swarm optimization algorithm: An overview. *Soft Comput.* **2018**, *22*, 387–408. [CrossRef]
25. Hakli, H.; Uguz, H. A novel particle swarm optimization algorithm with Levy flight. *Appl. Soft Comput.* **2014**, *23*, 333–345. [CrossRef]
26. Jensi, R.; Jiji, G.W. An enhanced particle swarm optimization with Levy flight for global optimization. *Appl. Soft Comput.* **2016**, *43*, 248–261. [CrossRef]
27. Tarkhaneh, O.; Shen, H. Training of feedforward neural networks for data classification using hybrid particle swarm optimization, Mantegna Levy flight and neighborhood search. *Heliyon* **2019**, *5*, e01275. [CrossRef]
28. Georgioudakis, M.; Plevris, V. On the Performance of Differential Evolution Variants in Constrained Structural Optimization. *Procedia Manuf.* **2020**, *44* 371–378. [CrossRef]
29. Leon, M.; Xiong, N. Adapting Differential Evolution Algorithms For Continuous Optimization via Greedy Adjustment of Control Parameters. *J. Artif. Intell. Soft Comput. Res.* **2016**, *6*, 103–118. [CrossRef]
30. Storn, R.; Price, K. Differential evolution—A simple and efficient heuristic for global optimization over continuous spaces. *J. Glob. Optim.* **1997**, *11*, 341–359. [CrossRef]
31. Chalup, S.K.; Wiklendt, L. Variations of the two-spiral task. *Connect. Sci.* **2007**, *19*, 183–199. [CrossRef]
32. Cangelosi, A.; Parisi, D.; Nolfi, S. Cell division and migration in a genotype for neural networks. *Netw. Comput. Neural Syst.* **1994**, *5*, 497–515. [CrossRef]
33. Nolfi, S.; Parisi, D. Growing neural networks. In *Artificial Life III*; Langton, C.G., Ed.; Addison-Wesley: Boston, MA, USA, 1992; pp. 431–434.
34. de Campos, L.M.L.; Oliveira, R.C.L.D.; Roisenberg, M. Evolving Artificial Neural Networks through L-system and evolutionary computation. In Proceedings of the International Joint Conference on Neural Networks (IJCNN), Killarney, Ireland, 12–17 July 2015. [CrossRef]
35. Gomez, A.B.; Saez, Y.; Vinuela, P.I. Evolutionary convolutional neural networks: An application to handwriting recognition. *Neurocomputing* **2018**, *283*, 38–52.
36. Kitano, H. Designing neural networks using genetic algorithms with graph generation system. *Complex Syst.* **1990**, *4*, 461–476.
37. Assuncao, F.; Lourenco, N.; Machado, P.; Ribeiro, B. Denser: Deep evolutionary network structured representation. *Genet. Program. Evolvable Mach.* **2018**, *20*, 5–35. [CrossRef]
38. Assuncao, F.; Lourenco, N.; Ribeiro, B.; Machado, P. Incremental Evolution and Development of Deep Artificial Neural Networks. *arXiv* **2020**, arXiv:2004.00302.
39. Assuncao, F.; Lourenco, N.; Ribeiro, B.; Machado, P. Fast-DENSER: Fast Deep Evolutionary Network Structured Representation. *SoftwareX* **2021**, *14*, 100694. [CrossRef]
40. Lima, R.H.R.; Magalhaes, D.; Pozo, A.; Mendiburu, A.; Santana, R. A grammar-based GP approach applied to the design of deep neural networks. *Genet. Program. Evolvable Mach.* **2022**. [CrossRef]
41. Rodrigues, N.M.; Silva, S.; Vanneschi, L. A study of generalization and fitness landscapes for neuroevolution. *IEEE Access* **2020**, *8*, 108216–108234. [CrossRef]
42. Koutnik, J.; Gomez, F.; Schmidhuber, J. Evolving Neural Networks in Compressed Weight Space. In Proceedings of the Genetic and Evolutionary Computation Conference 2010, Portland, OR, USA, 7–11 July 2010.
43. Gauci, J.; Stanley, K. Generating large-scale neural networks through discovering geometric regularities. In Proceedings of the Genetic and Evolutionary Computation Conference, London, UK, 7–11 July 2007.
44. Fernando, C.; Banarse, D.; Reynolds, M.; Besse, F.; Pfau, D.; Jaderberg, M.; Lanctot, M.; Wierstra, D. Convolution by evolution: Differentiable pattern producing networks. In Proceedings of the 2016 Genetic and Evolutionary Computation Conference, Denver, CO, USA, 20–24 July 2016.
45. Pugh, J.K.; Stanley, K.O. Evolving multimodal controllers with hyperneat. In Proceedings of the 15th Annual Conference on Genetic and Evolutionary Computation. Amsterdam, The Netherlands, 6–10 July 2013.
46. Schrum, J.; Lehman, J.; Risi, S. Using Indirect Encoding of Multiple Brains to Produce Multimodal Behavior. *arXiv* **2016**, arXiv:1604.07806.
47. Tenstad, A.; Haddow, P.C. DES-HyperNEAT: Towards Multiple Substrate Deep ANNs. In Proceedings of the 2021 IEEE Congress on Evolutionary Computation (CEC), Krakow, Poland, 28 June–1 July 2021; pp. 2195–2202. [CrossRef]
48. Buk, Z.; Koutnik, J.; Snorek, M. NEAT in HyperNEAT substituted with genetic programming. In Proceedings of the Adaptive and Natural Computing Algorithms, Kuopio, Finland, 23–25 April 2009; pp. 243–252.
49. Gruau, F. Neural network Synthesis Using Cellular Encoding And The Genetic Algorithm. Ph.D. Thesis, Ecole Normale Superieure de Lyon, Lyon, France, 1994.
50. Ritchie, M.D.; Motsinger, A.A.; Bush, W.S.; Coffey, C.S.; Moore, J.H. Genetic Programming Neural Networks: A Powerful Bioinformatics Tool for Human Genetics. *Appl. Soft Comput.* **2017**, *7*, 471–479. [CrossRef]

51. Rodriguez-Coayahuitl, L.A.; Escalante, H.J.; Morales-Reyes, A. *Deep Representation Learning with Genetic Programming*; Technical Report No. CCC-17-009; National Institute for Astrophysics Optics and Electronics Tonantzintla: Puebla, Mexico, 2018.
52. Luke, S.; Spector, L. Evolving Graphs and Networks with Edge Encoding: Preliminary Report. In Proceedings of the Late Breaking Papers at the Genetic Programming 1996 Conference, Stanford, CA, USA, 28–31 July 1996.
53. Mwaura, J.; Keedwell, E. Evolving Robotic Neuro-Controllers Using Gene Expression Programming. In Proceedings of the 2015 IEEE Symposium Series on Computational Intelligence, Cape Town, South Africa, 7–10 December 2015.
54. Broni-Bediako, C.; Murata, Y.; Mormille, L.H.; Atsumi, M. Evolutionary NAS with Gene Expression Programming of Cellular Encoding. *arXiv* **2020**, arXiv:2005.13110.
55. Praczyk, T. Assembler Encoding with Evolvable Operations. *Comput. Methods Sci. Technol.* **2015**, *21*, 123–139. [CrossRef]
56. Praczyk, T. Using augmenting modular neural networks to evolve neuro-controllers for a team of underwater vehicles. *Soft Comput.* **2014**, *18*, 2445–2460. [CrossRef]
57. Rempis, C.; Pasemann, F. Search space restriction of neuro-evolution through constrained modularization of neural networks. In Proceedings of the 6th International Workshop on Artificial Neural Networks and Intelligent Information Processing (ANNIIP), Funchal, Madeira, Portugal, 17–18 June 2010.
58. Gaier, A.; Asteroth, A.; Mouret, J.-B. Data-efficient Neuroevolution with Kernel-Based Surrogate Models. *arXiv* **2018**, arXiv:1804.05364.
59. Stork, J.; Zaefferer, M.; Bartz-Beielstein, T. Improving NeuroEvolution Efficiency by Surrogate Model-based Optimization with Phenotypic Distance Kernels. *arXiv* **2019**, arXiv:1902.03419.
60. Available online: https://github.com/hav4ik/tinyai (accessed on 8 June 2022).
61. Available online: http://www.cs.utexas.edu/users/ai-lab/?cosyne (accessed on 8 June 2022).
62. Available online: https://oceanai.mit.edu/ivpman/pmwiki/pmwiki.php?n=IvPTools.USimMarine (accessed on 8 June 2022).

electronics

MDPI

Review

Review of Collision Avoidance and Path Planning Algorithms Used in Autonomous Underwater Vehicles

Rafał Kot

Faculty of Mechanical and Electrical Engineering, Polish Naval Academy, 81-127 Gdynia, Poland; r.kot@amw.gdynia.pl

Abstract: The rapid technological development of computing power and system operations today allows for increasingly advanced algorithm implementation, as well as path planning in real time. The objective of this article is to provide a structured review of simulations and practical implementations of collision-avoidance and path-planning algorithms in autonomous underwater vehicles (AUVs). The novelty of the review paper is to consider not only the results of numerical research but also the newest results of verifying collision-avoidance and path-planning algorithms in real applications together with a comparison of the difficulties encountered during simulations and their practical implementation. Analysing the last 20 years of AUV development, it can be seen that experiments in a real environment are dominated by classical methods. In the case of simulation studies, artificial intelligence (AI) methods are used as often as classical methods. In simulation studies, the APF approach is most often used among classical methods, whereas among AI algorithms reinforcement learning and fuzzy logic methods are used. For real applications, the most used approach is reactive behaviors, and AI algorithms are rarely used in real implementations. Finally, this article provides a general summary, future works, and a discussion of the limitations that inhibit the further development in this field.

Keywords: collision avoidance; path planning; obstacle detection; autonomous underwater vehicle; artificial intelligence; autonomous navigation

Citation: Kot, R. Review of Collision Avoidance and Path Planning Algorithms Used in Autonomous Underwater Vehicles. *Electronics* **2022**, *11*, 2301. https://doi.org/10.3390/electronics11152301

Academic Editor: Zhiyun Lin

Received: 4 July 2022
Accepted: 21 July 2022
Published: 23 July 2022

1. Introduction

In recent years, a significant increase in the number of unmanned underwater, ground, and air vehicles (UxVs) in various environments has been observed. Most often, these vehicles are autonomous with different levels of autonomy. When speaking of autonomous vehicles, most will refer to air or ground vehicles. Less widespread and commercialised, but equally important are autonomous underwater vehicles (AUVs). This type of vehicle can be used for ocean exploration, performance of various industrial operations, or military missions. Due to the limited adaptability of humans, diving to very deep depths may be impossible or require a lot of time and preparation. In some cases, like observing the seafloor or looking for damage to pipes running along the bottom of a water reservoir, using AUV can bring many benefits to humans. To perform underwater operations properly and safely, it is necessary to equip the AUVs with all the essential systems, such as obstacle detection systems, motion- or path-planning systems, collision-avoidance systems, trajectory-planning systems, mapping systems, mission planners, path-tracking systems, etc. The underwater environment poses many difficulties for the vehicles' movement, i.e., high signal attenuation, limited possibilities of spatial orientation in relation to reference points, the presence of sea currents, limited access to light, and disturbing signal reflections from the bottom and surface of the water. Therefore, some motion-planning solutions implemented for robots or other groups of autonomous vehicles may not be appropriate for AUVs. The rapid technological development of computing power and system operations today allow for more and more advanced algorithm implementation as well as path planning in real time. As one study [1] indicated, increasing the computing power through hardware can significantly increase the data-processing speed and system efficiency compared to the implementation of more efficient software. Reliable performance is considered a crucial condition for AUV design. Although simulation analysis of collision-avoidance algorithms and path planning usually brings satisfactory results, real vehicle implementation has caused many difficulties in past studies [2]. As scientific and technical knowledge advance, researchers attempt to systematise current achievements in this field. In [3], motion-planning approaches for robots from the years 1980–2010 were analysed, indicating how the number of heuristic algorithms changed in individual decades compared to conventional algorithms. The basics of classical and heuristics-based methods for robots were discussed in [4]. The authors of [5] analysed 80 articles from 2011–2015 on motion planning in a dynamic environment for robots and systematised individual solutions in terms of parameters such as smooth path, safety, path length, run-time, accuracy, stability, control, computation cost, and efficiency. The study [6] reviewed 3D path-planning algorithms for robots, focusing on the universality of individual algorithms implementation in aerial, ground, and underwater robots. Reference [7] elaborated on heuristic algorithms for robots, showing examples of their efficient and inefficient operation. Reference [8] reviewed sampling-based motion-planning algorithms for robots, and [9] focused on probabilistic roadmap approaches in dealing with dynamic collision, narrow passages, multi-targets, and manipulation planning for deformable linear objects. Other studies also summarized the achievements in the field of quadrupedal robots [10], AGVs (usually road vehicles) [11,12], coverage path planning for robots [13], and searching methods for environmental monitoring [14]. Together with single robots, multi-robot performance has also been studied. Reference [15] discussed navigation techniques for the single and multi-robot systems in both static and dynamic environments. In addition, reference [16] analysed the techniques of movement of the robot group depending on the shape of the formation (cluster/line flocking). The authors of [17] reviewed the available literature in terms of the multi-robot movement in normal and faulty situations, and [18] focused on methods determining desired trajectories for multi-robots. In the path-planning studies, some researchers have been focusing on solutions designed for specific types of vehicles depending on the environment in which the vehicle is moving. In [19,20], the algorithms designed for unmanned aerial systems (UASs) were analysed, and in [21–23] algorithms for unmanned aerial vehicles (UAVs) were discussed. Reference [24] discussed vision-based algorithms for UAVs, and [25] for multiple UAVs. Many reviews in the field of underwater robotics have also been published. In [26], the technological development and limitations that are encountered in marine robotics are analysed. The authors of [27] define the trends in construction, instrumentation, and systems for unmanned underwater vehicles (UUVs). The UUVs can be remotely operated vehicles (ROVs) and AUVs. In [28], the main systems for both a single AUV and multi-AUV are described, as well as the possible directions of AUVs' further development. In terms of motion control of a UUV, reference [29] reviews the development of control methods and the problems encountered during the research. In turn, reference [30] provides detailed literature that analyses path-planning-related solutions for AUVs. The main surveys connected with motion planning for robots are presented in Table 1.

The objective of this article is to provide a structured review of simulations and practical implementations of collision-avoidance and path-planning algorithms in AUVs together with a comparison of the difficulties encountered during simulations and their practical implementation. The paper is organised as follows. In Section 2, the most essential path-planning and collision-avoidance algorithms for AUVs are explained. Section 3 includes a summary of simulation and practical research in this area. In Section 4, the general summary, future works, and limitations that inhabit the further development in this field are discussed. Final conclusions are presented in Section 5.

Table 1. Surveys connected with motion planning for robots.

Review	Field of Analisys	Year	Main Focus
[7]	Robots	2005	This review elaborated on heuristic algorithms for robots, showing examples of their efficient and inefficient operation.
[9]	Robots	2006	This review focused on probabilistic foadmap approaches in dealing with dynamic collision, narrow passages, multi-targets, and manipulation planning for deformable linear objects.
[8]	Robots	2008	This study reviewed sampling-based motion planning algorithms for robots.
[3]	Robots	2012	This study reviewed motion planning approaches for robots from the years 1980–2010 indicating how the number of heuristic algorithms changed in individual decades compared to conventional algorithms
[13]	Robots	2013	This study reviewed coverage path planning issues for robots
[4]	Robots	2015	This study presented the basics of classical and heuristics-based methods for robots.
[5]	Robots	2015	This review analysed 80 articles from 2011–2015 on motion planning in a dynamic environment for robots and systematised individual solutions in terms of parameters such as smooth path, safety, path length, run-time, accuracy, stability, control, computation cost, efficiency.
[6]	Robots	2016	This study reviewed 3D path-planning algorithms for robots, focusing on the universality of individual algorithms implementation in aerial, ground, and underwater robots.
[14]	Robots	2017	This study reviewed searching methods for environmental monitoring
[15]	Robots (Single and multi-robot)	2019	This study discussed navigation techniques for the single and multi-robot systems in both static and dynamic environments.
[17]	Robots (Multi-robot)	2020	This study reviewed the available literature in terms of the multi-robot movement in normal and faulty situations.
[18]	Robots (Multi-robot)	2021	This study focused on methods of determining desired trajectories for multi-robots.
[16]	Robots (Multi-robot)	2021	This study analysed the techniques of movement of the robot group depending on the shape of the formation.
[10]	Robots	2021	This study summarized the achievements in the field of quadrupedal robots
[11]	Ground Robots (AV)	2015	This study presented relevant works in the overtaking and obstacle avoidance maneuvers and also makes an overview of the state-of-the-art implementation of motion planning techniques for automated driving.
[12]	Ground Robots (AV)	2021	This study reviewed existing approaches on motion and behavior planning for AVs in terms of their feasibility, capability in handling dynamic constraints and obstacles, and optimality of motion for comfort.
[21]	Aerial Robots (UAVs)	2010	This study presented existing motion-planning algorithms and recent developments in the robotics and aerospace guidance and control fields. The review also analysed perspectives and practical examples from UAV guidance approaches.

Table 1. *Cont.*

Review	Field of Analisys	Year	Main Focus
[19]	Aerial Robots (UAS)	2015	This study presented a comparison between different motion-planning algorithms for a multi-rotor UAS with a multi-joint robotic arm.
[24]	Aerial Robots (UAVs)	2018	This study discussed vision-based algorithms for UAVs.
[25]	Aerial Robots (multi-UAV)	2019	This study presented a state-of-the-art overview of various approaches for multi-UAV collision avoidance.
[20]	Aerial Robots (multi-UAV)	2019	This study presented the most relevant Deep Learning techniques for autonomous collision avoidance as well as their application to UAV systems.
[23]	Aerial Robots (UAV)	2020	This study presented an extensive review of UAVs including physical characteristics, classification, standardization, battery charging, security challenges, and solutions. The research also discussed the perspective of using UAVs in 5G networks, Internet of Things Networks, and different applications.
[22]	Aerial Robots (UAV)	2021	This study presented the methodology for spatial orientation angle correction for UAV.
[29]	Marine robots(UUV)	2009	This study reviewed the development of control methods and the problems encountered during the research connected with motion control of UUVs.
[26]	Marine robots	2018	This study analysed the technological development and limitations that are encountered in marine robotics.
[27]	Marine robots (UUV)	2021	This study defined the trends in construction, instrumentation, and systems for UUVs.
[28]	Marine robots (single and multi-AUV)	2021	This study described the main systems for both a single AUV and multi-AUV as well as the possible directions of AUV's further development.
[30]	Marine robots (AUV)	2021	This study provides detailed literature that analyses path planning related solutions for AUVs.
This review	**Marine robots (AUV)**	**2022**	**This study provides a structured review of simulations and practical implementations of collision avoidance and path planning algorithm in AUV together with a comparison of the difficulties encountered during simulations and their practical implementation.**

2. Fundamentals of Path-Planning Algorithms

This section presents the main path-planning algorithms for AUVs. The description of each method includes an explanation of the fundamental principles. It also analyses the advantages and disadvantages, environmental limitations, and potential difficulties in real applications. Each approach contains several application examples from different fields of robotics.

2.1. A Algorithm*

In 1986, the heuristic method [31] A* was proposed, which consists of dividing the known area into individual cells and calculating the total cost of reaching the target for each of them. Given the total distance from the starting point to the currently analysed cell and the distance from the analysed cell to the target, the total path cost is calculated with the following formula:

$$F(n) = G(n) + H(n), \tag{1}$$

where:

$F(n)$ is the total path cost;

$G(n)$ is the cost of reaching from the starting point to the analysed cell; and

$H(n)$ is the cost of reaching from the analysed cell to the target.

The A* algorithm focuses on finding the shortest path to a destination in the presence of static obstacles assuming that both the environment and the location of the obstacles are known. In this method, however, an increased amount of computation is needed for large areas analysis or areas with many obstacles. This significantly increases the path planning time. For the marine vehicles, AUVs and autonomous surface vehicles (ASV), the A* algorithm method is in most cases used in combination with the visibility graph algorithm [32–34]. In [35], simulations of two grid-based methods in the 3D environment were compared. Compared to the Dijkstra algorithm, the multi-directional A* proved to be more efficient in terms of the number of nodes and the total length of the path. The practical implementation of this method is associated with the problem of variability of the underwater environment and the lack of precise knowledge about the location of obstacles. Another challenge for practical implementation of this method in AUV may be the unforeseen effect of sea currents.

2.2. *Artificial Potential Field*

The artificial potential field (APF) method has its origins in 1986 [36] and assumes the presence of a repulsive field around the obstacle and an attractive area around the target that affect a moving vehicle (e.g., AUV). As a result of the interaction of these virtual fields, the resultant force is determined according to which way the vehicle is moving. In this method, prior knowledge of the environment and the location of obstacles is not needed. It can be used regardless of whether the obstacles in the environment in which the AUV is moving are static or dynamic and whether they have regular or irregular shapes. The crucial condition for this algorithm to be efficient is accurate obstacle detection. The advantage of the APF method is the ease of implementation and low computational requirements, which makes it possible to control AUVs and avoid collisions in close to real time. Despite the many abovementioned advantages, the possibility of local minima or trap situations is considered as a significant disadvantage [37]. Under certain conditions, e.g., a U-shaped obstacle, when the resultant force acting on the AUV is balanced, the algorithm will control the movement of the AUV in a closed infinite loop without reaching the goal. In addition, the passage of the AUV between closely spaced obstacles may not be possible or cause oscillation due to the alternating influence of force fields from opposing obstacles [37,38]. Also, the AUV tends to demonstrate unstable movements when passing around obstacles. In order to solve the local minima problem, APF is combined with other methods, such as [39,40]. In the simulation-based study [41] for AUVs a solution was proposed based on the introduction of a virtual obstacle in the place where the local minimum occurs. Another way to avoid a trap situation can be by using random movements to lead the AUV out of the adverse area (randomised potential field) [42]. Despite the limitations of this method, it is used to control a swarm of robots [43–46]. Simulation-based studies also prove the possibility of controlling multi-AUV formations [47] and its application in mission planners [48]. In [49], the potential field-based method was used in the practical implementation in NPS ARIES AUV in a real-world environment. To avoid the negative impact of APF method limitations in real implementations, it is necessary to use the global path-planner module based on, e.g., heuristic methods.

2.3. *Rapidly Exploring Random Tree*

Introduced in 1998 [50], the motion-planning algorithm named rapid-exploring random tree (RRT) is a sampling-based method. Any node in the X_{rand} spatial is randomly determined from the initial point X_{init} (Figure 1). Then, depending on the given direction of movement and the maximum length of the section from the analysed point, the intermediate X_{near} node is determined. The longer the sections' connecting nodes, the higher the risk of moving in the wrong direction for a long time and encountering obstacles on the designated path. If any obstacle is detected between waypoints, further route calculation in that direction is ignored. If no obstacles are detected, a random point in the X_{rand} spatial is determined, followed by a new intermediate point X_{new}.

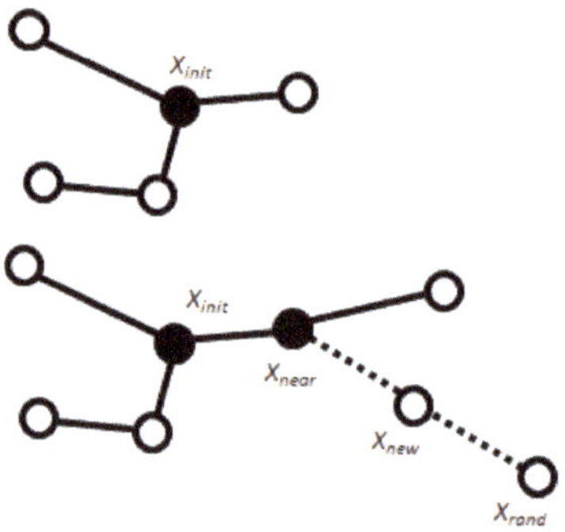

Figure 1. An illustration of the RRT mechanism.

As a result of subsequent iterations (time steps), new edges and path points are determined. The method is easy to process and ensures the finding of a collision-free path (if there is one) in an unknown environment, both 2D and 3D. RRT can be used in both static and dynamic environments [51]. A modification of this method was used in the SPARUS-II AUV and tested in a real-world environment [52]. The study [53] shows the validity of the modified RRT* algorithm for mapping in a 3D environment for multiple AUVs. In [54], an RRT-based approach was used to solve the problem of local minima in the fast warm starting module in the Aqua2 AUV. It was noted that this method is not always efficient in an environment where the path to the target leads through a narrow opening or gap. Another limitation of this method is the need to provide information about large areas of the environment, which is not always possible in practical implementations due to the technical limitations of sensors [55–57]. Additionally, the calculated path is suboptimal [58] which requires the use of additional optimisation algorithms.

2.4. *Artificial Neural Network*

An artificial neural network (ANN) is a machine learning method based on the mathematical mapping of information processing by the human brain. The general structure consists of three layers: input layer, hidden layers, and output layers (Figure 2). The neurons in each layer are connected to all the neurons in the neighboring layer and process the inputs based on the weights in between. The data from each neuron in the input layer multiplied by weights is sent as input to hidden layers where each neuron is assigned a value (bias). Depending on the activation function and bias value, only neurons with a value higher than the threshold value are activated. The output from each layer is also the input for the next layer. The data transfer between the layers is performed only by active neurons. The ANN method requires learning, i.e., indicating the accuracy of the output data. Based on the determination of the expected results, the neural network adjusts the weights between the individual neurons in each iteration in such a way that the output data is as close to the expectations as possible.

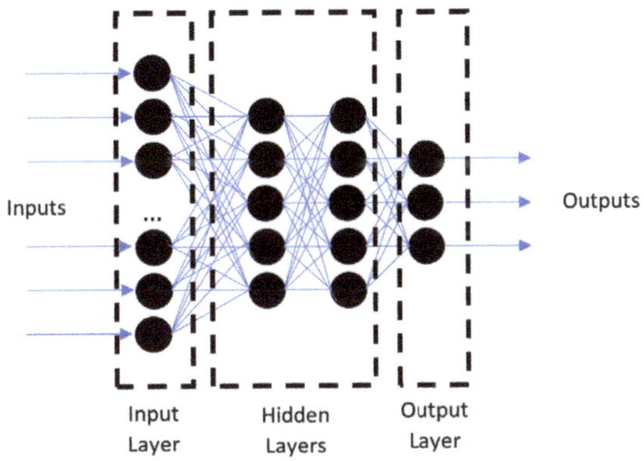

Figure 2. Neural network architecture.

The method has learning ability and is applicable in systems that implement complex functions, supporting many outputs based on data from multiple sensors. The algorithm is also efficient for the systems where the input data is incomplete or distorted, or in cases in which there poorly modelled, nonlinear dynamical systems [59–61]. The main drawback of this algorithm is the need for long-term training to achieve satisfactory results [62]. In the case of practical implementation for AUV, online learning does not bring satisfactory results due to slow learning speed and long training time [63]. Therefore, offline training for controllers is essential before the AUV can be used in a real-world environment. Reference [64] presents a collision-avoidance controller for static and dynamic obstacles based on neural networks that do not require training. The resulting behaviour of the AUV was defined by neuron weights. It should be noted, however, that this type of controller is suitable for simple AUV operational cases. In [65], the neural network method was used in real implementation in AUV-UPCT. The vehicle required prior learning in ROV mode. In [66], the neural network algorithm was used in a simulation study to control and avoid collisions for multiple AUVs in a 3D environment. In [67], the neural network was used to process real images of the underwater environment (simple, coloured monocular camera) to determine the free space enabling the escape of the AUV from the cluttered environment.

2.5. Genetic Algorithm

Genetic algorithm (GA) refers to research and optimisation methods inspired by the natural evolution process where only the fittest organisms have a chance for survival. In general, the algorithm starts looking for a solution to a problem by generating a random population of possible solutions. Depending on the applied function, a selection is made during which the least suitable solutions are eliminated [68]. For path planning, the main criterion is the energy cost required to run each path [69]. Then, through an operation called crossover, further potential solutions to the problem (offspring) are created, combining the best solutions from the previous generation (parents). Additionally, in the mutation process, random modifications of the best solutions are created. Then, the GA runs the selection again, adding to the population's successive possible solutions that are closer and closer to the correct result until a specific final condition is reached, e.g., the number of generations. The main advantage of the GA method is the possibility of fast and global stochastic searching for optimal solutions [70]. In addition, the algorithm is easy to implement and can be used to solve complex problems, such as determining the optimal AUV path in the presence of static and dynamic obstacles in the underwater environment. Due to the random nature of searching for solutions, the algorithm reduces the risk of the local minima problem [71]. In general, the method does not require large numbers of calculations to solve the path-planning problem. However, the route may be suboptimal if too few generations are executed. As the number of generations increases, the route becomes closer and closer to the optimal one due to the constant elimination of the least optimal solutions in the population. It entails an increase in computational costs. Similarly, in a rapidly changing environment or when the environment is very extensive, the amount of computation needed to determine a solution increases significantly. In [69], a modification of the GA was presented in order to determine the energy's optimal path. The modification involved the introduction of iterations consisting of additional runs of the algorithm with different initial conditions and the operator based on the random immigrants' mechanism, which sets the level of randomness of the developing population. Reference [72] discussed the framework of the collision-avoidance system based on the GA. The simulation test proved the proper functioning of the method for static and dynamic obstacles. Improved GA was also used in a simulation-based study [73] to optimise energetic routing in a complex 3D environment in the presence of static and dynamic obstacles. The algorithm with improved crossover and mutation probability and the modified fitness function was tested in simulation [70] in comparison with the traditional GA method. The simulation confirmed that the improved GA allows a shorter and smoother path with fewer generations. The method worked very well in the energy optimisation of routing. It should be noted that the efficiency of this method in real testing depends on the correct detection and determination of the obstacle location as well as the target.

2.6. Fuzzy Logic

The fuzzy logic (FL) method [74,75] is based on the evaluation of the input data depending on the fuzzy rules which can be determined by using the knowledge and experience of experts. In AUV obstacle avoidance systems, the fuzzy controller sensor processes data containing information about the surrounding environment and makes decisions based on it. Then, appropriate signals are transmitted to the actuators. The first stage of the algorithm's operation is fuzzification, i.e., assigning the input data to the appropriate membership function. Each of the functions is based on a descriptive classification of the input data, e.g., low, medium or high collision risk. Then, in the fuzzy inference process, a data evaluation is performed based on "if prerequisite then result" statements rule. Ultimately, the defuzzification process determines specific system output values (e.g., actuator control signal values). The method can be used for both static and dynamic obstacles. However, when an AUV operates in an unknown environment, the usability of the algorithm directly depends on the implemented rules, and therefore also on the knowledge and experience of experts. The main advantage of the algorithm is its usability in the case of incomplete information about the environment, even containing noises or errors [76]. The method is easy to implement and provides satisfactory results in real-time processing but requires a precise definition of membership functions and fuzzy rules [77]. Additionally, as the number of inputs increases, the amount of input data necessary for the system to process increases. An important issue affecting the effectiveness of the algorithm is also the AUV speed and the complexity of the environment. A fuzzy system usually has at least two inputs. In a very dynamic environment, the use of additional inputs can increase the efficiency of the controller [78]. In some cases, the necessity of performing complex maneuvers to avoid a collision may cause the AUV to be diverted far from the optimal path. For this reason, it is necessary to use additional algorithms that control the path in terms of energy. A simulation study [71] showed that the use of GA to optimise fuzzy logic path planner allows one to achieve greater efficiency and reduce cross-track errors and total traveled path. A modification of this method was also used in a practical implementation in a study [79] where the Bandler and Kohout product of fuzzy relations was used for preplanning in horizontal plane maneuvers. The fuzzy logic method was also used in [80] to control a single unmanned surface vehicle (USV) and in the simulation [81] to control virtual AUVs in the leader–follower formation of AUVs.

2.7. Reinforcement Learning

The reinforcement learning (RL) method is based on research about the observation of animal behaviour. As a result of strengthening the pattern of behavior by receiving a stimulus by the animal, it increases the tendency to choose actions [82]. Machine learning based on this idea was very intensively developed in the second half of the 19th century. Depending on the requirements of the environment model, the simplicity of data processing, and the iterative nature of calculations, several methods of solving RL problems have been developed, e.g., dynamic programming, Monte Carlo methods, and temporal difference learning. The most crucial method of solving RL problems is the temporal difference method. Depending on the occurrence of the policy function, it is divided into several algorithms, e.g., Actor-Critic, Q-learning or Sarsa method [82]. The actor-critic method consists of autonomous learning of the correct problem solving depending on the received reward or punishment. By performing action, the agent influences the environment by observing the effects of the action. Depending on the implemented reward function, different ways of influencing the environment are assessed. The objective of this method is to learn how to perform such actions in order to get the greatest reward. In the case of path planning, the agent will receive the greatest reward if moving toward the given destination. In general, an agent includes a combination of neural network's actor and critic. The interaction between them is a closed-loop learning situation [83]. The actor chooses the action for performing to improve the current policy. The critic observes the effects of this action and tries to assess them. The assessment is then compared with the reward function. Based on the error rate, the critic network is updated to predict actor network behaviour in the future better. The abovementioned approach was used in [84] to control four robots. Each of them was able to avoid collisions with other robots and with obstacles. In the study [85], the RL approach was used in a two-dimensional simulation of cooperative pursuit of unauthorized UAV by using a team of UAVs in an urbanized environment. In [86], a simulation of smart ship control algorithm was presented. Moreover, in simulation studies [87,88] the RL approach was used for path planning and motion control of AUVs. The method can be used for both static and dynamic obstacles in an unknown environment. It is characterised by a strong focus on problem solving and shows high environmental adaptability. The RL algorithm can learn to solve very complex problems as long as the correct reward function is used. The inability to manually change the parameters of the learned network is considered as the main drawback of this method. To change the operation of the algorithm, the network must be redesigned, and a long learning process must be performed. Also, checking this method's operation by simulation does not guarantee its correct operation in a real-world environment.

2.8. Comments

Among the classical and heuristic methods discussed above, it is difficult to find one that will work efficiently in various conditions and at the same time meet all the requirements for complete control and for a collision-avoidance system for the AUV. In an ideal scenario, the implemented method should assure high efficiency, real-time response, safety, path optimisation in terms of energy, low complexity, high accuracy, the shortest possible path, and the ability to operate properly in an unknown environment in relation to both static and dynamic obstacles. Each of the methods has its advantages and disadvantages depending on the given conditions. A situation in which the algorithm works very well and one in which the effectiveness of the algorithm will be very low can be found for each of the discussed methods [7]. Researchers still work on improving the parameters of individual methods by modifying them. In order to improve efficiency and eliminate defects, combinations of various algorithms are also used.

3. Chronological State of the-Art

After analysing the available literature in the field of path planning and collision avoidance for AUVs, a detailed classification of the studies was carried out in terms of chronology and in terms of the technique of testing the effectiveness of the presented algorithms. In each of the 4 intervals (until 2010, 2011–2015, 2016–2020 and from 2021 onward), the studies were divided into 2 groups, as follows.

- The first group includes algorithms validated in numerical research containing new methods, modified methods, or a combination of existing methods that have been validated in a simulation environment. Some of the methods presented have been developed to be applied to real AUVs. This group also includes algorithms whose effectiveness has been verified in hardware-in-loop simulations.
- The second group includes algorithms in real applications where AUV control methods are presented, the effectiveness of which has been tested in a real environment in the presence of both real and virtual obstacles

In each of the abovementioned groups, the research was ordered from oldest to newest in the analyzed period of time.

3.1. Until 2010

The need for a collision-avoidance system for AUVs has been recognised in the last century. At that time, many researchers proved algorithms that could detect and avoid collisions for static and moving obstacles [64,89]. Studies carried out in early 2000 showed a very intensive development of AUV collision-avoidance and path-planning algorithms. The capabilities of most of them, however, were tested only with simulation-based methods. The list of algorithms validated in numerical research until 2010 is presented in Table 2. At that time, intensive research on obstacle detection and classification were carried and analysed for physical sensor use and in simulation-based methods. In some cases, the algorithms were explicitly created for practical implementation. In [90] a virtual force field (VFF)-based system for the RAIS AUV was introduced and tested by using numerical simulations and hardware-in-loop simulation methods. In [91], researchers proposed a collision-avoidance algorithm through the pitch angle change. This algorithm was intended for physical implementation in the Taipan AUV. The authors of [92] proposed a collision avoidance algorithm based on the prediction of gradient lines (static obstacle) and deformed safety zone (for dynamic obstacles) intended foran AUV called DeepC. The approach was based on determining the trajectory by using the graph method combined with the A* method. The Redermor test platform equipped with forward-looking sonar (FLS), ten echosounder, and side scan sonar in the [93] study was used to collect sonar data during underwater tests. The authors have been able to classify the obstacles and prove that it is possible to define the boundary level around the obstacle. The implementation of more advanced hardware equipped with FPGAs allowed [1] to significantly increase the speed of data processing and power efficiency. At that time, solutions were also designed to control AUVs moving in formation. The accuracy of the proposed method was tested only in simulation-based analysis. In the study [94], the method based on the potential function and behaviour rules was proven in simulations for 3 AUVs moving in formation. Also, the authors of [81] proposed a well-functioning algorithm based on FL to control the velocity of the AUV formation. An algorithm based on the artificial potential field method [95] was also developed, which has been proven, under simulation conditions, to work efficiently for static obstacles.

Table 2. List of algorithms validated in numerical research until 2010.

Research	Algorithm	Main Characteristics
[89]	Modified Potential Field method	• Static and dynamic obstacles • Local path planning • Simulated clustering sonar returns
[64]	Neural Network	• Static and dynamic obstacles • Local path planning • Not necessarily adaptive training, neuron weights defined by the desired AUV response
[96]	Increase or decrease velocity, heading control	• Dynamic obstacles • Sonar data analyse in 2D (horizontal plane)
[97]	Deformable Virtual Zone	• Static obstacles • Local path planning • 25 ultrasonic sensors to scan front space in 25 directions
[2]	Reinforcement Learning based on the stochastic real value function	• Local path planning • For AUV named Twin Burger 2 • In the real application, only detection was achieved. Avoidance behaviors were not obtained due to the complexity of the implemented method.
[90]	Virtual Force Field (VFF)	• For AUV named RAIS • Numerical and Hardware in Loop simulation
[98]	Nonlinear programming technique	• Local path planning • 2D environment • Tested on ROV
[91]	Changing pitch angle	• Static obstacles (underwater cliffs) • Local path planning • For AUV named Taipan
[99]	Fuzzy Logic	• 2D environment • For AUV named REMUS
[32]	Two-layer control Path planning by Graph method with A* method Reactive control based on Deformable Safety Zone	• Static and dynamic obstacles • For AUV named DeepC • Simulations in 2D and 3D environment
[72]	Genetic Algorithm	• Static and dynamic obstacles • 2D simulation • Smooth collision-free path • Real-time algorithm
[100]	Markov decision process	• Uncertain static obstacles and vortex • 2D environment • Real-time algorithm
[71]	Genetic Algorithm tuned Fuzzy Logic planner	• Static obstacles • Global Path Planning • Better results than conventional fuzzy logic planners
[101]	Virtual Potential Method	• 2D environment • Static obstacles
[102]	Direct-method-based procedure	• Horizontal and vertical path planning • Static and dynamic obstacles
[103]	Graph algorithm based on time-varying ocean flow	• Static and dynamic obstacles • For AUV named SLOCUM
[104]	Fuzzy Logic	• Static obstacles • 5 possible behaviours for AUV

An example of practical implementation of collision-avoidance algorithms can be found in [49]. In this study, AUV avoided the sunken barge by changing the altitude by using a method based on the artificial Pptential field. Another practical implementation [38] was also performed with MEREDITH. 2D vector field histogram (VFH) based method functionality was tested in a real-world environment. The vehicle was able to avoid a large obstacle which was a breakwater. Also, in 2001–2009, a project code-named Autosub was carried out. At that time, several vehicles from the Autosub brand were made and practically tested during many under-ice missions. The collision-avoidance capability has been proven practically in this mission. In [105], a collision-avoidance module was triggered, and the vehicle correctly managed to avoid the obstacle. At that time, the algorithm worked only in the vertical plane. Over the years, the project's team gained a lot of experience and gradually improved their AUVs capability. Reference [106] presented an improved collision-avoidance system capable of turning in a horizontal plane. The upgraded Autosub-I AUV, after detecting an obstacle, performs a maneuver to change the depth and then turn and move in the horizontal plane. Then, the AUV retreats and tries to avoid the obstacle. After the maneuver, it returns to the previously taken path. As the authors claimed, the algorithm copes well with large obstacles. However, small-sized obstacles and a complex environment pose a significant challenge to the vehicle. The same algorithm was also implemented in another vehicle from the Autosub group—Autosub3. The vehicle operated for four days under the ice and covered 510 km during the mission, proving practical adaptation to even difficult operations [107]. In Table 3 the list of algorithms in real applications until 2010 is presented.

Table 3. List of algorithms in real applications until 2010.

Research	Algorithm	Main Characteristics
[49]	Artificial Potential Field	• Collision avoidance realised by changing depth • Equipped with FLS and light underwater camera
[38]	Vector Field Histogram +	• Static and dynamic obstacles • Only 2D heading
[106,107]	Reactive behaviours	• Static and dynamic obstacles • The simplified collision avoidance system • Efficient for large obstacles

3.2. 2011–2016

In the past decade, researchers focused more often on analysing the combinations of algorithms. In Table 4 the list of algorithms validated in numerical research between 2011–2016 is presented. In the simulation study [77], the particle swarm optimization (PSO) algorithm was used to regulate the membership function in the FL-based method in the AUV motion controller system. In the study [108], the method based on the combination of FL for pre-planning and reactive techniques is simulated. The authors consider collision avoidance only in the horizontal plane. The study [109] analyses the combination of potential field and edge detection to avoid collisions (only in the vertical plane). The method is designed for applications in which the AUV is responsible for inspecting the area (by using side scan sonar), where a straight course line should be kept. In [34], a combination of graph methods with the A* algorithm was used to determine the path. However, as concluded, that combination does not allow one to determine the optimal path in a completely unknown environment. In [110], the APF-based algorithm was used for local collision avoidance, and the velocity synthesis algorithm was used to optimise the AUV path. Also, modifications of the already existing algorithms were developed. The multi-point potential field method was analysed in [111]. The vehicle is steered in the direction of the field's lowest resultant value, which is generated in many directions. Additionally, in that time, more and more focus was given to path planning and global collision avoidance. Several control systems and collision-avoidance and path-planning studies have also been developed for multi-AUV formations. In a simulation study [47], a multi-layer region control concept was used to control and avoid both static and dynamic obstacles, and a PD-based regulator controlled the shape of the formation. The authors in [66] used neural networks to simulate the multi-AUV control system in the presence of static obstacles. Other issues closely related to collision avoidance and path planning by AUV were investigated in studies at that time. The authors of [111] performed a simulation in laboratory conditions of AUV with hardware-in-loop and software-in-loop methods with real sensor data (ultrasonic sensor- for distance to the obstacle, pan and tilt angle inputs by the user) and real implementation of actuators. This method is beneficial and effective for verifying the operation of algorithms in real time.

Table 4. List of algorithms validated in numerical research between 2011–2016.

Research	Algorithm	Main Characteristics
[112]	Fuzzy interface controller with A/B (accelerate/break) module	• Static and dynamic obstacles • More human-like module's decision making confirmed by simulation
[77]	Fuzzy-PSO controller with A/B (accelerate/break) module	• Static and dynamic obstacles • Improved navigation performance with great real-time
[108]	Potential Field and Edge Detection methods	• Increased precision compared to other methods • Horizontal plane only • Intended for the AUV called Blue Fin 12
[109]	Potential Field and Edge detection methods	• Vertical plane only • The algorithm strongly focuses on safety
[34]	Graph method with A* path find algorithm	• Not allowed to determine the optimal path in a completely unknown environment • Simulation conducted for static obstacles
[111]	Multi-Point Potential Field	• Static and dynamic obstacles • HIL and SIL simulation
[113]	Modified balance points of motion	• Static and dynamic obstacles • Attempt to combine local and global planning
[114]	Vector Polar Histogram (VPH)	• Uses real sonar data for the simulation • Memory function for trap environment
[110]	Velocity synthesis algorithm and Artificial Potential Field method	• APF method used to avoid obstacles effectively • Velocity synthesis algorithm to achieve the optimal path

In [65], researchers successfully tested a AUV-UPCT control system based on the neural network method in Menor Lagoon. The vehicle has the ability to move in 3D, and obstacle avoidance is implemented only in the horizontal plane. In the study [115], the successful maneuver to avoid a group of four obstacles was accomplished by making a 90° turn and bypassing the obstacle, which was the island and the points around it. However, one test resulted in AUV looping due to the local planner not being fully integrated with mission planner. The study [52] tested the possibilities of the SPARUS-II AUV in a real-world environment. In the simulation conditions, a computation time equal to 1 s was assumed, whereas in real-world conditions, it was necessary to change this to 1.5 s. This indicates that the method works quickly but is not intended for a dynamically changing environment. The study [116] proposed an algorithm based on reactive behaviours that allows it to correctly navigate and avoid obstacles in both vertical and horizontal planes. Due to the fact that the collision avoidance in both planes is not parallel, there may be cases of a non-optimal AUV trajectory. The study discusses the situation in which the AUV, to avoid an obstacle, gradually ascends to the minimum depth and then avoids the obstacle by using horizontal plane behaviours. the list of algorithms in real applications between 2011–2016 is presented in Table 5.

Table 5. List of algorithms in real applications between 2011–2016.

Research	Algorithm	Main Characteristics
[65]	Neural Network	• Sonar and camera • Requires learning in ROV mode • Moving in 3D but avoiding obstacles only in the horizontal plane • Can track the seafloor
[115]	Combination of the vector field, reactive algorithm and object mapping algorithm	• Static and dynamic obstacles • Data from 3 sonars • 2D maneuvers
[52]	Optimal Rapidly exploring Random Tree (RRT*)	• 2D workspace • Static obstacles
[116]	Reactive behaviours	• Test performed only for static obstacles • Sonar • Vertical and horizontal plane reactive maneuvers

3.3. 2016–2020

In the next five years (2016–2020), there was an increased interest in collision avoidance and control systems for single AUV and multi-AUV formations. Researchers focused their simulation studies on modifying existing methods to improve efficiency. List of algorithms validated in numerical research between 2016 and 2020 is presented in Table 6. In the survey [117], an APF algorithm modification was introduced, consisting of the disappearance of the repulsive force effect on AUV in the case of increasing the distance from the obstacle. For this method, an obstacle is detected only when at less than 80 m distance. In the study [118], the path optimisation determined by the neural network was solved using the Radau pseudospectral method. The authors of [73] presented a solution to modify the mutation operator through a grey wolf optimiser for energy optimisation in path planning. In the modified APF-based method in [119], the amount of data necessary for the correct operation of the algorithm was reduced to provide information only about the distance between the vehicle and the obstacle. The study [78] introduced a three-input FL controller that can provide greater efficiency for fast-moving obstacles than the generally used two-input controllers. In the survey [70], a modified GA is presented by adaptively adjusted crossover and mutation probability and by assessing the diversity of the new population. The collision-avoidance and controls systems for AUV formations were also points of focus at that time. In [120], authors used the fuzzy artificial potential function and leader–follower algorithm for flocking control of group AUV. Another solution based on the neural network method for controlling AUVs and APF for collision detection and avoidance was simulated in the study [121]. Due to hardware and environmental limitations, real-time in-system simulation was performed. The results proved to be similar to the simulation run in the MATLAB environment without using AUVs. Researchers also modified well-known routing optimisation algorithms for multiple AUVs and their energy consumption [122]. Also, many other studies have been focusing on the control and avoidance of collisions for multiple AUV formations, e.g., [123–125], which shows that in the time period considered in this paragraph, there was a significant increase in interest not only in collision avoidance and path planning algorithms for individual AUVs but also a lot of focus has been given to the research of AUV-formation control algorithms. In [126], the authors developed a practical implementation of the 2D control method based on the Dijkstra algorithm with its testing on the ROV. The authors claimed that the method can also be easily implemented in AUVs. In the study [127], a real-world test of the biomimetic AUV (BAUV) was carried out. The study, however did not bring satisfactory results due to the insufficient accuracy of the navigation system. Also, some false obstacles were detected due to reflections from testing the BAUV in shallow water. The examples of BAUV and AUV are shown in Figure 3. Another study [128] analysed the possibilities of determining the optimal and collision-free path for the Inspection-AUV (I-AUV) with a manipulator in cluttered environments. The authors of [129] confirmed that it is possible to build a collision-avoidance controller for AUVs by using a hydroacoustic obstacle-detection system. Reference [130] analyses the collision avoidance problem for various types of underwater objects (usually vehicles) and various environmental conditions in cases of one-way communication.

Figure 3. BAUV examples made at the Polish Naval Academy [own source].

Table 6. List of algorithms validated in numerical research between 2016 and 2020.

Research	Algorithm	Main Characteristics
[41]	Artificial Potential Field	• 2D static environment • Local minima problem solved by introducing a virtual obstacle in local minimum position
[131]	Dynamic Window	• Horizontal plane collision avoidance • Simulation conducted using nonlinear model of AUV named HUGIN 1000 • The algorithm is prone to trapping in the local minima
[76]	Fuzzy reactive architecture for different forward speed	• Simulation specifically for Guanay II AUV • Various obstacles changing location and shape
[67]	Deep Learning	• Simulation-based on real camera images • Finding a collision free space and determining the direction of AUV escape
[132]	Safety spheres	• Relates to I-AUV • The obstacles are represented by Octrees obtained from the Octomap implementation
[117]	Improved APF	• Static obstacles • APF modification consists in determining the distance of 80 m between the obstacle and the AUV beyond which the repulsive force disappears
[118]	Radial Basis Function (RBF) Neural Networks	• Horizontal plane only • Path planning simulation conducted with dynamic obstacle

Table 6. *Cont.*

Research	Algorithm	Main Characteristics
[133]	Fuzzy Logic	• Higher degree security and safety • The Infrared Sensor module for obstacle distance information
[134]	Rapid random search tree algorithm	• Static and dynamic obstacles Real-time path planning
[73]	Improved Interfered Fluid Dynamical System, Improved Genetic Algorithm for energy optimal path obtaining	• Static and dynamic obstacles • 3D path planning with complex obstacles and the ocean flow
[135]	Machine Learning-based algorithm	• Static obstacles • Real-time path planning
[87]	Image segmentation and Reinforcement Learning	• Static and dynamic obstacles • Estimates path dynamically
[119]	Modified Potential Field	• Static and dynamic obstacles • No need to know the exact position of the AUV, only the distance between the obstacle and the AUV
[78]	Three inputs Fuzzy Logic	• Static and dynamic obstacles • Increased efficiency compared to 2 inputs Fuzzy Logic System
[136]	Improved histogram-based EDA (LFHH-Learning Fixed-Height Histogram)	• Static and dynamic obstacles • 2D and 3D environments
[70]	Improved Genetic Algorithm	• Static obstacles • Improved crossover and mutation probability • Modified fitness function
[137]	Reactive behaviours	• Vertical plane only • The ability to maintain a constant or varying distance from an obstacle above or below the vehicle
[88]	Reinforcement learning	• Static obstacles • 2D environment
[138]	APF based method	• Improves the suboptimal solution of the traditional trajectory planning algorithm • Selection of cost function is optimised by potential field intensity
[35]	3D Multi direction A*	• Static environment • More satisfying results compared to Dijkstra Algorithm
[139]	Evolutionary Neural Network	• Static obstacles (AUVs operate in mountainous and underwater areas) • An algorithm designed for BAUV

The list of algorithms in real applications between 2016–2020 is presented in Table 7. In the study performed in [79], horizontal maneuvers are preferred over vertical ones. If it is impossible to calculate the path by using horizontal maneuvers, the vertical approach is activated. This approach allows one to avoid collisions in 3D, although in some cases the path on which the AUV moves may not be optimal or is far from optimal. The authors [140] tested AUV in a real-world environment with simulated obstacles in the form of a labyrinth. Therefore, obstacle detection and sensor input processing have been simplified. Nonetheless, the algorithm recalculated the path in around 5 s, which means the AUV might not be able to react appropriately in an emergency. The study [141] discussed an obstacle-avoidance algorithm, changing the AUV altitude when an obstacle appears in the scanning area. When performing evasive maneuvers, the AUV ascends vertically, and a downward-looking camera and laser monitor the obstacle, memorising the position of the obstacle's peak of the obstacles. After executing the maneuver, the AUV returns to the before-avoidance altitude. In the study [142], the reactive algorithm enforces turning when an obstacle is detected. If not, the AUV is controlled parallel to a wall, keeping a constant distance from it. In [143], the reactive algorithm uses a combination of behaviours depending on the situation: gain altitude, reduce altitude, move forward, move backwards, and escape from hot water. The study includes photographing the bottom of the test reservoir containing hydrothermal springs. When avoiding collisions with seafloor elements, the AUV makes a vertical plane movements only. In [144], extensive simulation and practical tests (under various conditions, including regular and irregularly shaped obstacles) were conducted to confirm the correct operation of the collision-avoidance algorithm implemented on the SPARUS-II AUV. The research proved the possibility of an effective operation of the algorithm in the field of online planning with full 3D control. The authors [54] tested the Aqua2 AUV in a pool with static obstacles. Although offline path planning was successfully accomplished in 3D, and online planning was only realized in the horizontal plane.

Table 7. List of algorithms in real applications between 2016–2020.

Research	Algorithm	Main Characteristics
[79]	Two-layer algorithm Fuzzy Logic for preplanning, potential field and edge detection to reactive maneuvers	• Horizontal maneuvers, if no routing possible, then the vertical approach activated • Strongly focused on safety
[140]	RRT based algorithm	• Experiment with artificially imposed obstacles • Around 5 s to calculate the path
[141]	Reactive behaviours	• Cameras, line lasers (image processing techniques) • Change the altitude when obstacle detected
[142]	Reactive behaviours	• Forward Looking Sonar • AUV moving in 3D, collision avoidance only in 2D
[143]	Reactive behaviours	• Forward Looking Sonar, Forward Looking Camera-Laser • Vertical plane collision avoidance maneuvers • Improved the existing system with Forward Looking Sonar, backward movement and thermometer (high-temperature water detection—hydrothermal vent fields.
[144]	Optimal RRT* based algorithm	• Static and dynamic obstacles • Online path planning in 3D unknown environment
[54]	Sequential convex optimisation	• The experiment in the pool with static obstacles • Offline path planning realised in 3D • Online path planning—only horizontal plane maneuvers

3.4. 2021–Now

In the last two years, simulation methods for route planning and collision avoidance are still being developed. Another aspect that is becoming more important is accurate navigation, together with algorithms for reducing the navigation errors and also methods of tracking the previously determined path [63,145–148]. There is still continuing interest in control and collision avoidance systems for multi-AUV formations [149–151]. To improve the efficiency of simulation algorithms, combinations or improvements to the existing methods are implemented. The list of algorithms validated in numerical research from 2021 is presented in Table 8.

Table 8. List of algorithms validated in numerical research from 2021.

Research	Algorithm	Main Characteristics
[152]	Improved Deep Deterministic Policy Gradient (DDPG)	• Static and dynamic obstacles • APF method to set continuous rewards • 2D environment
[153]	HA* (Hybrid A*)	• Static and dynamic obstacles • 3D path planning
[154]	Deep Reinforcement Learning	• Static and dynamic obstacles • 2D environment
[155]	Reinforcement Learning—RRT*	• Static obstacles • Validated in an unknown virtual maze
[156]	(SDEQPSO) algorithm Selective Differential Evolution Quantum-behaved Particle Swarm Optimisation	• Static and dynamic obstacles • HIL test validated • MOOS-IvP framework
[157]	Dynamic Virtual AUV principle and Barrier Lyapunov function	• Static and dynamic obstacles • Vertical plane avoidance maneuvers

In [158], a real-world environment test was conducted in which the AUV, after detecting a fishing net, was to make a turn and then return to the original path. In a result, 13 out of 16 attempts were successful. For the three unsuccessful attempts, the vehicle did not react accurately due to the too high detection threshold to exclude false detections. The MRF-net was used to detect obstacles in real time. The method is suitable for detection of a single static obstacle. The study [159] examined the capabilities of the PI AUV, which is intended for damage inspection of underwater pipes. It was tested whether the AUV correctly detects a leak from the pipe. Also, the anti-collision system operation was tested in which the avoidance expert system takes over the control if an obstacle is detected at a distance of 80 m from the AUV. Depending on the obstacle in relation to the AUV, the appropriate maneuver (turn left/right or bypass left/right) is selected in the horizontal plane. The position of dynamic obstacles was predicted by using the Kalman filter. The authors [160] introduced the path-planning and collision-avoidance systems for the spherical underwater robot SUR IV. The combination of ant colony and PSO approaches allowed the researchers to avoid collisions in 3D, which was confirmed by an experiment in a real water environment. To detect obstacles in this study, a camera was used. It must be noted that placing the SUR IV in a real water environment with low transparency could interfere with the correct detection of the environment and obstacles and as a result, malfunction of the entire system. The list of algorithms in real applications from 2021 is presented in Table 9.

Table 9. List of algorithms in real applications from 2021.

Research	Algorithm	Main Characteristics
[158]	Reactive behaviours	• Static obstacles • MRF-net to obstacle detection
[159]	Reactive behaviours (APF based)	• Static and dynamic obstacles • Kalman filters for the motion prediction of dynamic obstacles • Multibeam echosounder and FLS
[160]	Merged Ant Colony and PSO algorithms	• Static and dynamic obstacles • 3D path planning • Spherical Underwater Robot

4. Discussion

This section summarises the results of the literature analysis, containing statistics on the use of path planning and collision-avoidance methods. Quantitative statistics include a division into simulation-tested and real-environment validated methods from 2001 to 2020. It also contains classification according to classical and artificial intelligence methods. The next part of this chapter discusses future works and limitations that inhabit the further development in this field.

4.1. Summary

After analysing the literature on motion planning and collision avoidance for AUVs, it is clear that a significant increase in publications in this area occurred only in the last decade (Figure 4). The rise in simulation tests of new, improved, or combined algorithms for path planning and collision avoidance is accompanied by an increase in the number of practical implementations in AUV motion controllers and tests in a real environment. However, the practical implementations of the control and collision-avoidance algorithms are not developed as fast as the simulation methods. A practical approach to the study requires dealing with design and technological problems, which are often time-consuming and costly. Additionally, performing tests in a real environment requires thorough preparation and logistical organisation. Analysing the last 20 years of AUV development, it can also be noticed that experiments in the physical environment are dominated by classical methods (Figure 5). In the case of simulation studies, artificial intelligence methods are used as often as classical methods. Although artificial intelligence algorithms have origins in the previous century and allow for solving very complex problems, their use by researchers has not been dominated in further studies on the development of AUV. Figure 6 shows that in simulation studies, the APF approach is most often used among classic methods, whereas among artificial intelligence algorithms RL and FL methods predominate (Figure 7). For real applications, the most used approach is reeactive behaviors (Figure 8), as can be seen in Figure 9. Artificial intelligence algorithms are rarely used in real implementations. This probably results from the fact that these methods need a lots of data to train the artificial intelligence system. Summarizing the above analysis, it can be stated that in the field of collision avoidance and path planning in underwater environment algorithms, researchers still show great interest in classical methods. Thanks to the research to date, many disadvantages of classic methods have been eliminated, which, combined with the ease of implementation, is still an alternative to the often complex methods of artificial intelligence.

The currently designed AUVs are diverse in terms of intended use. Local collision-avoidance algorithms are usually connected to a global system that analyses and optimises the path. This creates the need to develop many side areas of studies, such as the selection of the path-planning method, preplanning, algorithms and drivers for tracking the designated path, navigation and mapping related to determining the most accurate position of obstacles and the AUV. Moreover, in recent years the development of I-AUVs has also been noticeable. I-AUVs are usually equipped with an arm with several degrees of freedom and are designed to perform inspections and repairs, for example of gas and oil installations [161]. When carrying out specific underwater missions, I-AUVs must have collision-avoidance algorithms for the correct task execution and safe movement. For the I-AUV with several arms, it is also necessary to use collision-avoidance algorithms between arms. Implementing that component to the control system, however, significantly increases the complexity of the controller system. Another type of AUV is a spherical underwater robot (SUR). All control elements, electronic devices, and power sources are within that spherical shape [162–164]. Due to the shape and center of gravity, SURs are characterised by high stability and maneuverability, allowing for a tiny turning radius (even 0 degrees). The obstacle-avoidance and control systems for the multiple AUV formations are also studied more often recently. In such formations, the control is, however, very complex due to the need to simultaneously control multiple vehicles, their communication and cooperation, as well as to perform specific maneuvers when avoiding obstacles and then returning all formations to the previous positions. Currently, AUV control systems mostly use artificial intelligence methods to define the environment and the situation they are in, as well as to find solutions in the event of encountering problems or obstacles. Due to the limitations of individual methods in their basic versions, to increase the efficiency of the AUV motion control system, modifications or combinations of methods are used. That approach in some cases allows one to use the advantages to eliminate the disadvantages of the chosen methods.

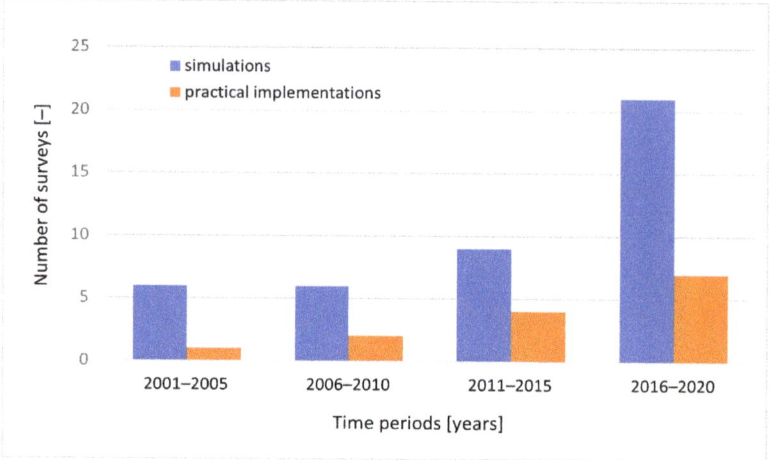

Figure 4. Number of publications in the field of simulation methods and practical implementations of control and collision avoidance systems for AUV in the last 20 years.

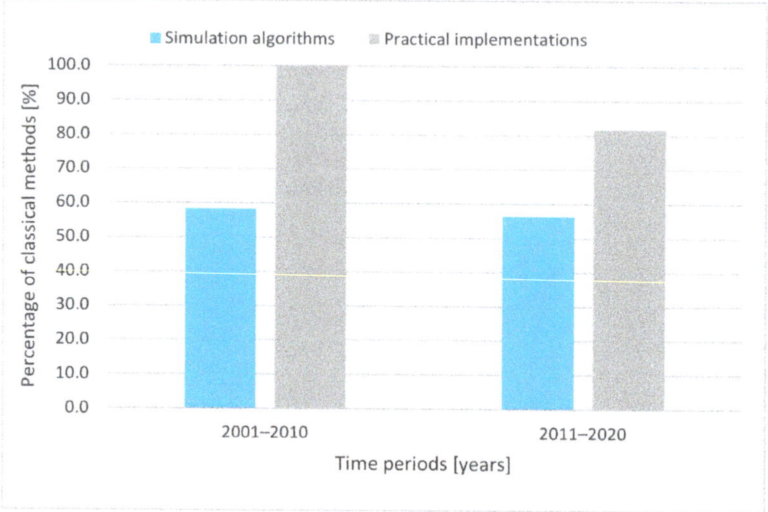

Figure 5. Percentage of classical methods used in simulations studies and real applications in the last 20 years.

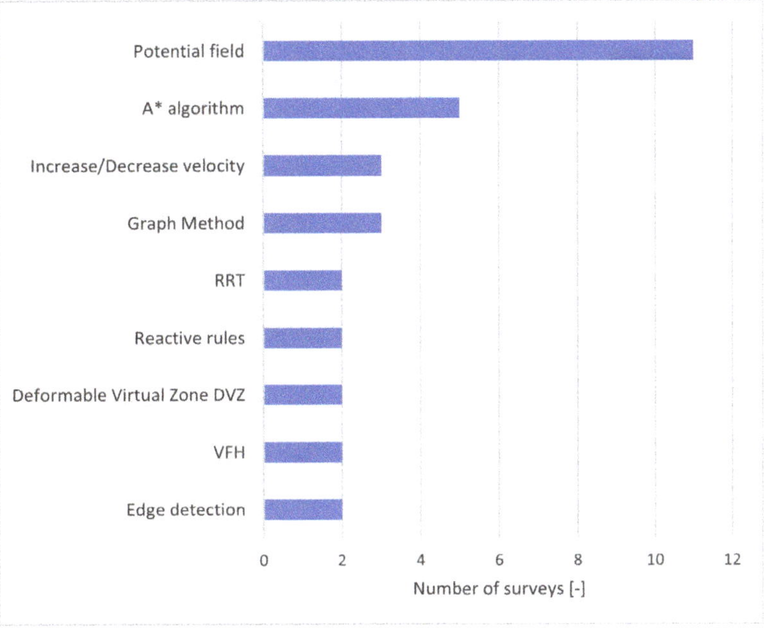

Figure 6. Classical methods validated in numerical research.

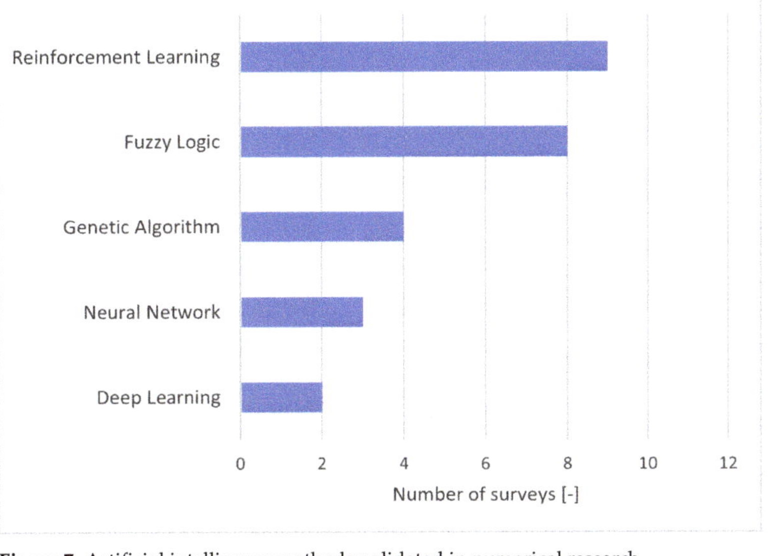

Figure 7. Artificial intelligence methods validated in numerical research.

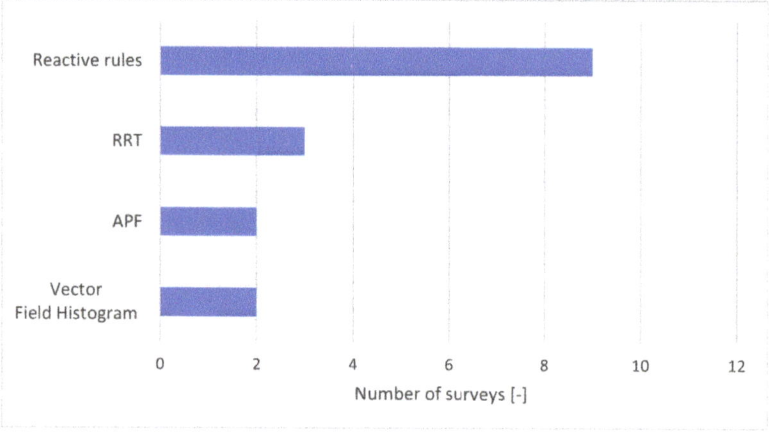

Figure 8. Classical methods in real applications.

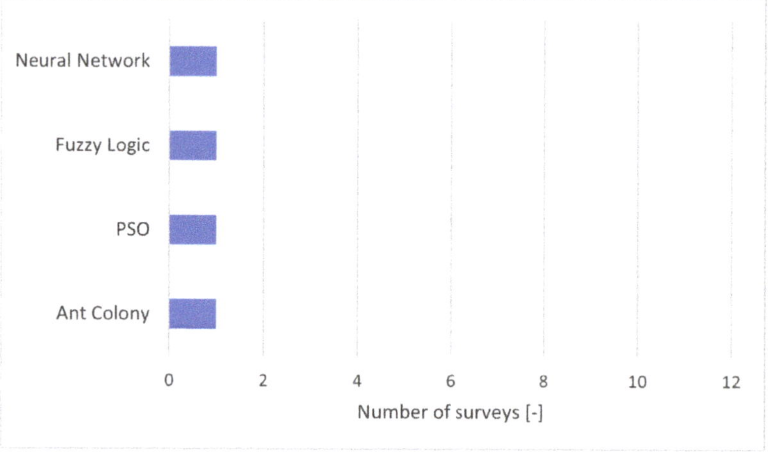

Figure 9. Artificial intelligence methods in real applications.

4.2. Future Works

Over the last few decades, multiple path-planning methods and collision-avoidance methods have been proposed and simulated in the field of AUV technology. Nevertheless, there are only a few practical AUV implementations with an effective and efficient system of optimal path planning and collision avoidance. In the coming years, it will be necessary to focus more on practical implementations by using the existing knowledge gained in simulation studies. Technological development provides better possibilities for more accurate detection, and more precise navigation or increasing the computing speed. The appropriate use of evolving technology is crucial in the further development of collision avoidance and path-planning systems for AUVs. Many problems in this field have already been resolved. However, there are still issues that require extensive research. AUVs are still not able to achieve high speeds during collision-avoidance maneuvers. Additionally, despite the developing technology, the navigational uncertainties still have to be compensated with algorithms that approximate the exact location. Recently, research on the methods of following a designated path and compensating for navigational error has been developing very intensively. Another aspect that requires more analysis and further work is navigation of the AUV in shallow water. This applies particularly to AUVs that use obstacle-detection sonars, where ground and water surface reflections create fictional obstacles.

4.3. Bottlenecks of Future Development

Artificial intelligence methods usually require a large amount of computation when processing 3D data in a complex environment. The vehicle's speed depends on how quickly the system makes decisions when detecting an obstacle or a group of obstacles. Increasing the computing power allows for more efficient operation of the control system and faster data processing and consequently making faster decisions on how to avoid collisions. Ensuring high computing power, however, requires increasing the vehicle's size, which limits the vehicle's maneuvering parameters. Furthermore, highly efficient systems that provide high computing power consume more energy and require more financial outlay. The solution to this problem could be, for example, the creation of wireless charging stations for underwater vehicles; however, there are still limitations related to the specificity of the underwater environment, which inhibits the development of such infrastructure [23]. Thus, the optimal selection of the computing power for path planning and collision-avoidance drivers is crucial. Another aspect that limits the development of AUV in collision avoidance and path planning is the rough conditions in the underwater environment. The greater the operating depth, the higher the technical requirements for the AUV design. The lack of reference points, poor visibility and high signal attenuation make it challenging to navigate AUVs. Moreover, sea currents make it difficult to predict the environmental impact on AUVs. Another issue is the nonlinear dynamics of AUVs. Methods based on machine learning can achieve the ability to correctly control taking into account the dynamics of the vehicle. However, this requires a very long training time and does not guarantee correct vehicle control in situations that differ slightly from those covered by the training. For this reason, vehicle dynamics plays a vital role in practical implementations.

5. Conclusions

This article presented quantitative and qualitative comparisons of path-planning and collision-avoidance systems verified by using numerical tests and also real applications. We performed a literature review and analysis, summarizing the achievements, future works, and aspects limiting the further development of AUVs, especially in terms of path planning and collision avoidance.

Although since the 1990's many of the AUV's collision avoidance-related issues, such as the local minima problem, mapping the optimal path in an unknown environment, 3D path planning, re-planning and real-time decision-making, have been addressed and resolved by using mainly numerical studies, there are still many issues, which have not been entirely resolved, e.g., moving in a complex environment at high speed, nonlinear dynamics of the AUV, detection disruptions related to shallow water and a limited environment, or accurate navigation in the underwater environment. Moreover, it is difficult to indicate the best method or group of methods for collision avoidance, based on the state-of-the-art analysis. Depending on the circumstances, various methods were used with success for different cases. Quite often, hybrid methods showed good efficiency. Each method has disadvantages that reduce its effectiveness in a challenging underwater environment. The presented results show that the most frequently used algorithms tested in the simulated environment are APF, RL, and FL. These algorithms appear to be the most suitable for use in a real environment. Currently, in real implementations, the most frequently used method is the application of behavioural rules based on reacting to changes in the environment in a specific way. Although artificial intelligence methods have long dominated the field of collision avoidance for robots [3], in AUV, they are used in numerical research as often as classical methods. However, in real applications, mainly classical methods are still used.

Funding: This research received no external funding.

Institutional Review Board Statement: Not applicable.

Informed Consent Statement: Not applicable.

Data Availability Statement: Not applicable.

Conflicts of Interest: The authors declare no conflict of interest.

Abbreviations

The following abbreviations are used in this manuscript:

AGVs	Autonomous Ground Vehicles
AI	Artificial Inteligence
ANN	Artificial Neural Network
APF	Artificial Potential Field
ASV	Autonomous Surface Vehicle
AUVs	Autonomous Underwater Vehicles
BAUV	Biomimetic AUV
FL	Fuzzy Logic
FLS	Forward Looking Sonar
GA	Genetic Algorithm
I-AUV	Inspection AUV
PSO	Particle Swarm Optimization
RL	Reinfircement Learning
ROV	Remotely Operated Vehicles
RRT	Rapidly-exploring Random Tree
SUR	Spherical Underwater Robot
UAS	Unmanned Aerial Systems
UAV	Unmanned Aerial Vehicle
USV	Unmanned Surface Vehicle
UUV	Unmanned Underwater Vehicle
UxVs	Unmanned Underwater, Ground and Air Vehicles
VFF	Virtual Force Field
VFH	Vector Field Histogram

References

1. Karabchevsky, S.; Kahana, D.; Guterman, H. Acoustic real-time, low-power FPGA based obstacle detection for AUVs. In Proceedings of the 2010 IEEE 26th Convention of Electrical and Electronics Engineers in Israel, Eilat, Israel, 17–20 November 2010; pp. 655–659.
2. Sayyaadi, H.; Ura, T.; Fujii, T. Collision avoidance controller for AUV systems using stochastic real value reinforcement learning method. In Proceedings of the SICE 2000 IEEE 39th SICE Annual Conference, Iizuka, Japan, 28 July 2000; pp. 165–170.
3. Tang, S.H.; Khaksar, W.; Ismail, N.; Ariffin, M. A review on robot motion planning approaches. *Pertanika J. Sci. Technol.* **2012**, *20*, 15–29.
4. Atyabi, A.; Powers, D. Review of classical and heuristic-based navigation and path planning approaches. *Int. J. Adv. Comput. Technol. (IJACT)* **2013**, *5*, 14.
5. Kamil, F.; Tang, S.; Khaksar, W.; Zulkifli, N.; Ahmad, S. A review on motion planning and obstacle avoidance approaches in dynamic environments. *Adv. Robot. Autom.* **2015**, *4*, 134–142.
6. Yang, L.; Qi, J.; Song, D.; Xiao, J.; Han, J.; Xia, Y. Survey of robot 3D path planning algorithms. *J. Control Sci. Eng.* **2016**, *2016*, 7426913. [CrossRef]
7. Ferguson, D.; Likhachev, M.; Stentz, A. A guide to heuristic-based path planning. In Proceedings of the International Workshop on Planning under Uncertainty for Autonomous Systems, International Conference on Automated Planning and Scheduling (ICAPS), Monterey, CA, USA, 5–10 June 2005; pp. 9–18.
8. Sánchez López, A.; Zapata, R.; Osorio Lama, M.A. Sampling-based motion planning: A survey. *Comput. Sist.* **2008**, *12*, 5–24.
9. Saha, M. *Motion Planning with Probabilistic Roadmaps*; ProQuest: Cambridge, UK, 2006.
10. Biswal, P.; Mohanty, P.K. Development of quadruped walking robots: A review. *Ain Shams Eng. J.* **2021**, *12*, 2017–2031. [CrossRef]
11. González, D.; Pérez, J.; Milanés, V.; Nashashibi, F. A review of motion planning techniques for automated vehicles. *IEEE Trans. Intell. Transp. Syst.* **2015**, *17*, 1135–1145. [CrossRef]
12. Sharma, O.; Sahoo, N.C.; Puhan, N.B. Recent advances in motion and behavior planning techniques for software architecture of autonomous vehicles: A state-of-the-art survey. *Eng. Appl. Artif. Intell.* **2021**, *101*, 104211. [CrossRef]
13. Galceran, E.; Carreras, M. A survey on coverage path planning for robotics. *Robot. Auton. Syst.* **2013**, *61*, 1258–1276. [CrossRef]
14. Bayat, B.; Crasta, N.; Crespi, A.; Pascoal, A.M.; Ijspeert, A. Environmental monitoring using autonomous vehicles: A survey of recent searching techniques. *Curr. Opin. Biotechnol.* **2017**, *45*, 76–84. [CrossRef]
15. Patle, B.; Pandey, A.; Parhi, D.; Jagadeesh, A. A review: On path planning strategies for navigation of mobile robot. *Def. Technol.* **2019**, *15*, 582–606. [CrossRef]

16. Beaver, L.E.; Malikopoulos, A.A. An overview on optimal flocking. *Annu. Rev. Control* **2021**, *51*, 88–99. [CrossRef]
17. Kamel, M.A.; Yu, X.; Zhang, Y. Formation control and coordination of multiple unmanned ground vehicles in normal and faulty situations: A review. *Annu. Rev. Control* **2020**, *49*, 128–144. [CrossRef]
18. Madridano, Á.; Al-Kaff, A.; Martín, D.; de la Escalera, A. Trajectory planning for multi-robot systems: Methods and applications. *Expert Syst. Appl.* **2021**, *173*, 114660. [CrossRef]
19. Ragel, R.; Maza, I.; Caballero, F.; Ollero, A. Comparison of motion planning techniques for a multi-rotor UAS equipped with a multi-joint manipulator arm. In Proceedings of the 2015 IEEE Workshop on Research, Education and Development of Unmanned Aerial Systems (RED-UAS), Cancún, México, 23–25 November 2015; pp. 133–141.
20. Fraga-Lamas, P.; Ramos, L.; Mondéjar-Guerra, V.; Fernández-Caramés, T.M. A review on IoT deep learning UAV systems for autonomous obstacle detection and collision avoidance. *Remote Sens.* **2019**, *11*, 2144. [CrossRef]
21. Goerzen, C.; Kong, Z.; Mettler, B. A survey of motion planning algorithms from the perspective of autonomous UAV guidance. *J. Intell. Robot. Syst.* **2010**, *57*, 65–100. [CrossRef]
22. Naus, K.; Szymak, P.; Piskur, P.; Niedziela, M.; Nowak, A. Methodology for the Correction of the Spatial Orientation Angles of the Unmanned Aerial Vehicle Using Real Time GNSS, a Shoreline Image and an Electronic Navigational Chart. *Energies* **2021**, *14*, 2810. [CrossRef]
23. Mohsan, S.A.H.; Islam, A.; Khan, M.A.; Mahmood, A.; Rokia, L.S.; Mazinani, A.; Amjad, H. A Review on Research Challenges, Limitations and Practical Solutions for Underwater Wireless Power Transfer. *Int. J. Adv. Comput. Sci. Appl.* **2020**, *11*, 554–562. [CrossRef]
24. Al-Kaff, A.; Martin, D.; Garcia, F.; de la Escalera, A.; Armingol, J.M. Survey of computer vision algorithms and applications for unmanned aerial vehicles. *Expert Syst. Appl.* **2018**, *92*, 447–463. [CrossRef]
25. Huang, S.; Teo, R.S.H.; Tan, K.K. Collision avoidance of multi unmanned aerial vehicles: A review. *Annu. Rev. Control* **2019**, *48*, 147–164. [CrossRef]
26. Zereik, E.; Bibuli, M.; Mišković, N.; Ridao, P.; Pascoal, A. Challenges and future trends in marine robotics. *Annu. Rev. Control* **2018**, *46*, 350–368. [CrossRef]
27. Neira, J.; Sequeiros, C.; Huamani, R.; Machaca, E.; Fonseca, P.; Nina, W. Review on unmanned underwater robotics, structure designs, materials, sensors, actuators, and navigation control. *J. Robot.* **2021**, *2021*, 5542920. [CrossRef]
28. Orlowski, M. Directions of development of the autonomous unmanned underwater vehicles. A review. *Parameters* **2021**, *2*, 24. [CrossRef]
29. Yildiz, Ö.; Gökalp, R.B.; Yilmaz, A.E. A review on motion control of the underwater vehicles. In Proceedings of the 2009 IEEE International Conference on Electrical and Electronics Engineering-ELECO 2009, Bursa, Turkey, 5–8 November 2009; pp. II-337–II-341.
30. Cheng, C.; Sha, Q.; He, B.; Li, G. Path planning and obstacle avoidance for AUV: A review. *Ocean Eng.* **2021**, *235*, 109355. [CrossRef]
31. Hart, P.E.; Nilsson, N.J.; Raphael, B. A formal basis for the heuristic determination of minimum cost paths. *IEEE Trans. Syst. Sci. Cybern.* **1968**, *4*, 100–107. [CrossRef]
32. Eichhorn, M. An obstacle avoidance system for an autonomous underwater vehicle. In Proceedings of the 2004 IEEE International Symposium on Underwater Technology (IEEE Cat. No. 04EX869), Taipei, Taiwan, 20–23 April 2004; pp. 75–82.
33. Casalino, G.; Turetta, A.; Simetti, E. A three-layered architecture for real time path planning and obstacle avoidance for surveillance USVs operating in harbour fields. In Proceedings of the Oceans 2009 IEEE-Europe, Bremen, Germany, 11–14 May 2009; pp. 1–8.
34. Li, J.H.; Lee, M.J.; Park, S.H.; Kim, J.G. Real time path planning for a class of torpedo-type AUVs in unknown environment. In Proceedings of the 2012 IEEE/OES Autonomous Underwater Vehicles (AUV), Southampton, UK, 24–27 September 2012; pp. 1–6.
35. Li, M.; Zhang, H. AUV 3D path planning based on A* algorithm. In Proceedings of the 2020 IEEE Chinese Automation Congress (CAC), Shanghai, China, 6–8 November 2020; pp. 11–16.
36. Khatib, O. Real-time obstacle avoidance for manipulators and mobile robots. In *Autonomous Robot Vehicles*; Springer: New York, NY, USA, 1986; pp. 396–404.
37. Koren, Y.; Borenstein, J. Potential field methods and their inherent limitations for mobile robot navigation. In Proceedings of the ICRA, Sacramento, CA, USA, 9–11 April 1991; Volume 2, pp. 1398–1404.
38. Teo, K.; Ong, K.W.; Lai, H.C. Obstacle detection, avoidance and anti collision for MEREDITH AUV. In Proceedings of the OCEANS IEEE 2009, Biloxi, MS, USA, 26–29 October 2009; pp. 1–10.
39. Qing, L.; Li-jun, W.; Bo, C.; Zhou, Z.; Yi-xin, Y. An improved artificial potential field method with parameters optimization based on genetic algorithms. *J. Univ. Sci. Technol.* **2012**, *34*, 202–206.
40. Barraquand, J.; Latombe, J.C. Robot motion planning: A distributed representation approach. *Int. J. Robot. Res.* **1991**, *10*, 628–649. [CrossRef]
41. Solari, F.J.; Rozenfeld, A.F.; Villar, S.A.; Acosta, G.G. Artificial potential fields for the obstacles avoidance system of an AUV using a mechanical scanning sonar. In Proceedings of the 2016 3rd IEEE/OES South American International Symposium on Oceanic Engineering (SAISOE), Buenos Aires, Arhentina, 14–17 June 2016; pp. 1–6.
42. Youakim, D.; Ridao, P. Motion planning survey for autonomous mobile manipulators underwater manipulator case study. *Robot. Auton. Syst.* **2018**, *107*, 20–44. [CrossRef]

43. Leonard, N.E.; Fiorelli, E. Virtual leaders, artificial potentials and coordinated control of groups. In Proceedings of the 40th IEEE Conference on Decision and Control (Cat. No. 01CH37228), Orlando, FL, USA, 4–7 December 2001; Volume 3, pp. 2968–2973.
44. Gazi, V. Swarm aggregations using artificial potentials and sliding-mode control. *IEEE Trans. Robot.* **2005**, *21*, 1208–1214. [CrossRef]
45. Olfati-Saber, R. Flocking for multi-agent dynamic systems: Algorithms and theory. *IEEE Trans. Autom. Control* **2006**, *51*, 401–420. [CrossRef]
46. Meng, Z.; Lin, Z.; Ren, W. Leader–follower swarm tracking for networked Lagrange systems. *Syst. Control Lett.* **2012**, *61*, 117–126. [CrossRef]
47. Hou, S.P.; Cheah, C.C. Can a simple control scheme work for a formation control of multiple autonomous underwater vehicles? *IEEE Trans. Control Syst. Technol.* **2010**, *19*, 1090–1101. [CrossRef]
48. Sorbi, L.; De Capua, G.P.; Toni, L.; Fontaine, J.G. Target detection and recognition: A mission planner for Autonomous Underwater Vehicles. In Proceedings of the IEEE OCEANS'11 MTS/IEEE KONA, Waikoloa, HI, USA, 19–22 September 2011; pp. 1–5.
49. Horner, D.; Healey, A.; Kragelund, S. AUV experiments in obstacle avoidance. In Proceedings of the IEEE OCEANS 2005 MTS/IEEE, Washington, DC, USA, 17–23 September 2005; pp. 1464–1470.
50. LaValle, S.M. Rapidly-Exploring Random Trees: A New Tool for Path Planning. Available online: https://www.cs.csustan.edu/~xliang/Courses/CS4710-21S/Papers/06 (accessed on 1 July 2022).
51. Tan, C.S.; Sutton, R.; Chudley, J. An integrated collision avoidance system for autonomous underwater vehicles. *Int. J. Control* **2007**, *80*, 1027–1049. [CrossRef]
52. Hernández, J.D.; Vidal, E.; Vallicrosa, G.; Galceran, E.; Carreras, M. Online path planning for autonomous underwater vehicles in unknown environments. In Proceedings of the 2015 IEEE International Conference on Robotics and Automation (ICRA), Seattle, WA, USA, 26–30 May 2015; pp. 1152–1157.
53. Cui, R.; Li, Y.; Yan, W. Mutual information-based multi-AUV path planning for scalar field sampling using multidimensional RRT. *IEEE Trans. Syst. Man Cybern. Syst.* **2015**, *46*, 993–1004. [CrossRef]
54. Xanthidis, M.; Karapetyan, N.; Damron, H.; Rahman, S.; Johnson, J.; O'Connell, A.; O'Kane, J.M.; Rekleitis, I. Navigation in the presence of obstacles for an agile autonomous underwater vehicle. In Proceedings of the 2020 IEEE International Conference on Robotics and Automation (ICRA), Paris, France, 31 May–31 August 2020; pp. 892–899.
55. Tan, C.S. A Collision Avoidance System for Autonomous Underwater Vehicles. Available online: https://pearl.plymouth.ac.uk/handle/10026.1/2258 (accessed on 1 July 2022).
56. Piskur, P.; Szymak, P.; Jaskólski, K.; Flis, L.; Gąsiorowski, M. Hydroacoustic system in a biomimetic underwater vehicle to avoid collision with vessels with low-speed propellers in a controlled environment. *Sensors* **2020**, *20*, 968. [CrossRef]
57. Piskur, P.; Gasiorowski, M. Digital Signal Processing for Hydroacoustic System in Biomimetic Underwater Vehicle. *NAŠE MORE Znan. časopis Za More I Pomor.* **2020**, *67*, 14–18. [CrossRef]
58. Tan, C.S.; Sutton, R.; Chudley, J. An incremental stochastic motion planning technique for autonomous underwater vehicles. *IFAC Proc.* **2004**, *37*, 483–488. [CrossRef]
59. Zhao, S.; Lu, T.F.; Anvar, A. Multiple obstacles detection using fuzzy interface system for auv navigation in natural water. In Proceedings of the 2010 IEEE 5th Conference on Industrial Electronics and Applications, Taichung, Taiwan, 15–17 June 2010; pp. 50–55.
60. Piskur, P.; Szymak, P.; Flis, L.; Sznajder, J. Analysis of a Fin Drag Force in a Biomimetic Underwater Vehicle. *NAŠE MORE Znan. časopis Za More I Pomor.* **2020**, *67*, 192–198. [CrossRef]
61. Piskur, P.; Szymak, P.; Przybylski, M.; Naus, K.; Jaskólski, K.; Żokowski, M. Innovative energy-saving propulsion system for low-speed biomimetic underwater vehicles. *Energies* **2021**, *14*, 8418. [CrossRef]
62. Szymak, P.; Piskur, P.; Naus, K. The Effectiveness of Using a Pretrained Deep Learning Neural Networks for Object Classification in Underwater Video. *Remote Sens.* **2020**, *12*, 3020. [CrossRef]
63. Meng, F.; Liu, A.; Jing, S.; Zu, Y. FSM trajectory tracking controllers of OB-AUV in the horizontal plane. In Proceedings of the 2021 IEEE International Conference on Intelligence and Safety for Robotics (ISR), Tokoname, Japan, 4–6 March 2021; pp. 204–208.
64. DeMuth, G.; Springsteen, S. Obstacle avoidance using neural networks. In Proceedings of the IEEE Symposium on Autonomous Underwater Vehicle Technology, Washington, DC, USA, 5–6 June 1990; pp. 213–215.
65. Guerrero-González, A.; García-Córdova, F.; Gilabert, J. A biologically inspired neural network for navigation with obstacle avoidance in autonomous underwater and surface vehicles. In Proceedings of the OCEANS 2011 IEEE, Santander, Spain, 6–9 June 2011; pp. 1–8.
66. Ding, G.; Zhu, D.; Sun, B. Formation control and obstacle avoidance of multi-AUV for 3-D underwater environment. In Proceedings of the 33rd IEEE Chinese Control Conference, Nanjing, China, 28–30 July 2014; pp. 8347–8352.
67. Gaya, J.O.; Gonçalves, L.T.; Duarte, A.C.; Zanchetta, B.; Drews, P.; Botelho, S.S. Vision-based obstacle avoidance using deep learning. In Proceedings of the 2016 XIII IEEE Latin American Robotics Symposium and IV Brazilian Robotics Symposium (LARS/SBR), Recife, Brazil, 8–12 October 2016; pp. 7–12.
68. Jurczyk, K.; Piskur, P.; Szymak, P. Parameters identification of the flexible fin kinematics model using vision and Genetic Algorithms. *Pol. Marit. Res.* **2020**, *27*, 39–47. [CrossRef]
69. Alvarez, A.; Caiti, A.; Onken, R. Evolutionary path planning for autonomous underwater vehicles in a variable ocean. *IEEE J. Ocean. Eng.* **2004**, *29*, 418–429. [CrossRef]

70. Yan, S.; Pan, F. Research on route planning of auv based on genetic algorithms. In Proceedings of the 2019 IEEE International Conference on Unmanned Systems and Artificial Intelligence (ICUSAI), Xi'an, China, 22–24 November 2019; pp. 184–187.
71. Wu, X.; Feng, Z.; Zhu, J.; Allen, R. Line of sight guidance with intelligent obstacle avoidance for autonomous underwater vehicles. In Proceedings of the OCEANS 2006 IEEE, Singapore, 16–19 May 2006; pp. 1–6.
72. Chang, Z.H.; Tang, Z.D.; Cai, H.G.; Shi, X.C.; Bian, X.Q. GA path planning for AUV to avoid moving obstacles based on forward looking sonar. In Proceedings of the 2005 IEEE International Conference on Machine Learning and Cybernetics, Waikoloa, HI, USA, 10–12 October 2005; Volume 3, pp. 1498–1502.
73. Yao, P.; Zhao, S. Three-dimensional path planning for AUV based on interfered fluid dynamical system under ocean current (June 2018). *IEEE Access* **2018**, *6*, 42904–42916. [CrossRef]
74. Zadeh, L. Fuzzy sets. *Inf. Control* **1965**, *8*, 338–353. [CrossRef]
75. Zadeh, L. Fuzzy algorithms. *Inf. Control* **1968**, *12*, 94–102. [CrossRef]
76. Galarza, C.; Masmitja, I.; Prat, J.; Gomaríz, S. Design of obstacle detection and avoidance system for Guanay II AUV. In Proceedings of the 2016 IEEE 24th Mediterranean Conference on Control and Automation (MED), Athens, Greece, 21–24 June 2016; pp. 410–414.
77. Zhu, D.; Yang, Y.; Yan, M. Path planning algorithm for AUV based on a Fuzzy-PSO in dynamic environments. In Proceedings of the 2011 IEEE Eighth International Conference on Fuzzy Systems and Knowledge Discovery (FSKD), Shanghai, China, 26–28 July 2011; Volume 1, pp. 525–530.
78. Li, X.; Wang, W.; Song, J.; Liu, D. Path planning for autonomous underwater vehicle in presence of moving obstacle based on three inputs fuzzy logic. In Proceedings of the 2019 IEEE 4th Asia-Pacific Conference on Intelligent Robot Systems (ACIRS), Nagoya, Japan, 13–15 July 2019; pp. 265–268.
79. Braginsky, B.; Guterman, H. Obstacle avoidance approaches for autonomous underwater vehicle: Simulation and experimental results. *IEEE J. Ocean. Eng.* **2016**, *41*, 882–892. [CrossRef]
80. Szymak, P. *Zorientowany na Sterowanie Model Ruchu oraz Neuro-Ewolucyjno-Rozmyta Metoda Sterowania bezzałOgowymi Jednostkami Pływającymi*; Politechnika Krakowska: Krakow, Poland, 2015.
81. Huang, W.; Fang, H.; Li, L. Obstacle avoiding policy of multi-AUV formation based on virtual AUV. In Proceedings of the 2009 IEEE Sixth International Conference on Fuzzy Systems and Knowledge Discovery, Tianjin, China, 14–16 August 2009; Volume 4, pp. 131–135.
82. Sutton, R.S.; Barto, A.G. Introduction to Reinforcement Learning. Introduction to Reinforcement Learning. Available online: https://login.cs.utexas.edu/sites/default/files/legacy_files/research/documents/1%20intro%20up%20to%20RL%3ATD.pdf (accessed on 1 July 2022).
83. Szepesvári, C. Algorithms for reinforcement learning. *Synth. Lect. Artif. Intell. Mach. Learn.* **2010**, *4*, 1–103.
84. Arai, Y.; Fujii, T.; Asama, H.; Kaetsu, H.; Endo, I. Collision avoidance in multi-robot systems based on multi-layered reinforcement learning. *Robot. Auton. Syst.* **1999**, *29*, 21–32. [CrossRef]
85. Du, W.; Guo, T.; Chen, J.; Li, B.; Zhu, G.; Cao, X. Cooperative pursuit of unauthorized UAVs in urban airspace via Multi-agent reinforcement learning. *Transp. Res. Part C Emerg. Technol.* **2021**, *128*, 103122. [CrossRef]
86. Chen, C.; Chen, X.Q.; Ma, F.; Zeng, X.J.; Wang, J. A knowledge-free path planning approach for smart ships based on reinforcement learning. *Ocean Eng.* **2019**, *189*, 106299. [CrossRef]
87. Gore, R.; Pattanaik, K.; Bharti, S. Efficient Re-Planned Path for Autonomous Underwater Vehicle in Random Obstacle Scenario. In Proceedings of the 2019 IEEE 5th International Conference for Convergence in Technology (I2CT), Bombay, India, 29–31 March 2019; pp. 1–5.
88. Li, W.; Yang, X.; Yan, J.; Luo, X. An obstacle avoiding method of autonomous underwater vehicle based on the reinforcement learning. In Proceedings of the 2020 IEEE 39th Chinese Control Conference (CCC), Shenyang, China, 27–29 July 2020; pp. 4538–4543.
89. Williams, G.N.; Lagace, G.E.; Woodfin, A. A collision avoidance controller for autonomous underwater vehicles. In Proceedings of the IEEE Symposium on Autonomous Underwater Vehicle Technology, Washington, DC, USA, 5–6 June 1990; pp. 206–212.
90. Antonelli, G.; Chiaverini, S.; Finotello, R.; Schiavon, R. Real-time path planning and obstacle avoidance for RAIS: An autonomous underwater vehicle. *IEEE J. Ocean. Eng.* **2001**, *26*, 216–227. [CrossRef]
91. Creuze, V.; Jouvencel, B. Avoidance of underwater cliffs for autonomous underwater vehicles. In Proceedings of the IEEE/RSJ International Conference on Intelligent Robots and Systems, Lausanne, Switzerland, 30 September–4 October 2002; Volume 1, pp. 793–798.
92. Healey, A.J. Obstacle avoidance while bottom following for the REMUS autonomous underwater vehicle. *IFAC Proc.* **2004**, *37*, 251–256.
93. Quidu, I.; Hétet, A.; Dupas, Y.; Lefèvre, S. AUV (Redermor) obstacle detection and avoidance experimental evaluation. In Proceedings of the IEEE OCEANS 2007-Europe, Aberdeen, Scotland, 18–21 June 2007; pp. 1–6.
94. Jia, Q.; Li, G. Formation control and obstacle avoidance algorithm of multiple autonomous underwater vehicles (AUVs) based on potential function and behavior rules. In Proceedings of the 2007 IEEE International Conference on Automation and Logistics, Jinan, China, 18–21 August 2007; pp. 569–573.
95. Fan, S.; Feng, Z.; Lian, L. Collision free formation control for multiple autonomous underwater vehicles. In Proceedings of the OCEANS'10 IEEE Sydney, Sydney, Australia, 24–27 May 2010; pp. 1–4.

96. Conte, G.; Zanoli, S. A sonar based obstacle avoidance system for AUVs. In Proceedings of the IEEE Symposium on Autonomous Underwater Vehicle Technology (AUV'94), Cambridge, MA, USA, 19–20 July 1994; pp. 85–91.
97. Zapata, R.; Lepinay, P. Collision avoidance and bottom following of a torpedo-like AUV. In Proceedings of the OCEANS 96 MTS/IEEE Conference Proceedings, The Coastal Ocean-Prospects for the 21st Century, Fort Lauderdale, FL, USA, 23–26 September 1996; Volume 2, pp. 571–575.
98. Petillot, Y.; Ruiz, I.T.; Lane, D.M. Underwater vehicle obstacle avoidance and path planning using a multi-beam forward looking sonar. *IEEE J. Ocean. Eng.* **2001**, *26*, 240–251. [CrossRef]
99. Fodrea, L.R.; Healey, A.J. Obstacle avoidance control for the REMUS autonomous underwater vehicle. *IFAC Proc.* **2003**, *36*, 103–108. [CrossRef]
100. Kawano, H. Real-time obstacle avoidance for underactuated autonomous underwater vehicles in unknown vortex sea flow by the mdp approach. In Proceedings of the 2006 IEEE/RSJ International Conference on Intelligent Robots and Systems, Beijing, China, 9–13 October 2006; pp. 3024–3031.
101. Barišić, M.; Vukić, Z.; Mišković, N. A kinematic virtual potentials trajectory planner for AUV-s. *IFAC Proc. Vol.* **2007**, *40*, 90–95. [CrossRef]
102. Yakimenko, O.A. Real-Time Computation of Spatial and Flat Obstacle Avoidance Trajectories for UUVs. *IFAC Proc.* **2008**, *41*, 202–207. [CrossRef]
103. Eichhorn, M. A new concept for an obstacle avoidance system for the AUV "SLOCUM Glider" operation under ice. In Proceedings of the IEEE Oceans 2009-Europe, Bremen, Germany, 11–14 May 2009; pp. 1–8.
104. Xu, H.; Feng, X. An AUV fuzzy obstacle avoidance method under event feedback supervision. In Proceedings of the IEEE Oceans 2009, Biloxi, MS, USA, 26–29 October 2009; pp. 1–6.
105. Wadhams, P.; Wilkinson, J.P.; McPhail, S. A new view of the underside of Arctic sea ice. *Geophys. Res. Lett.* **2006**, *33*. [CrossRef]
106. Pebody, M. Autonomous underwater vehicle collision avoidance for under-ice exploration. *Proc. Inst. Mech. Eng. Part M J. Eng. Marit. Environ.* **2008**, *222*, 53–66. [CrossRef]
107. McPhail, S.D.; Furlong, M.E.; Pebody, M.; Perrett, J.; Stevenson, P.; Webb, A.; White, D. Exploring beneath the PIG Ice Shelf with the Autosub3 AUV. In Proceedings of the IEEE Oceans 2009-Europe, Bremen, Germany, 11–14 May 2009; pp. 1–8.
108. Braginsk, B.; Karabchevsk, S.; Guterma, H. Two layers obstacle avoidance algorithm for autonomous underwater vehicle. In Proceedings of the 2012 IEEE 27th Convention of Electrical and Electronics Engineers in Israel, Eilat, Israel, 14–17 November 2012; pp. 1–5.
109. Karabchevsky, S.; Braginsky, B.; Guterman, H. AUV real-time acoustic vertical plane obstacle detection and avoidance. In Proceedings of the 2012 IEEE/OES Autonomous Underwater Vehicles (AUV), Southampton, UK, 24–27 September 2012; pp. 1–6.
110. Cheng, C.; Zhu, D.; Sun, B.; Chu, Z.; Nie, J.; Zhang, S. Path planning for autonomous underwater vehicle based on artificial potential field and velocity synthesis. In Proceedings of the 2015 IEEE 28th Canadian Conference on Electrical and Computer Engineering (CCECE), Halifax, NS, Canada, 3–6 May 2015; pp. 717–721.
111. Subramanian, S.; George, T.; Thondiyath, A. Hardware-in-the-Loop verification for 3D obstacle avoidance algorithm of an underactuated flat-fish type AUV. In Proceedings of the 2012 IEEE International Conference on Robotics and Biomimetics (ROBIO), Guangzhou, China, 11–14 December 2012; pp. 545–550.
112. Yang, Y.; Zhu, D. Research on dynamic path planning of auv based on forward looking sonar and fuzzy control. In Proceedings of the 2011 IEEE Chinese Control and Decision Conference (CCDC), Mianyang, China, 23–25 May 2011; pp. 2425–2430.
113. Sun, Y.; Wu, H.; Zhang, Y.; Li, C. Obstacle avoidance of autonomous underwater vehicle based on improved balance of motion. In Proceedings of the 2013 IEEE 2nd International Conference on Measurement, Information and Control, Harbin, China, 16–18 August 2013; Volume 2, pp. 788–792.
114. Wang, H.; Wang, L.; Li, J.; Pan, L. A vector polar histogram method based obstacle avoidance planning for AUV. In Proceedings of the 2013 MTS/IEEE OCEANS-Bergen, Bergen, Norway, 10–14 June 2013; pp. 1–5.
115. Millar, G. An obstacle avoidance system for autonomous underwater vehicles: A reflexive vector field approach utilizing obstacle localization. In Proceedings of the 2014 IEEE/OES Autonomous Underwater Vehicles (AUV), Oxford, MS, USA, 6–9 October 2014; pp. 1–4.
116. Xu, H.; Gao, L.; Liu, J.; Wang, Y.; Zhao, H. Experiments with obstacle and terrain avoidance of autonomous underwater vehicle. In Proceedings of the Oceans 2015-MTS/IEEE Washington, Washington, DC, USA, 19–22 October 2015; pp. 1–4.
117. Yan, Z.; Li, J.; Jiang, A.; Wang, L. An Obstacle Avoidance Algorithm for AUV Based on Obstacle's Detected Outline. In Proceedings of the 2018 IEEE 37th Chinese Control Conference (CCC), Wuhan, China, 25–27 July 2018; pp. 5257–5262.
118. Chu, Z.; Zhu, D. Obstacle Avoidance Trajectory Planning and Trajectory Tracking Control for Autonomous Underwater Vehicles. In Proceedings of the 2018 IEEE 13th World Congress on Intelligent Control and Automation (WCICA), Changsha China, 4–8 July 2018; pp. 450–454.
119. Haoran, L.; Hongjun, D.; Yakun, Z. Avoidance for AUV with mobile obstacles based on current interference. In Proceedings of the 2019 3rd International Symposium on Autonomous Systems (ISAS), Shanghai, China, 29–31 May 2019; pp. 137–142. [CrossRef]
120. Sahu, B.K.; Subudhi, B. Flocking control of multiple AUVs based on fuzzy potential functions. *IEEE Trans. Fuzzy Syst.* **2017**, *26*, 2539–2551. [CrossRef]
121. Li, X.; Zhu, D. An adaptive SOM neural network method for distributed formation control of a group of AUVs. *IEEE Trans. Ind. Electron.* **2018**, *65*, 8260–8270. [CrossRef]

122. Wu, J.; Song, C.; Fan, C.; Hawbani, A.; Zhao, L.; Sun, X. DENPSO: A distance evolution nonlinear PSO algorithm for energy-efficient path planning in 3D UASNs. *IEEE Access* **2019**, *7*, 105514–105530. [CrossRef]
123. Hou, X.; Pan, X.; Bi, M. Research on Non-collision Coordination Control Technology for Formation Autonomous Underwater Vehicle. In Proceedings of the 2019 IEEE 3rd Advanced Information Management, Communicates, Electronic and Automation Control Conference (IMCEC), Chongqing, China, 11–13 October 2019; pp. 1178–1183.
124. Zhu, Y. A multi-AUV searching algorithm based on neuron network with obstacle. In Proceedings of the 2019 IEEE 3rd International Symposium on Autonomous Systems (ISAS), Shanghai, China, 29–31 May 2019; pp. 131–136.
125. Ma, X.; Yanli, C.; Bai, G.; Liu, J. Multi-AUV collaborative operation based on time-varying navigation map and dynamic grid model. *IEEE Access* **2020**, *8*, 159424–159439. [CrossRef]
126. Grefstad, Ø.; Schjølberg, I. Navigation and collision avoidance of underwater vehicles using sonar data. In Proceedings of the 2018 IEEE/OES Autonomous Underwater Vehicle Workshop (AUV), Porto, Portugal, 6–9 November 2018; pp. 1–6.
127. Praczyk, T.; Szymak, P.; Naus, K.; Pietrukaniec, L.; Hożyń, S. Report on research with biomimetic autonomous underwater vehicle—Navigation and autonomous operation. *Zesz. Nauk. Akad. Mar. Wojennej* **2018**, *59*. [CrossRef]
128. Yu, H.; Lu, W.; Liu, D. A unified closed-loop motion planning approach for an I-AUV in cluttered environment with localization uncertainty. In Proceedings of the 2019 IEEE International Conference on Robotics and Automation (ICRA), Montreal, QC, Canada, 20–24 May 2019; pp. 4646–4652.
129. Felski, A.; Jaskolski, K.; Zwolak, K.; Piskur, P. Analysis of Satellite Compass Error's Spectrum. *Sensors* **2020**, *20*, 4067. [CrossRef]
130. Ferreira, F.; Petroccia, R.; Alves, J. Underwater/surface collision avoidance using underwater acoustic communications-a preliminary analysis. In Proceedings of the IEEE Oceans 2019-Marseille, Marseille, France, 17–20 June 2019; pp. 1–10.
131. Eriksen, B.O.H.; Breivik, M.; Pettersen, K.Y.; Wiig, M.S. A modified dynamic window algorithm for horizontal collision avoidance for AUVs. In Proceedings of the 2016 IEEE Conference on Control Applications (CCA), Buenos Aires, Argentina, 19–22 September 2016; pp. 499–506.
132. Simoni, R.; Rodríguez, P.R.; Cieślak, P.; Youakim, D. A novel approach to obstacle avoidance for an I-AUV. In Proceedings of the 2018 IEEE/OES Autonomous Underwater Vehicle Workshop (AUV), Porto, Portugal, 6–9 November 2018; pp. 1–6.
133. Kappagantula, S.; Ramadass, G.; Adlinge, S.D. Design of a Biomimetic Robot Fish for Realization of Coefficient of Drag with Control Architecture and Fuzzy Logic Algorithm for Autonomous Obstacle Avoidance. In Proceedings of the 2018 IEEE 3rd International Conference for Convergence in Technology (I2CT), Pune, India, 6–8 April 2018; pp. 1–9.
134. Yan, Z.; Li, J.; Wu, Y.; Yang, Z. A novel path planning for AUV based on objects' motion parameters predication. *IEEE Access* **2018**, *6*, 69304–69320. [CrossRef]
135. Vibhute, S. Adaptive dynamic programming based motion control of autonomous underwater vehicles. In Proceedings of the 2018 IEEE 5th International Conference on Control, Decision and Information Technologies (CoDIT), Thessaloniki, Greece, 10–13 April 2018; pp. 966–971.
136. Liu, R.D.; Chen, Z.G.; Wang, Z.J.; Zhan, Z.H. Intelligent path planning for AUVs in dynamic environments: An EDA-based learning fixed height histogram approach. *IEEE Access* **2019**, *7*, 185433–185446. [CrossRef]
137. Fanelli, F.; Fenucci, D.; Marlow, R.; Pebody, M.; Phillips, A.B. Development of a multi-platform obstacle avoidance system for autonomous underwater vehicles. In Proceedings of the 2020 IEEE/OES Autonomous Underwater Vehicles Symposium (AUV), St Johns, NL, Canada, 30 September–2 October 2020; pp. 1–6.
138. Liu, Y.; Liu, L.; Yu, X.; Wang, C. Optimal Path Planning Algorithm of AUV State Space Sampling Based on Improved Cost Function. In Proceedings of the 2020 IEEE 39th Chinese Control Conference (CCC), Shenyang, China, 27–29 July 2020; pp. 3747–3752.
139. Praczyk, T. Neural collision avoidance system for biomimetic autonomous underwater vehicle. *Soft Comput.* **2020**, *24*, 1315–1333. [CrossRef]
140. McMahon, J.; Plaku, E. Mission and motion planning for autonomous underwater vehicles operating in spatially and temporally complex environments. *IEEE J. Ocean. Eng.* **2016**, *41*, 893–912. [CrossRef]
141. Okamoto, A.; Sasano, M.; Seta, T.; Inaba, S.; Sato, K.; Tamura, K.; Nishida, Y.; Ura, T. Obstacle avoidance method appropriate for the steep terrain of the deep seafloor. In Proceedings of the 2016 IEEE Techno-Ocean (Techno-Ocean), Kobe, Japan, 6–8 October 2016; pp. 195–198.
142. McEwen, R.S.; Rock, S.P.; Hobson, B. Iceberg wall following and obstacle avoidance by an AUV. In Proceedings of the 2018 IEEE/OES Autonomous Underwater Vehicle Workshop (AUV), Porto, Portugal, 6–9 November 2018; pp. 1–8.
143. Okamoto, A.; Sasano, M.; Seta, T.; Hirao, S.C.; Inaba, S. Deployment of the auv hobalin to an active hydrothermal vent field with an improved obstacle avoidance system. In Proceedings of the 2018 IEEE OCEANS-MTS/IEEE Kobe Techno-Oceans (OTO), Kobe, Japan, 28–31 May 2018; pp. 1–6.
144. Hernández, J.D.; Vidal, E.; Moll, M.; Palomeras, N.; Carreras, M.; Kavraki, L.E. Online motion planning for unexplored underwater environments using autonomous underwater vehicles. *J. Field Robot.* **2019**, *36*, 370–396. [CrossRef]
145. Zheng, J.; Song, L.; Liu, L.; Yu, W.; Wang, Y.; Chen, C. Fixed-time sliding mode tracking control for autonomous underwater vehicles. *Appl. Ocean Res.* **2021**, *117*, 102928. [CrossRef]
146. Chen, Y.L.; Hu, X.Y.; Ma, X.W.; Bai, G.Q. Adaptive location correction and path re-planning based on error estimation method in underwater sensor networks. *Ocean Eng.* **2022**, *252*, 111257. [CrossRef]
147. Elhaki, O.; Shojaei, K.; Mehrmohammadi, P. Reinforcement learning-based saturated adaptive robust neural-network control of underactuated autonomous underwater vehicles. *Expert Syst. Appl.* **2022**, *197*, 116714. [CrossRef]

148. Heshmati-Alamdari, S.; Nikou, A.; Dimarogonas, D.V. Robust trajectory tracking control for underactuated autonomous underwater vehicles in uncertain environments. *IEEE Trans. Autom. Sci. Eng.* **2020**, *18*, 1288–1301. [CrossRef]
149. Chen, B.; Ma, H.; Kang, H.; Liang, X. Multi-agent Distributed Formation Control Based on Improved Artificial Potential Field and Neural Network for Connectivity Preservation. In Proceedings of the 2021 IEEE International Conference on Unmanned Systems (ICUS), Beijing, China, 15–17 October 2021; pp. 455–460.
150. Zhu, D.; Zhou, B.; Yang, S.X. A Novel Algorithm of Multi-AUVs Task Assignment and Path Planning Based on Biologically Inspired Neural Network Map. *IEEE Trans. Intell. Veh.* **2021**, *6*, 333–342. [CrossRef]
151. Zhang, J.; Liu, M.; Zhang, S.; Zheng, R.; Dong, S. Multi-AUV Adaptive Path Planning and Cooperative Sampling for Ocean Scalar Field Estimation. *IEEE Trans. Instrum. Meas.* **2022**, *71*, 1–14. [CrossRef]
152. Sun, Y.; Luo, X.; Ran, X.; Zhang, G. A 2D Optimal Path Planning Algorithm for Autonomous Underwater Vehicle Driving in Unknown Underwater Canyons. *J. Mar. Sci. Eng.* **2021**, *9*, 252. [CrossRef]
153. Scharff Willners, J.; Gonzalez-Adell, D.; Hernández, J.D.; Pairet, È.; Petillot, Y. Online 3-dimensional path planning with kinematic constraints in unknown environments using hybrid A* with tree pruning. *Sensors* **2021**, *21*, 1152. [CrossRef]
154. Yuan, J.; Wang, H.; Zhang, H.; Lin, C.; Yu, D.; Li, C. AUV obstacle avoidance planning based on deep reinforcement learning. *J. Mar. Sci. Eng.* **2021**, *9*, 1166. [CrossRef]
155. Qiu, X.; Feng, C.; Shen, Y. Obstacle avoidance planning combining reinforcement learning and RRT* applied to underwater operations. In Proceedings of the OCEANS 2021: San Diego–Porto IEEE, San Diego, France, 20–23 September 2021; pp. 1–6.
156. Lim, H.S.; King, P.; Chin, C.K.; Chai, S.; Bose, N. Real-time implementation of an online path replanner for an AUV operating in a dynamic and unexplored environment. *Appl. Ocean Res.* **2022**, *118*, 103006. [CrossRef]
157. Liu, J.; Zhao, M.; Qiao, L. Adaptive barrier Lyapunov function-based obstacle avoidance control for an autonomous underwater vehicle with multiple static and moving obstacles. *Ocean Eng.* **2022**, *243*, 110303. [CrossRef]
158. Qin, R.; Zhao, X.; Zhu, W.; Yang, Q.; He, B.; Li, G.; Yan, T. Multiple receptive field network (MRF-Net) for autonomous underwater vehicle fishing net detection using forward-looking sonar images. *Sensors* **2021**, *21*, 1933. [CrossRef] [PubMed]
159. Zhang, H.; Zhang, S.; Wang, Y.; Liu, Y.; Yang, Y.; Zhou, T.; Bian, H. Subsea pipeline leak inspection by autonomous underwater vehicle. *Appl. Ocean Res.* **2021**, *107*, 102321. [CrossRef]
160. An, R.; Guo, S.; Zheng, L.; Hirata, H.; Gu, S. Uncertain moving obstacles avoiding method in 3D arbitrary path planning for a spherical underwater robot. *Robot. Auton. Syst.* **2022**, *151*, 104011. [CrossRef]
161. Ridao, P.; Carreras, M.; Ribas, D.; Sanz, P.J.; Oliver, G. Intervention AUVs: The next challenge. *Annu. Rev. Control* **2015**, *40*, 227–241. [CrossRef]
162. Yue, C.; Guo, S.; Shi, L. Hydrodynamic analysis of the spherical underwater robot SUR-II. *Int. J. Adv. Robot. Syst.* **2013**, *10*, 247. [CrossRef]
163. Li, Y.; Guo, S.; Wang, Y. Design and characteristics evaluation of a novel spherical underwater robot. *Robot. Auton. Syst.* **2017**, *94*, 61–74. [CrossRef]
164. Gu, S.; Guo, S.; Zhang, L.; Yao, Y. A hybrid propulsion device for the spherical underwater robot (SUR III). In Proceedings of the 2017 IEEE International Conference on Mechatronics and Automation (ICMA), Kagawa, Japan, 6–9 August 2017; pp. 387–392.

electronics

Article

A Concept of Autonomous Multi-Agent Navigation System for Unmanned Surface Vessels

Agnieszka Lazarowska [1,*,†] and Andrzej Żak [2,†]

1 Department of Ship Automation, Gdynia Maritime University, 81-225 Gdynia, Poland
2 Faculty of Mechanical and Electrical Engineering, Polish Naval Academy, 81-127 Gdynia, Poland
* Correspondence: a.lazarowska@we.umg.edu.pl
† These authors contributed equally to this work.

Abstract: The paper introduces a proposal of an Autonomous Navigation System for Unmanned Surface Vessels. The system architecture is presented with a special emphasis on collision avoidance and maneuver auto-negotiation. For the purpose of maneuver auto-negotiation, the concept of multi-agent systems has been applied. The algorithm developed for the task of collision avoidance is briefly described and the results of the simulation tests, confirming the effectiveness of applied method, are also given. Presented outcomes include solutions of test scenarios from the perspectives of different ships taking part in the considered situations, confirming the applicability of the collision avoidance algorithm in the process of maneuver auto-negotiation.

Keywords: autonomous navigation system; collision avoidance; maneuver auto-negotiation; multi-agent system; obstacle avoidance; real-time path planning; unmanned surface vehicle

Citation: Lazarowska, A.; Żak, A. A Concept of Autonomous Multi-Agent Navigation System for Unmanned Surface Vessels. *Electronics* **2022**, *11*, 2853. https://doi.org/10.3390/electronics11182853

Academic Editor: Arturo de la Escalera Hueso

Received: 28 June 2022
Accepted: 5 September 2022
Published: 9 September 2022

1. Introduction

Navigation is a vital task in the operation of every autonomously moving object. This relates to self-driving cars, autonomous mobile robots, drones and unmanned ships. Autonomous navigation of a craft has to cover perception, obstacle detection and avoidance and path planning and following. Marine units that move autonomously can be divided into autonomous merchant vessels and smaller crafts—Unmanned Surface Vehicles (USVs). The International Maritime Organization (IMO) introduced a term for autonomous merchant vessels—the Maritime Autonomous Surface Ship (MASS), which is used to express *a ship which, to a varying degree, can operate independent of human interaction* [1].

Over the last few years, many research and development projects dedicated to MASSs have been carried out. Examples of such projects, listed in Table 1 and in Figure 1, include the Maritime Unmanned Navigation through Intelligence in Networks (MUNIN) [2], ReVolt [3], Advanced Autonomous Waterborne Applications (AAWA) [4], Autosea [5], Autoferry [6], Yara Birkeland [7], Safer Vessel with Autonomous Navigation (SVAN) [8] and Mayflower Autonomous Ship (MAS) [9].

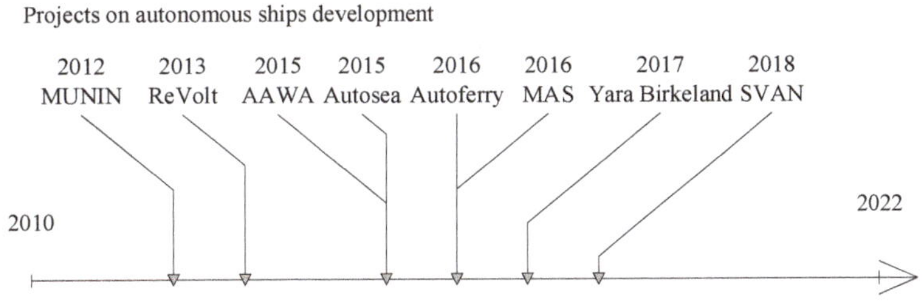

Figure 1. A timeline of recent projects on autonomous ships.

Table 1. Research projects on autonomous ships.

Project	Type of Vessel	Years	Aim	Type of Experiments/Analysis
MUNIN [2]	dry bulk carrier, 75.000 DWT, speed: 16 kn	2012–2015	deep-sea voyage autonomous navigation	feasibility study/theoretical analysis
ReVolt [3]	60 m, 1300 DWT, battery powered	2013	short-sea voyage	ship concept
AAWA [4]	not specified	2015–2018	remotely-controlled ship	theoretical analysis for a proof of concept demonstrator
Autosea [5]	autonomous passenger ferries	2015–2019	methods for guidance and navigation	full-scale experiments
Autoferry [6]	5 m, electric passenger ferry milliAmpère	2016–2019	urban water transport	full-scale prototype
Yara Birkeland [7]	79.5 m, fully electric container feeder, 120 TEU	2017–2022	short-sea voyage	model testing/vessel building and operation planned
SVAN [8]	53.8 m, ferry Finferries Falco	2018	short-sea voyage	fully autonomous ferry demonstration
MAS [9]	5 m trimaran Mayflower 400	2016–2022	oceanographic surveys and research	Atlantic Crossing planned for 2022

The technology of USVs has also developed dynamically in recent years. Examples of recently developed USVs are listed in Table 2. USVs application areas include defense area, such as surveillance, search and rescue, reconnaissance and strike missions, but also ocean surveying, such as collecting oceanographic data (bathymetry, pollution monitoring). A review of USVs with a special emphasis on the design aspects of the GNC system for these marine crafts can be found in [10].

Table 2. Examples of recently developed USVs.

USV	Country	Length [m]	Max Speed	Payload	Mission Endurance	Application
Katana [11]	Israel	11.9	60 kn	2200 kg	350 nm	Defense/military
Protector [12]	Israel	11 or 9	40 kn	-	>48 h	Naval & security missions
C-Target 9 [13]	USA	9.6	50 kn	-	-	Defense/military
C-Target 6 [14]	USA	6.5	35 kn	-	-	Defense/military
Edredon [15,16]	Poland	5.7	30 kn	1000 kg	8–130 h	Defense/military
C-Worker 7 [17]	USA	7.5	6 kn	500 kg	25 days @ 2 kn	Ocean Surveying
C-Worker 6 [18]	USA	5.8	6.5 kn	-	30 days	Ocean Surveying
Sounder [19]	Norway	8	12 kn	-	20 days @ 4 kn	Hydroacoustic Applications

Recently, different classification societies, maritime organizations and companies have been developing classifications of ships based on their degrees of autonomy [20]. The Maritime Safety Committee (MSC) of the IMO defined 4 degrees of ship autonomy, as shown in Table 3 [1].

Table 3. Degrees of ship autonomy according to IMO.

Degree of Autonomy	Description
One	Ship with automated processes and decision support
Two	Remotely controlled ship with seafarers on board
Three	Remotely controlled ship without seafarers on board
Four	Fully autonomous ship

The aim of this paper is to present the developed concept of the Autonomous Navigation System for the USV. The main contributions of the research presented in this paper are:

- An analysis of recent projects on autonomous ships and USVs with a special emphasis on ANSs and the collision avoidance feature;
- The definition of an USV Autonomous Navigation System structure based on the analyzed projects;

- The definition of the agent architecture and control flow in the autonomous multi-agent navigation system for USVs, on the basis of a general structure of an agent system;
- The development of an effective collision avoidance algorithm to be applied for the maneuver auto-negotiation in the proposed multi-agent system;
- Evaluation of the deterministic collision avoidance method for the purpose of the possibility to apply collaborative strategies to avoid collisions with other agents by the calculation of multiple trajectories for all USVs taking part in the considered navigational scenario.

Section 2 introduces the general structure of the Autonomous Navigation System for USVs. Section 3 presents the concept of the Multi-agent system for USV maneuver auto-negotiation. In Section 4, the collision avoidance algorithm developed for the application in the ANS for USVs has been briefly described. In Section 5, simulation tests results obtained with the use of the collision avoidance algorithm are shown, with a special emphasis of solutions achieved from the perspectives of different USVs taking part in the considered encounter situations. Discussion on the results is given in Section 6 and the paper is summarized in Section 7.

2. Autonomous Navigation System Structure

The Autonomous Navigation System is responsible for the navigation of the USV. The main tasks of the Autonomous Navigation System of every vehicle include: perception of the environment, path planning and path following (vehicle control).

One of the main subsystems of the ANS is the Collision Avoidance (CA) module. The CA module is responsible for the collision risk assessment on the basis of information obtained from the system that fuses data from various navigational sensors. The second task of the CA module is to ensure the safe navigation of the USV, both on the open sea and in restricted waters. The main element of the CA module is the collision avoidance algorithm, responsible for determining a safe maneuver or a safe trajectory for the vehicle, when a risk of collision has been detected. Besides the CA module, the route planning module is also applied. This subsystem is responsible for the calculation of a global path between the initial and final position, e.g., two harbors, including the craft's mission, but also weather conditions (therefore known also as weather routing). Figure 2 presents a general structure of the Autonomous Navigation System for USVs.

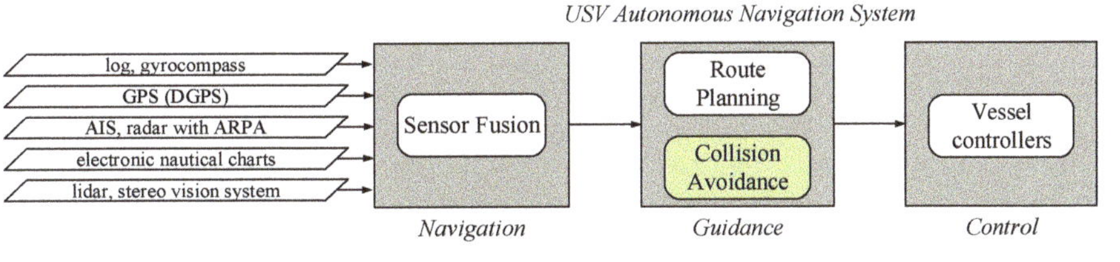

Figure 2. Autonomous Navigation System for USVs.

The module responsible for navigational data reception and fusion is the Advanced Sensor Module—ASM, also known as the Situation Awareness module or Sensor Fusion module. Applied sensors include: nautical charts, long-range radars, the Automatic Identification System (AIS), stereo vision systems, short-range radars, lidars and the Global Positioning System (GPS). In relation to the ANS of USVs, the term Obstacle Detection and Avoidance (ODA) system is also used, which reflects a combined version of the Situation Awareness and Collision Avoidance modules. Motion Control system is responsible for the calculation of appropriate control forces to steer the USV along the calculated trajectory or to execute a maneuver determined by the CA module. Table 4 presents a comparative analysis of ANSs proposed in different research projects on autonomous ships, listed in

Figure 1 and Table 1. The ANS structure applied in the research presented in this paper was developed based on the solutions proposed in the mentioned projects.

Table 4. ANSs in research projects on autonomous ships.

Project	Module for Collision Avoidance	Module for Data Perception and Data Fusion	Sensors	Autonomous Navigation System	Algorithms	Other Systems
MUNIN [2]	Collision Avoidance (CA) module	Advanced Sensor Module (ASM)	marine radar, AIS receiver, daylight and infrared cameras, nautical data	Deep Sea Navigation System, track pilot and rudder and engine control: path following, weather routing, collision avoidance	safe weather routing: A-star algorithm, collision avoidance algorithm: formalized description of COLREGs	Autonomous Engine and Monitoring Control system, Shore Control Centre
AAWA [4]	Collision Avoidance (CA) module	Situation Awareness (SA) system (sensor fusion)	visual and IR cameras, short range and long range radars, lidar, GPS, inertial sensors, electronic navigational charts	Route planning module, SA module, CA module, Ship state definition module	Velocity Obstacles (VO) method	Dynamic positioning system, Propulsion control system, Remote operator
Autosea [5]	Collision Avoidance (CA) module	Sensor Fusion (SF) module	AIS, radar, camera, charts	CA module (collision detection, avoidance, guidance), SF module (target tracking)	Model Predictive Control (MPC)	other systems not considered

3. Multi-Agent System for USV Maneuver Auto-Negotiation

Multi-agent systems are regarded as a powerful concept for solving real-world problems, especially in complex, dynamic environments, enabling to achieve increased autonomy within control systems [21]. Table 5 presents a summary of multi-agent systems proposed in the recent literature in relation to ships and different types of unmanned vehicles: USVs, UUVs (Unmanned Underwater Vehicles) and UAVs (Unmanned Aerial Vehicles). The comparison lists the control objects, the purpose of proposed systems, the type of tasks solved by the system, i.e., cooperation and/or competition, the type of applied method and information about the optimization of the task. The last column contains additional remarks, concerning advantages, limitations or other comments with regard to the relation between the mentioned methods and the approach described in this paper. Some of the proposed multi-agent systems are intended for solving a different task than collision avoidance, considered in this paper, such as formation control [22,23], training a team of USVs [24], searching water region by a team of UUV [25] and/or consider a different type of vehicle, such as UAV [26], UUV or a car [27].

The concept of the maneuver auto-negotiation system in relation to autonomous cars was proposed in [27], to UAV in [26] and to ships in [28–30]. In [28], collision avoidance and maneuver auto-negotiation is based on geometrical relationships. Two ships encounters are considered in this approach and the solutions might not be COLREGs compliant. In [29], a concept stage of the task has been presented. The authors assumed the application of evolutionary sets of safe trajectories, where the best set of reciprocal trajectories is calculated, but results of the collision avoidance approach are not presented in the paper. In [30], the author proposes a structure of the maneuver auto-negotiation system with the concentration on the data acquisition aspect. Safe trajectories are assumed to be calculated as evolutionary sets of safe trajectories. This approach has been presented in the author's previous works [x, y]. The author proposed a control flow, in which one of the vessels is a leader ship. The leader is responsible for data gathering, determination and optimization of maneuvers and distribution of the results among other participants of the encounter.

The literature analysis shows that the development of a multi-agent system for collision avoidance with maneuver auto-negotiation is an open research problem, as algorithms for the calculation of multiple trajectories are rare in the recent literature. There exists a demand in the industry for the development of an effective, robust and reliable multi-agent system, which will enable for the calculation of a set of safe trajectories for all ships or USVs taking part in an encounter. The development of such a solution is the aim of the research presented in this paper.

Table 5. Multi-agent systems for ships and unmanned vehicles.

Authors	Year	Control Object	Purpose	Type of Task	Method	Optimization	Remarks
Huang et al. [26]	2022	UAV	collision avoidance in UAV swarms	cooperation	multi-agent reinforcement learning	safety, energy consumption, response time	different control object
Xue and Wu [22]	2021	USVs + UAV	formation control	cooperation	leader-following consensus	APF path planning + sliding mode control	different task
Han et al. [24]	2019	USV	training a team of USVs	cooperation within the team, competition with other teams	genetic-based fuzzy rule training algorithm	optimizing agents' coordination decisions	different task
Wang et al. [23]	2019	USV	formation control	cooperation	leader-following consensus	sampled-data consensus protocol	different task
Żak [25]	2019	UUV	searching water region by a team of UUVs	cooperation	UUV operation algorithm	minimize the total time	different task
Visintainer et al. [27]	2016	car	maneuver negotiation	cooperation	algorithms for automated lane change, distance keeping	no optimization	concept stage
Szlapczynska [30]	2015	ship	maneuver auto-negotiation, collision avoidance	cooperation	evolutionary sets of safe trajectories	the best set of reciprocal trajectories	data acquisition
Hornauer and Hahn [29]	2013	ship	maneuver auto-negotiation, collision avoidance	cooperation	evolutionary sets of safe trajectories	the best set of reciprocal trajectories	concept stage
Hu et al. [28]	2008	ship	maneuver auto-negotiation, collision avoidance	cooperation	geometrical relationships	details not given	two ships encounter, might not be COLREGs compliant

Extending the ANS architecture for the purpose of maneuver auto-negotiation was based on multi-agent systems. Multi-agent systems are classified as an approach of Distributed Artificial Intelligence (DAI) [31]. A multi-agent system is composed of a number of agents, which interact with each other and the environment in order to achieve their goals. Such approach has been applied for modeling and solving problems by cooperation between local solvers and the design of complex systems. An agent is an object, being in a certain situation, that has the ability to perceive the environment and influence it through an autonomous action to achieve a defined goal. In the approach presented in this paper an agent is an USV, perceiving the environment through navigational equipment such as AIS, radar with ARPA, GPS, log, gyrocompass, lidar and stereovision systems. The USV also interacts with other USVs in the surroundings and acts on the environment by changing its course and/or speed.

Assumed agent architecture and control flow is shown in Figure 3 [25,32]. The algorithms related to detecting obstacles are included in the layer of reflex action. The planning layer implements obstacle avoidance algorithms [33–35] and these responsible for the control of the vehicle along a set trajectory. At the level of cooperation, the vehicle exchanges data with other vehicles and conducts negotiations about planned collision avoidance actions.

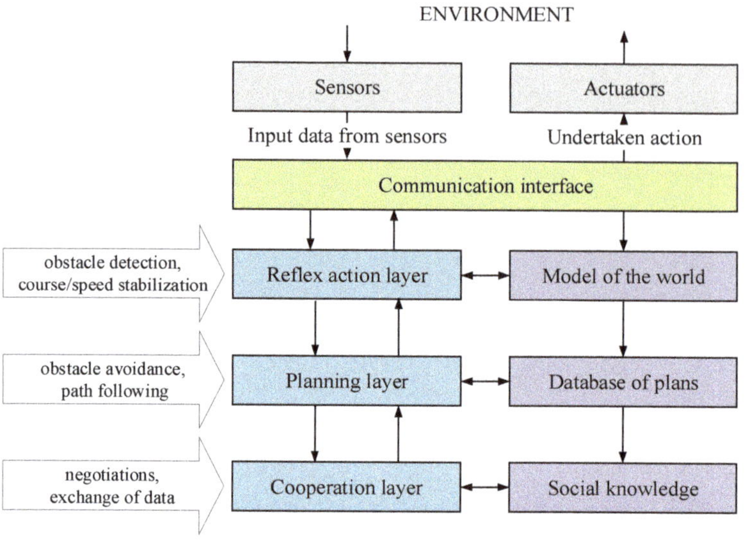

Figure 3. Agent architecture and control flow.

The agent architecture applied in this research is the InteRRaP model [36], which means Integration of Reactive behavior and Rational Planning. This architecture utilized the BDI model of reasoning, which comes from Beliefs, Desires (Goals) and Intentions [37,38]. In such an architecture, two types of interaction between different layers are possible. These are: bottom-up activation and top-down execution. In the first type, when a lower layer is not able to deal with the current situation, it passes control to a higher layer. In the second type of interaction, a higher layer uses the functionality assured by a lower layer in order to reach its goal. In the InteRRaP architecture, when new data from sensors arrive at the lowest reflex action layer, it evaluates whether it can deal with the situation. When the layer is not able to perform the task, the control will be passed in the bottom-up activation manner to the planning layer. This layer will also asses the task and when it will be able to deal with the it, after which it will apply a top-down execution. Otherwise, it will pass control to the highest cooperation layer. Such an architecture, in comparison with other layered architectures of multi-agent systems, has as an advantage that only the lowest layer has direct access to actuators. This excludes the occurrence of conflicting decisions between different layers. A different layered approach was proposed in [39], where appropriate filtering and suppression mechanisms have to be applied in order to prevent the system from conflicts among the actions of different layers.

Agents communicate with each other, while maintaining their autonomy of action and decision-making, which translates into actions taken by individual agents. Such interaction between agents leads to the modification of knowledge and actions of individual agents based on the behavior of other agents. A single agent is equipped with a strategy, leading it to achieve the assumed goal, which in this case is to reach a specific geographical position—the end point of the trajectory. However, as part of a multi-agent system, it must implement collaborative strategies to avoid collisions with other agents. The key feature of agents in a multi-agent system is their collaboration and competition. Therefore, agents exchange information on individual actions taken in order to avoid a collision.

The collision avoidance process with the use of the proposed autonomous multi-agent navigation system for USVs can be performed in one of the two following ways, both of which are possible with the use of the applied collision avoidance algorithm:

(a) A decentralized system, where collision avoidance calculations are performed by individual agents (USVs), these partial solutions are then used in the auto-negotiation process;
(b) One USV calculates the complete solution of the current navigational situation and then distributes the partial solutions to other agents.

4. USV Collision Avoidance Algorithm

Collision avoidance and safe path planning algorithm applied in the introduced concept of Autonomous Multi-agent Navigation System is a deterministic approach. The algorithm's operation is based on the search through a database with stored candidate trajectories in order to find the best safe path solving the currently considered situation; therefore, the approach has been called the Trajectory Base Algorithm (TBA). This algorithm has been chosen for further development due to its competitive results in relation to other deterministic and non-deterministic methods. A comparative analysis of different methods, along with the TBA, can be found in [40].

The main advantages of this approach are: the solution repeatability for every run of calculations with the same input data, achievement of COLREGs compliant solutions with minimal path length and run time of the algorithm enabling the application of this approach for real-time path planning.

The input variables, shown in Figure 4, are:

- Ψ—the course of an own ship (USV no. 0, for which the safe trajectory is currently calculated by the algorithm);
- V—the speed of an own ship;
- Ψ_j—the course of the j-th USV (also called target ship);
- V_j—the speed of the j-th USV;
- N_j—the bearing of the j-th USV;
- D_j—the distance of the j-th USV from an own ship (USV no.0);
- t—the number of a currently evaluated candidate trajectory retrieved from the database;
- t_{max}—the maximum number of candidate trajectories in the database.

The inputs from the database of trajectories are: candidate trajectories composed of a number of waypoints (x and y coordinates of the USV position). The output variables are: safe trajectories composed of a number of waypoints (x and y coordinates of the USV position) and the values of the USVs course at the consecutive parts of the trajectories.

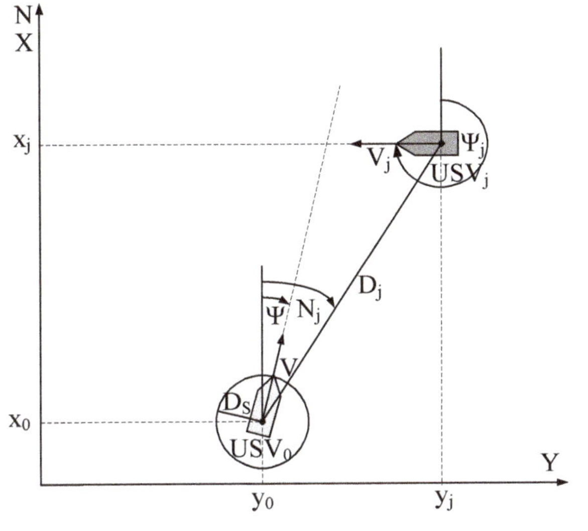

Figure 4. Parameters defining a navigational situation at sea.

Figure 5 presents a flowchart of the TBA. As mentioned above, input data are the courses, speeds, bearings and distances of all vessels taking part in an encounter. After the reception of input data describing the current navigational situation, the first candidate trajectory is retrieved from the database for the evaluation procedure. The evaluated candidate trajectory is divided into a number of steps. Afterwards, in every step, an own ship is moved into a new instantaneous position along the evaluated trajectory. Target ships are also moved into their next instantaneous positions, resulting from their motion parameters and a trajectory selected for the implementation. Then, the procedure of checking, whether the vessels positions do not lead to a collision, is carried out. When the whole trajectory is verified as a safe path, which means that it does not cause a collision with any of encountered ships during the vessel's movement along it, it becomes the final best solution. The optimization criterion applied in the collision avoidance algorithm, defined by Equation (1), is the minimal path length. It is calculated as a sum of the lengths of line segments composing the safe path, where $i = 1, \ldots, e$ are the waypoints composing the path:

$$f(t) = dist(t) = \sum_{i=1}^{e-1} \sqrt{(x_{i+1} - x_i)^2 + (y_{i+1} - y_i)^2} \rightarrow min \tag{1}$$

The information about the selected trajectory, as a part of cooperation activities between agents, is transferred to other agents in a multi-agent system to be used in their decision-making. Further calculations are terminated. A more detailed description of the TBA can be found in [41].

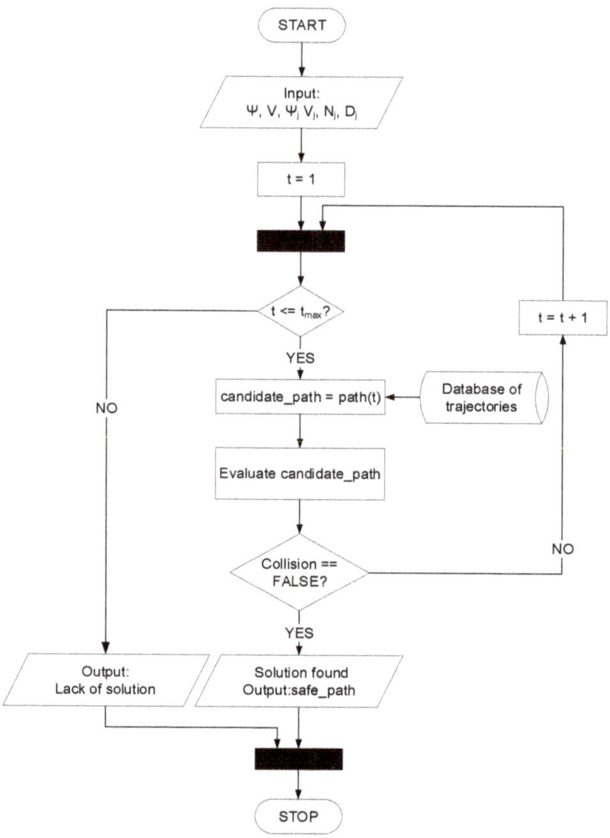

Figure 5. A flowchart of the Trajectory Base Algorithm.

5. Results of Collision Avoidance Algorithm

The TBA proposed in this paper for USVs collision avoidance has been tested comprehensively in order to validate its reliability and robustness. Various tests of this algorithm have been carried out, including:

- Simple encounter situations with one target ship considered in The International Regulations for Preventing Collisions at Sea (COLREGs) (head-on, crossing and overtaking);
- Encounter situations with static obstacle (lands, shallows);
- Complex encounter situations with real navigational data.

The results of these test can be found in [42].

One of the methods to assure a proper distance between the USVs or other marine crafts in a collision avoidance algorithm is to model other USVs (target ships) with the use of a ship domain. A ship domain is defined as an area around a ship, which a navigator wants to keep free from static and dynamic obstacles. In the presented approach, a ship domain with a hexagon shape, as proposed in [43], has been applied. The dimensions of the target ship domain used in the tests were: distance towards the bow: 1.3 nm, distance of amidships: 0.6 nm, distance towards the stern: 0.5 nm, distance towards the starboard side: 0.6 nm and distance towards the port side: 0.5 nm.

For the purpose of the algorithm's application in the USV Autonomous Multi-Agent Navigation System, the evaluation of solutions consistency from different ships' perspectives has been carried out. The results of these experiments are presented here. The algorithm was implemented in the MATLAB programming language and tested on a PC with Intel Core i7-10750H CPU, 32 GB RAM, 64-bit Windows 10 operating system. Out of many test cases, four scenarios have been chosen for the presentation in this paper. These test cases have been chosen as standard scenarios, which present the solutions compliance with the COLREGs. Similar test scenarios were presented in the cooperative path planning approach proposed in [44]. Input data of USVs, including courses in degrees, speeds in knots, bearings in degrees and distances between the marine crafts in nautical miles, are given in Tables 6–9. Tables 10–13 present output data, such as USV courses in degrees at consecutive parts of the calculated safe path, path lengths in nautical miles and run time of the algorithm in seconds. Figures 6–9 show graphical solutions of test cases 1–4. Initial positions of USVs are marked by digits representing the number of an USV (USV No. in tables). The numbers given along the consecutive positions of the USVs show the time of the USV arrival at a given point in minutes. Presented results have been discussed in the following section.

Table 6. Input data of test case 1—head-on.

USV No.	Course [°]	Speed [kn]	Bearing [°]	Distance [nm]
0	0	10	-	-
1	180	14	0	4

Table 7. Input data of test case 2.

USV No.	Course [°]	Speed [kn]	Bearing [°]	Distance [nm]
0	0	10	-	-
1	270	16	45	3
2	180	9	0	4

Table 8. Input data of test case 3.

USV No.	Course [°]	Speed [kn]	Bearing [°]	Distance [nm]
0	0	14	-	-
1	270	10	45	4
2	90	10	315	4

Table 9. Input data of test case 4.

USV No.	Course [°]	Speed [kn]	Bearing [°]	Distance [nm]
0	0	14	-	-
1	270	8	45	6
2	180	12	0	8
3	90	8	315	6

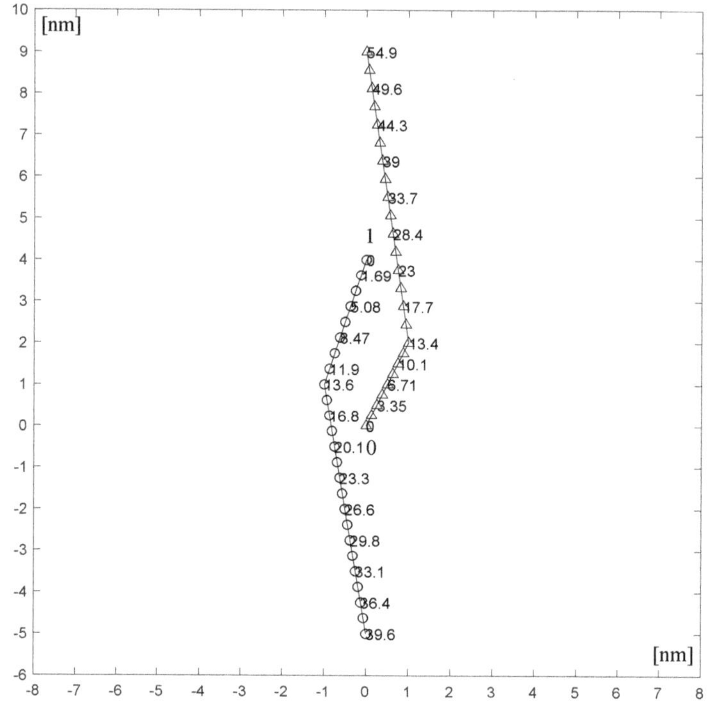

Figure 6. Solution of test case 1.

Table 10. Output data of test case 1.

USV No.	Course [°]	Path Length [nm]	Run Time [s]
0	27, 352	9.31	0.21
1	198, 171	9.25	0.12

Table 11. Output data of test case 2.

USV No.	Course [°]	Path Length [nm]	Run Time [s]
0	18, 351	9.25	0.16
1	315, 270, 262	9.49	0.39
2	198, 171	9.25	0.14

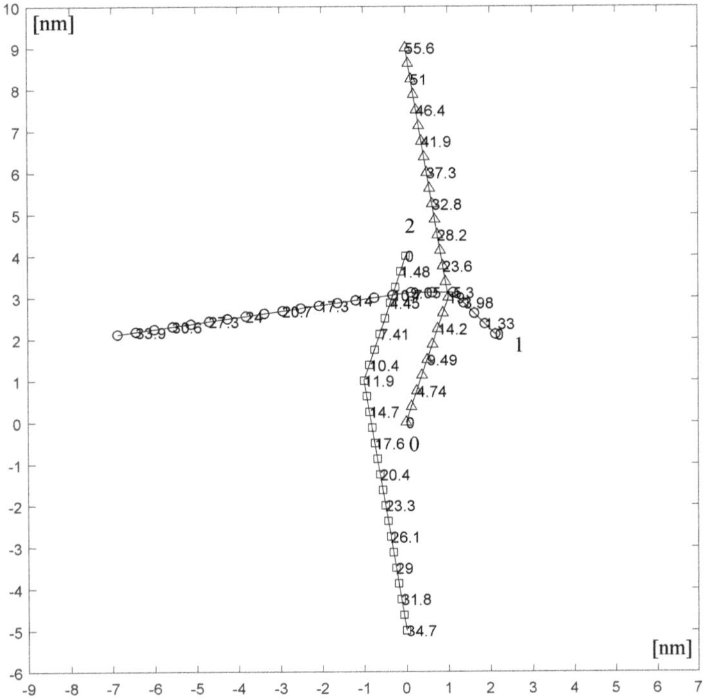

Figure 7. Solution of test case 2.

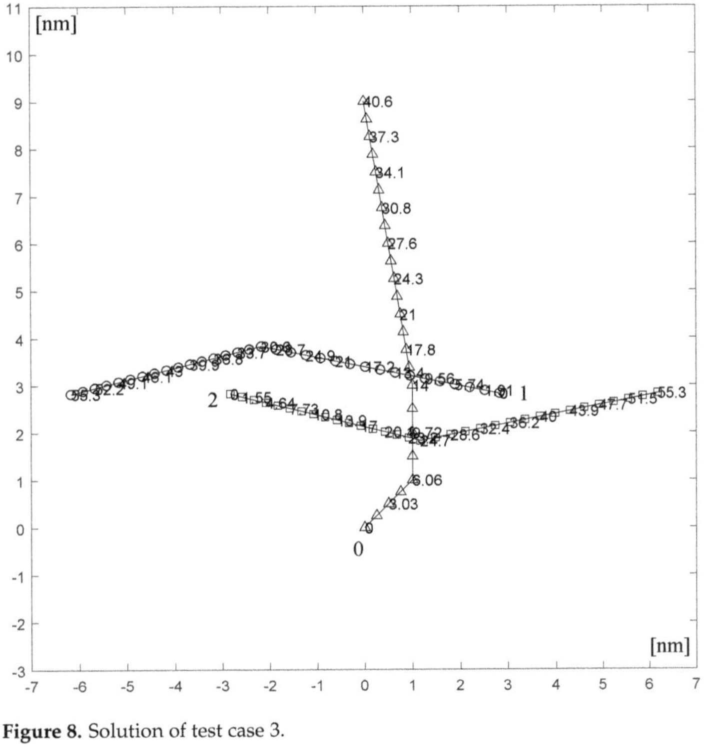

Figure 8. Solution of test case 3.

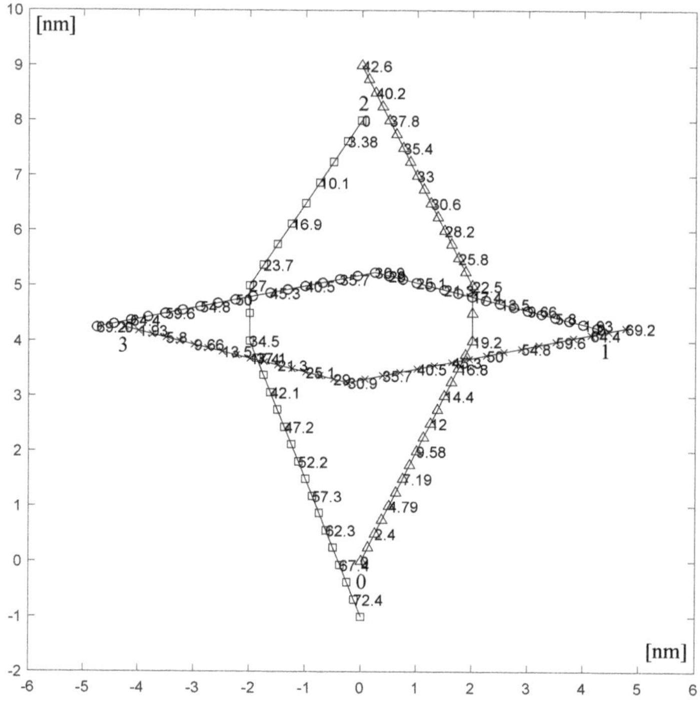

Figure 9. Solution of test case 4.

Table 12. Output data of test case 3.

USV No.	Course [°]	Path Length [nm]	Run Time [s]
0	45, 0, 351	9.5	0.47
1	281, 256	9.22	0.12
2	104, 79	9.22	0.11

Table 13. Output data of test case 4.

USV No.	Course [°]	Path Length [nm]	Run Time [s]
0	27, 0, 333	9.94	0.7
1	284, 259	9.22	0.09
2	214, 180, 158	9.99	0.8
3	104, 79	9.22	0.09

6. Discussion

6.1. Analysis of Results

Results of TBA applied in the presented concept for USV collision avoidance allow to state the following remarks:

- Safe paths calculated by the algorithm are compliant with COLREGs (especially rules 8b, 14 and 15);
- Paths calculated by the algorithm from the perspectives of different USVs taking part in the considered test case are consistent (do not lead to a collision between the crafts);
- The run time of the algorithms (less than a second) is acceptable for real-time path planning purposes.

It should be underlined here that results of collision avoidance algorithms, regarding calculation of multiple trajectories, are very rare in the literature on that topic. Examples of approaches other that the one presented in this paper can be found in [44,45].

Test case 1 presents a head-on situation. In this encounter situation, according to rule 14 of COLREGs, both vessels should alter her course to starboard side in order to pass on the port side of the other vessel. Figure 1 presents paths calculated by the algorithm for both vessels. As can be observed there, both USVs execute maneuvers to their starboard sides. Maneuvers are also compliant with rule 8b of COLREGs—they are large enough to be easily noticed and interpreted by the other vessel.

Test case 2 presents an encounter of three USVs and is composed of head-on (between vessels 0–2) and crossing situations (between vessels 0–1 and 1–2) between the different pairs of vessels. The behavior of vessels in a crossing situation is defined by rule 15 of COLREGs. According to this rule, the vessel that has the other vessel on her starboard side should keep out of the way and avoid crossing ahead of the other vessel. As can be observed in Figure 2, the USV no. 0 keeps out of the way of the USV no. 1 and does not cross ahead on the other vessel. In the same manner, the USV no. 1 keeps out of the way of the USV no. 2. The solution is also compliant with rule 14 of COLREGs—both USVs no. 0 and no. 2 alter their course to starboard, passing on the port side of the other vessel. Similarly, as in test case 1, maneuvers of all USVs also fulfill rule 8b.

In test case 3, a crossing situation occurs between vessels 0–1 and 0–2, whereas a head-on encounter takes place between vessels 1–2. Likewise, in test case 2, solutions are calculated by the TBA for all of the USVs fulfill rules 8b, 14 and 15.

Test case 4 is an encounter between four USVs, where crossing situations occur between the vessels 0–1, 1–2, 2–3 and 3–0, and head-on situations take place between vessels 0–2 and 1–3. Analyses of safe paths returned by the TBA enable to state that the solutions are compliant with COLREGs rules 8b, 14 and 15.

The obtained results confirm the prospect to apply the proposed collision avoidance algorithm in the Autonomous Multi-Agent Navigation System for Unmanned Surface Vessels, as it has been proven that the solutions calculated for different vessels in the same encounter situation constitute safe paths (do not lead to a collision between the marine crafts). Therefore, they can constitute a solution of maneuver auto-negotiation procedure.

6.2. Comparison with Other Collision Avoidance Algorithms

The collision avoidance algorithm has been compared with other approaches proposed for the application in maneuver auto-negotiation systems or presenting results from different ships' perspectives. The results of this analysis are shown in Tables 14 and 15. The comparison includes: applied method, type of the method, i.e., deterministic or non-deterministic, consideration of static navigational restrictions, dynamic obstacles and COLREGs, applied optimization criterion. The run time of the algorithm and repeatability of a solution for every run of calculations with the same input data are also specified in the tables. The adopted calculation approach has also been defined for every algorithm, whether the trajectories are calculated by only one vessel: the leader ship or by all of marine crafts taking part in the encounter. The main advantages of the approach proposed in this paper, in relation to other existing methods, are as follows:

- Deterministic nature of the algorithm, which guarantees repeatable solutions, similarly to [28,44];
- Near-real run time: up to a second, making it applicable for USVs;
- COLREGs compliant solution (rules 8b, 13–15);
- Consideration of static constraints (lands, shallows, buoys), as in [45];
- Possibility to calculate the solution from the leader USV perspective as well as all ships' perspectives.

Table 14. Comparison of collision avoidance algorithms applicable in maneuver auto-negotiation systems—part 1.

Authors	Year	Method	Type	Static Obstacles	Dynamic Obstacles	COLREGs
Tam and Bucknall [44]	2013	Cooperative Path Planning (CPP)	deterministic	static point-based obstructions (buoys)	simple and complex encounters	determined priority based on COLREGs
Szapczynski and Szlapczynska [45]	2012	Evolutionary Sets of Safe Ship Trajectories (ESoSST)	non-deterministic	shorelines, shallows	simple and complex encounters	COLREGs-violation penalties
Tam and Bucknall [46]	2010	evolutionary algorithm	non-deterministic	ship with 0 speed	simple and complex encounters	COLREGs-influenced area (CA)
Hu et al. [28]	2008	Collision-Avoidance Negotiation FramewOrk (CANFO)	deterministic	not considered	two ship encounters	might not be COLREGs compliant
This approach	2022	Trajectory Base Algorithm (TBA)	deterministic	shorelines, shallows, point-based	simple and complex encounters	COLREGs enforced by ship domain shape and size

Table 15. Comparison of collision avoidance algorithms applicable in maneuver auto-negotiation systems—part 2.

Authors	Year	Optimization Criteria	Run Time	Repeatability	Perspective
Tam and Bucknall [44]	2013	course change of 30 degrees)	7 s for complex test cases	yes	consistency from all ships' perspectives
Szapczynski and Szlapczynska [45]	2012	min. way losses of trajectories in a set;	maximum 1 min	no—small differences possible	leader ship perspective
Tam and Bucknall [46]	2010	path length, avg. speed, travelling time, engine adjustment;	200–800 s	no	consistency from all ships' perspectives
Hu et al. [28]	2008	details not given	not given	yes	both ships' perspectives
This approach	2022	minimal path length	less than 1 s	yes	leader ship/all ships perspectives

7. Conclusions

The main contributions of the research presented in this paper in relation to the previous works of the authors are the application of the developed deterministic collision avoidance algorithm within the framework of a multi-agent system for the purpose of USV maneuver auto-negotiation and also the carried out tests of the algorithm in order to assess the possibility to apply collaborative strategies to avoid collisions with other agents in the proposed multi-agent system by the calculation of multiple trajectories for all USVs taking part in the considered navigational scenario.

A concept of the Autonomous Navigation System for Unmanned Surface Vessels, with a special emphasis on the task of collision avoidance, was presented first. A general structure of the ANS system was developed, based on the idea of multi-agent systems. An agent is an autonomous entity that perceives the environment using sensors and acts on the environment through effectors (actuators). The agent's interaction can be based on cooperation or competition. The goal of the USV, modeled as an agent, is to reach the end point of the trajectory. However, in order to achieve this objective, it has to apply collaborative strategies to avoid collisions with other agents. For this purpose, agents exchange information on planned actions. Proposed collision avoidance algorithm was validated by simulation experiments with a special emphasis on the verification of its applicability for the purpose of maneuver auto-negotiation. The achieved results prove that the trajectories obtained for different agents do not lead to a collision; therefore, the algorithm is suitable for the application in the Autonomous Multi-Agent Navigation System for the USV.

Further research include the development of the communication protocol between agents, allowing them to exchange the information efficiently. After that, experiments in real operating conditions are planned to be carried out.

Author Contributions: Conceptualization, A.L. and A.Ż.; methodology, A.L. and A.Ż.; software, A.L.; validation, A.L. and A.Ż.; formal analysis, A.L. and A.Ż.; investigation, A.L. and A.Ż.; resources, A.L. and A.Ż.; data curation, A.L. and A.Ż.; writing—original draft preparation, A.L. and A.Ż.; writing—review and editing, A.L. and A.Ż.; visualization, A.L. and A.Ż.; supervision, A.Ż.; project administration, A.L. and A.Ż.; funding acquisition, A.L. and A.Ż. All authors have read and agreed to the published version of the manuscript.

Funding: This research was funded within the framework of the program of the Minister of Science and Higher Education under the name "Regional Excellence Initiative" in the years 2019–2022; project number 006/RID/2018/19; the amount of financing: 11 870 000 PLN.

Conflicts of Interest: The authors declare no conflict of interest.

Abbreviations

The following abbreviations are used in this manuscript:

AIS	Automatic Identification System
ANS	Autonomous Navigation System
ARPA	Automatic Radar Plotting Aid
ASM	Advanced Sensor Module
CA	Collision Avoidance module
COLREGs	The International Regulations for Preventing Collisions at Sea
DAI	Distributed Artificial Intelligence
GPS	Global Positioning System
IMO	The International Maritime Organization
MASS	Maritime Autonomous Surface Ship
MPC	Model Predictive Control
SA	Situation Awareness module
SF	Sensor Fusion module
TBA	Trajectory Base Algorithm
UAV	Unmanned Aerial Vehicle
USV	Unmanned Surface Vessel (Vehicle)
UUV	Unmanned Underwater Vehicle
VO	Velocity Obstacles method

References

1. The Maritime Safety Committee of the International Maritime Organization, Outcome of the Regulatory Scoping Exercise for the Use of Maritime Autonomous Surface Ships (MASS), MSC.1/Circ.1638. June 2021. Available online: https://www.imo.org (accessed on 3 February 2022).
2. MUNIN—Maritime Unmanned Navigation through Intelligence in Networks. Available online: http://www.unmanned-ship.org/munin/about/ (accessed on 28 February 2022).
3. DNV GL. The ReVolt. A New Inspirational Ship Concept. Available online: https://www.dnvgl.com/technology-innovation/revolt/index.html (accessed on 28 February 2022).
4. Rolls-Royce. Remote and Autonomous Ship—The Next Step. Available online: https://www.rolls-royce.com/~/media/Files/R/Rolls-Royce/documents/customers/marine/ship-intel/aawa-whitepaper-210616.pdf (accessed on 28 February 2022).
5. NTNU. Autosea—Sensor Fusion and Collision Avoidance for Autonomous Surface Vehicles. Available online: https://www.ntnu.edu/autosea/ (accessed on 28 February 2022).
6. NTNU. Autoferry—Autonomous All-Electric Passenger Ferries for Urban Water Transport. Available online: https://www.ntnu.edu/autoferry (accessed on 28 February 2022).
7. Kongsberg. YARA Birkeland—Autonomous Ship Project. Available online: https://www.kongsberg.com/maritime/support/themes/autonomous-ship-project-key-facts-about-yara-birkeland/ (accessed on 28 February 2022).
8. Rolls-Royce 2018. SVAN—Safer Vessel with Autonomous Navigation. Available online: https://breakingwaves.fi/wp-content/uploads/2019/06/SVAN-presentation.pdf (accessed on 28 February 2022).

9. Promare. Mayflower Autonomous Ship (MAS). Available online: https://mas400.com (accessed on 28 February 2022).
10. Campbell, S.; Naeem, W.; Irwin, G.W. A review on improving the autonomy of unmanned surface vehicles through intelligent collision avoidance manoeuvres. *Annu. Rev. Control* **2012**, *36*, 267–283. [CrossRef]
11. Katana—USV System. Available online: https://www.iai.co.il/p/katana (accessed on 2 March 2022).
12. Rafael Advanced Defense Systems Ltd. Protector USV. Available online: https://https://www.rafael.co.il/worlds/naval/usvs/ (accessed on 2 March 2022).
13. Available online: https://www.l3harris.com/all-capabilities/c-target-9-asv (accessed on 2 March 2022).
14. Available online: https://www.l3harris.com/all-capabilities/c-target-6-asv (accessed on 2 March 2022).
15. Kalinowski, A.; Małecki, J. Polish USV 'EDREDON' and non-European USV: A comparative sketch. *J. Mar. Eng. Technol.* **2017**, *16*, 416–419. [CrossRef]
16. Kitowski, Z.; Soliński, R. Application of Domestic Unmanned Surface Vessels in the Area of Internal Security and Maritime Economy—Capacities and Directions for Development. *Sci. J. Pol. Nav. Acad.* **2016**, *206*, 67–83. [CrossRef]
17. Available online: https://www.l3harris.com/all-capabilities/c-target-7-asv (accessed on 2 March 2022).
18. Available online: https://www.unmannedsystemstechnology.com/company/autonomous-surface-vehicles-ltd/ (accessed on 2 March 2022).
19. Kongsberg Maritime Ltd. Sounder USV. Available online: https://www.kongsberg.com/maritime/products/marine-robotics/uncrewed-surface-vehicle-sounder/ (accessed on 18 May 2022).
20. Zubowicz, T.; Arminski, K.; Witkowska, A.; Smierzchalski, R. Marine autonomous surface ship-control system configuration. *IFAC-PapersOnLine* **2019**, *52*, 409–415. [CrossRef]
21. Tweedale, J.W. Using Multi-agent Systems to Pursue Autonomy with Automated Components. *Procedia Comput. Sci.* **2013**, *22*, 1369–1378. [CrossRef]
22. Xue, K.; Wu, T. Distributed Consensus of USVs under Heterogeneous UAV-USV Multi-Agent Systems Cooperative Control Scheme. *J. Mar. Sci. Eng.* **2021**, *9*, 1314. [CrossRef]
23. Wang, L.; Yue, W.; Zhang, R. Consensus for Multiple Unmanned Surface Vehicle (Musv) Systems with Markov Switching Topologies. *Pol. Marit. Res.* **2019**, *26*, 145–152. [CrossRef]
24. Han, W.; Zhang, B.; Wang, Q.; Luo, J.; Ran, W.; Xu, Y. A Multi-Agent Based Intelligent Training System for Unmanned Surface Vehicles. *Appl. Sci.* **2019**, *9*, 1089. [CrossRef]
25. Zak, A. Control of Unmanned Underwater Vehicle as a Member of Vehicles Team Performing a Given Task. *Trans. Marit. Sci.* **2019**, *8*, 18–25. [CrossRef]
26. Huang, S.; Zhang, H.; Huang, Z. Multi-UAV Collision Avoidance using Multi-Agent Reinforcement Learning with Counterfactual Credit Assignment. *arXiv* **2022**, arXiv:2204.08594.
27. Visintainer, F.; Altomare, L.; Toffetti, A.; Kovacs, A.; Amditis, A. Towards Manoeuver Negotiation: AutoNet2030 Project from a Car Maker Perspective. *Transp. Res. Procedia* **2016**, *14*, 2237–2244. [CrossRef]
28. Hu, Q.; Yang, C.; Chen, H.; Xiao, B. Planned Route Based Negotiation for Collision Avoidance Between Vessels. *TransNav Int. J. Mar. Navig. Saf. Sea Transp.* **2008**, *2*, 363–368.
29. Hornauer, S.; Hahn, A. Towards Marine Collision Avoidance Based on Automatic Route Exchange. *IFAC Proc. Vol.* **2013**, *46*, 103–107. [CrossRef]
30. Szłapczyńska, J. Data Acquisition in a Manoeuver Auto-negotiation System. *TransNav Int. J. Mar. Navig. Saf. Sea Transp.* **2015**, *9*, 343–348.
31. Weiss, G. *Multiagent Systems. A Modern Approach to Distributed Artificial Intelligence*; The MIT Press: Cambridge, MA, USA, 1999.
32. Zak, A. Controlling a team of unmanned underwater vehicles performing water region search tasks. *Pol. Marit. Res.* **2013**, *20*, 82–89. [CrossRef]
33. Lisowski, J. Comparison of dynamic games in application to safe ship control. *Pol. Marit. Res.* **2014**, *21*, 3–12. [CrossRef]
34. Lisowski, J. Game Control Methods Comparison when Avoiding Collisions with Multiple Objects Using Radar Remote Sensing. *Remote Sens.* **2020**, *12*, 1573. [CrossRef]
35. Lisowski, J.; Mohamed-Seghir, M. Comparison of computational intelligence methods based on fuzzy sets and game theory in the synthesis of safe ship control based on information from a radar ARPA system. *Remote Sens.* **2019**, *11*, 82. [CrossRef]
36. Müller, J.P. The agent architecture interrap. In *The Design of Intelligent Agents. Lecture Notes in Computer Science*; Springer: Berlin/Heidelberg, Germany, 1996; Volume 1177, pp. 45–123.
37. Tweedale, J.; Ichalkaranje, N.; Sioutis, C.; Jarvis, B.; Consoli, A.; Phillips-Wren, G. Innovations in multi-agent systems. *J. Netw. Comput. Appl.* **2007**, *30*, 1089–1115. [CrossRef]
38. Tweedale, J.W. Enhancing the Degree of Autonomy by Creating Automated Components within a Multi-Agent System Framework. In *Knowledge-Based Information Systems in Practice. Smart Innovation, Systems and Technologies*; Weedale, J., Jain, L., Watada, J., Howlett, R., Eds.; Springer: Cham, Switzerland, 2015; Volume 30, pp. 251–276.
39. Ferguson, I. Towards an architecture for adaptive, rational, mobile agents. In *Decentralized AI 3—Proceedings of the Third European Workshop on Modelling Autonomous Agents in a Multi-Agent World (MAAMAW 1991)*; Werner, E., Demazeau, Y., Eds.; Elsevier Science Publishers: Amsterdam, The Netherlands, 1991; Volume 1177, pp. 249–262.
40. Lazarowska, A. Verification of Ship's Trajectory Planning Algorithms Using Real Navigational Data. *TransNav Int. J. Mar. Navig. Saf. Sea Transp.* **2019**, *13*, 559–564. [CrossRef]

41. Lazarowska, A. A new deterministic approach in a decision support system for ship's trajectory planning. *Expert Syst. Appl.* **2017**, *71*, 469–478. [CrossRef]
42. Lazarowska, A. *Safe Trajectory Planning for Maritime Surface Ships*, 1st ed.; Springer Nature Switzerland AG: Cham, Switzerland, 2022.
43. Śmierzchalski, R. Ships' domains as collision risk at sea in the evolutionary method of trajectory planning. In *Information Processing and Security Systems*; Saeed, K., Pejaś, J., Eds.; Springer: Boston, MA, USA, 2005; pp. 411–422.
44. Tam, C.; Bucknall, R. Cooperative path planning algorithm for marine surface vessels. *Ocean Eng.* **2013**, *57*, 25–33. [CrossRef]
45. Szlapczynski, R.; Szlapczynska, J. On evolutionary computing in multi-ship trajectory planning. *Appl. Intell.* **2012**, *37*, 155–174. [CrossRef]
46. Tam, C.; Bucknall, R. Path-planning algorithm for ships in close-range encounters. *J. Mar. Sci. Technol.* **2010**, *15*, 395–407. [CrossRef]

electronics

MDPI

Article

LTE and NB-IoT Performance Estimation Based on Indicators Measured by the Radio Module

Robert Burczyk, Agnieszka Czapiewska *, Malgorzata Gajewska and Slawomir Gajewski

Faculty of Electronics, Telecommunications and Informatics, Gdańsk University of Technology, Gabriela Narutowicza 11/12 Street, 80-233 Gdańsk, Poland

* Correspondence: agnieszka.czapiewska@pg.edu.pl

Abstract: Monitoring the operating parameters of power grids is extremely important for their proper functioning as well as for ensuring the security of the entire infrastructure. As the idea of the Internet of Things becomes more ubiquitous, there are tools for monitoring the state of the complex electrical grid and means to control it. There are also developed new measuring devices and transmission technologies allowing for the transfer of performed measurements from many places to the network management center. However, there are still no devices that act as data concentrators, which would integrate many transmission technologies and protocols in one device, supporting the communication between those different transmission technologies and which would realize edge computing to assist the management center by prioritizing and combining transmitted data. In this article, the authors present a device that meets the above-mentioned requirements. There are presented research results leading to the development of a decision algorithm, called Multilink—ML, dedicated to the presented device. This algorithm enables the selection between LTE and NB-IoT interfaces for packet transmission without the need to burden the communication system with additional transmissions.

Keywords: LTE; NB-IoT; reception quality indicators; data concentrator

Citation: Burczyk, R.; Czapiewska, A.; Gajewska, M.; Gajewski, S. LTE and NB-IoT Performance Estimation Based on Indicators Measured by the Radio Module. *Electronics* **2022**, *11*, 2892. https://doi.org/10.3390/electronics11182892

Academic Editor: Djuradj Budimir

Received: 18 August 2022
Accepted: 8 September 2022
Published: 13 September 2022

1. Introduction

In recent years, many products have been developed for AMI (Advanced Metering Infrastructure) [1], SCADA (Supervisory Control And Data Acquisition) [2], HAN (Home Area Network) [3] and IoT (Internet of Things) systems. However, there is a lack of devices working at the edge of those networks that concentrate data and create datasets for managing and monitoring electrical grids. At the Department of Radio Communication Systems and Networks, at the faculty of Electronics, Telecommunications and Informatics of the Gdańsk University of Technology is being realized the project entitled "Power data concentrator with innovative decision functionality and gate functionality in AMI, SCADA, HAN, IoT environments" with the acronym KODEŚ [4]. This project is realized with strong cooperation with DGT company which is a Polish producer and integrator of modern ICT systems. The aim of the project is the development of the data concentrator that enables the work of AMI systems at the interface of SCADA, HAN and IoT for energy networks. The KODEŚ concentrator will be equipped with elements capable of making independent decisions at the networks interface (edge computing), gateway functions in terms of DNP3/DLMS (Distributed Network Protocol version 3/Device Language Message Specification) and the innovative multilink algorithm enabling automatic choice of the radio communication link based on the transmission quality and the purpose of the interfaces. The device presented in this paper is prepared to work in electric networks; however, such a device can merge different technologies and data from other fields, e.g., in harbors or sea vessels where many different communication devices and technologies are used.

KODEŚ device is equipped with two commercially available radio modules. The first one is the NeoWay N27 module to cover NB-IoT (Narrow Band Internet of Things)

technology. The second one is Quectel RM500Q-GL which enables, among others, radio communication in LTE (Long Term Evolution) cellular network. The data concentrator will decide which communication interface should be used for particular data, based on the importance and the amount of this data. To make such a decision there must be also knowledge about the present quality of a particular communication link. Obviously, a communication link is strongly connected with available communication interfaces in the KODEŚ device. In this paper, the authors are interested in ways of forecasting if a packet of data can be sent via radio interfaces working in NB-IoT and LTE technologies. As KODEŚ is planned to be a commercial product, it was decided to integrate with KODEŚ available commercial radio modules, to make it less expensive. Hence, there is a lack of possibility to perform signal processing. Nevertheless, utilized radio modules measure standardized quality parameters of received signal called also Key Performance Indicators (KPI), such as RSRP (Reference Signal Received Power), RSRQ (Reference Signal Received Quality), SINR (Signal to Interference plus Noise Ratio), RSSI (Received Signal Strength Indicator), MCS (Modulation and Coding Scheme) and CQI (Channel Quality Indicator) [5,6].

In this paper, the authors proposed a decision algorithm (called Multilink—ML) for link quality monitoring which allows forecasting if a particular connection is working (packet sent through this connection will be successfully received). The ML algorithm is based on quasi-real-time measurements in a real cellular network (LTE and LTE NB-IoT).

The possibility of the transmission quality estimation is important for adjusting transmission parameters to maximize the effectiveness of resource utilization. KPI parameters are mainly utilized by a network to realize the handover procedure [7–10]. A different utilization of signal quality parameters is presented in [11] where authors compare a couple of machine learning algorithms to estimate throughput in the LTE system. Worth mentioning is that the resulting probability of correct estimation in [11] is similar to that one we achieve for the LTE system, while our algorithm needs less computational power. Another difference in approach in [11] presented in this article, emerges from the measurement stand. We use a commercial radio module, without specialized measurement applications and devices, which makes our solution more suitable for developed data concentrator KODEŚ. KPIs may also be used to estimate network load [12,13]. In [12], measurements were done by RF scanners. In [13], presented results gathered by a commercial smartphone and Nemo device. In the paper [14], authors cover the issue of cell reselection based on RSRP and RSRQ values measured by an RF scanner; however, compared to bars shown by iPhone 5. In [15], the authors present RSRP, RSRQ and SNR measurements for LTE for different cellular operators. Those measurements were done with the utilization of the G-NetTrack Pro app. Unfortunately, these measurements are not compared with real transmission quality. In [16], the authors used the build-in cellular modem signal parameters measurement mechanisms to research the correlation between achieved download speed, upload speed, latency, jitter and KPIs available in LTE. Despite interesting analysis, those results are insufficient to build our multilink algorithm.

NB-IoT is an emerging technology well suited for wireless IoT end devices [17,18]. There is a wide range of possible applications. Most common are those connected with the smart cities field [19], especially with energy grids [20] also for monitoring marine energy generation [21]. However, NB-IoT may be utilized in other areas of the economy for example in agriculture [22]. In literature are analyses of NB-IoT coverage [23,24] and performance in selected environments. Those research suggest that NB-IoT is very robust even in deep indoor, so the choice of NB-IoT for our data concentrator became obvious. Nevertheless, there was a need to conduct own measurements to implement mentioned before multilink algorithm.

This paper is organized as follows: in part 2 there is a description of the developed data concentrator KODEŚ. Part 3 covers the measurement stand. Consecutive parts present results accordingly for NB-IoT and LTE.

2. Realized Data Concentrator Device—KODEŚ

KODEŚ is an intelligent device, supposed to work as a sophisticated, specialized NB-IoT router and data concentrator realizing edge computing within the boundaries of the network. It is going to enable connection, scheduling, selection and optimization of the quality of data transmission between wireless and wired interfaces for AMI systems, SCADA, HAN and IoT. The proposed concentrator will be equipped with specialized functions. First of all, there will be a **decisive function** (DF) as part of edge computing whose aim would be the detection of malfunction and undesirable states of the energetic network, to improve the response time of service. This is especially important in a situation when there is no connection with the management center due to e.g., terrorist act. DF will make a decision based on information gathered from energy meters, controllers, disconnectors, sensors, predictions and known profiles of work of energetic network elements. Another important functionality would be a **gateway function** (GF) which would allow to convert data streams between different protocols according to the specification of a given network, act as a data concentrator, control the network and work as a specialized router for WWAN (Wireless Wide Area Network) interfaces. To support the last functionality, in KODEŚ must be implemented a **multilink functionality** (MF), a decisive algorithm to choose the most appropriate radio network for transmission based on the state of the network, the priority of transmission and required quality.

Figure 1 presents the block diagram of the KODEŚ device which was described in detail in article [4]. As was mentioned before, realized data concentrator needs to support interfaces that can work in different networks.

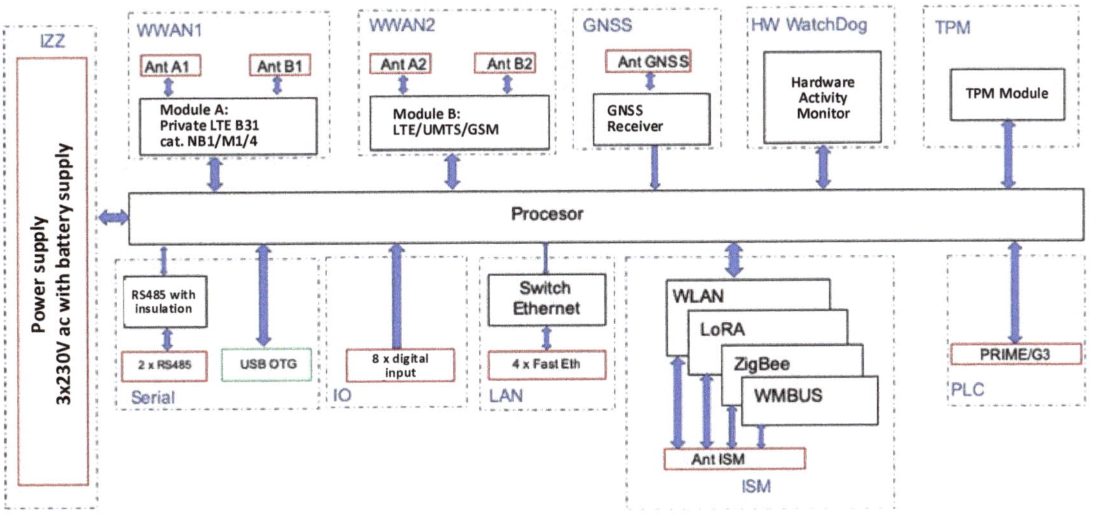

Figure 1. Block diagram of KODEŚ device [4].

For this reason, in the KODEŚ device, many interfaces are provided. Some of them can support a connection in different technologies e.g., ISM, Serial and WWAN. The data concentrator has to periodically analyze the state of all provided interfaces to decide which of them is the most suitable to send user data. The authors of this paper are mostly interested in WWAN interfaces, that support all fully functional cellular networks. As for purpose of machine-to-machine communication in a deep indoor environment, NB-IoT technology was developed therefore, the WWAN1 interface is intended to work in this technology. Whereas the WWAN2 may switch utilized technology between GSM, UMTS or LTE, depending on which one can realize transmission with the best quality for a specific application. Figure 2 shows the device in cover (b) and without cover (a). In Figure 2a can be seen two antenna ports, four Ethernet ports, digital input IO, two SIM card slots and a power supply connection.

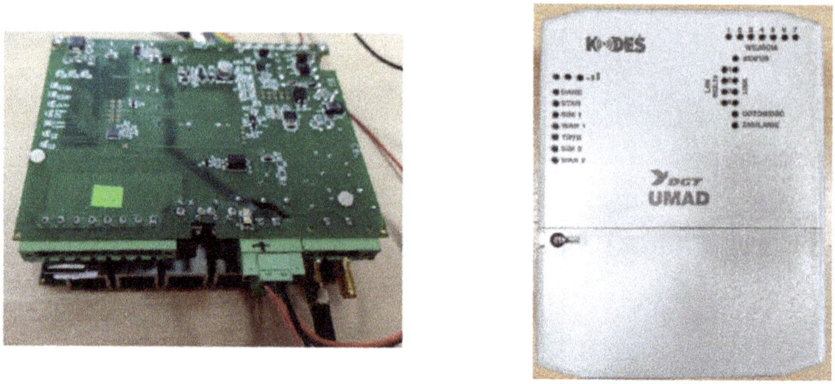

(**a**) (**b**)

Figure 2. KODEŚ device: (**a**) without cover; (**b**) with cover.

3. Measurements

To implement the Multilink (ML) algorithm a knowledge of the correlation between values available in particular cellular technology quality parameters and the quality of realized transmission is needed. To gain this knowledge a set of measurements was conducted. In the article [5], the authors presented results and analysis of measurements performed with the utilization of radiocommunication tester Rohde & Schwarz CMW500. As the propagation conditions were controlled during measurements, there could be analyzed the correlation between different quality parameters with BLER (BLock Error Rate). While in the present article authors would like to present results for measurements performed in a real cellular network. For this purpose, a laboratory stand was prepared, and which block diagram is presented in Figure 3. Presented KODEŚ device is a prototype and needs a PC where a TFTP server is set with data necessary to start the operating system on KODEŚ. It is realized by Ethernet connection. To get access to KODEŚ an RS232 connection is used. The data concentrator is connected via LAN with a Server. This Server has a public IP address and an Internet connection. To change the quality of the received signal for measurement purposes adjustable attenuators were put between WWAN antennas and WWAN interfaces in KODEŚ. Measurements are initialized by the Server in the following steps:

1. The server prepares UDP (User Datagram Protocol) packet and sends it via LAN to KODEŚ;
2. KODEŚ redirect this packet to WWAN1 and WWAN2;
3. KODEŚ reads and saves quality parameters from both radio interfaces;
4. The packet is sent through both WWAN interfaces to Server on different ports;
5. Server register time of packets reception;
6. Steps 1–5 are repeated until the required quantity of packets is sent.

User Datagram Protocol was used instead of Transmission Control Protocol (TCP) as it has not built-in mechanisms to guarantee data reception. Nevertheless, the block error rate could not be measured in described research as there was no access to the physical layer. Moreover, because of the upper layers of the ISO-OSI model, during all measurements, none of the corrupted packets was received. For this reason, the quality parameters' results of each cellular radio access technology were compared to information if a UDP packet was received in Server. To collect measurements for different environment conditions i.e., quality of the received signal, at the beginning of each measurement adjustable attenuators were set to 0 dB and then the attenuation was gradually slowly increased until a particular radio module lost connection. Then for this module attenuation was turned off (0 dB) and all procedure was repeated. For each measurement, a 3-second timeout was set.

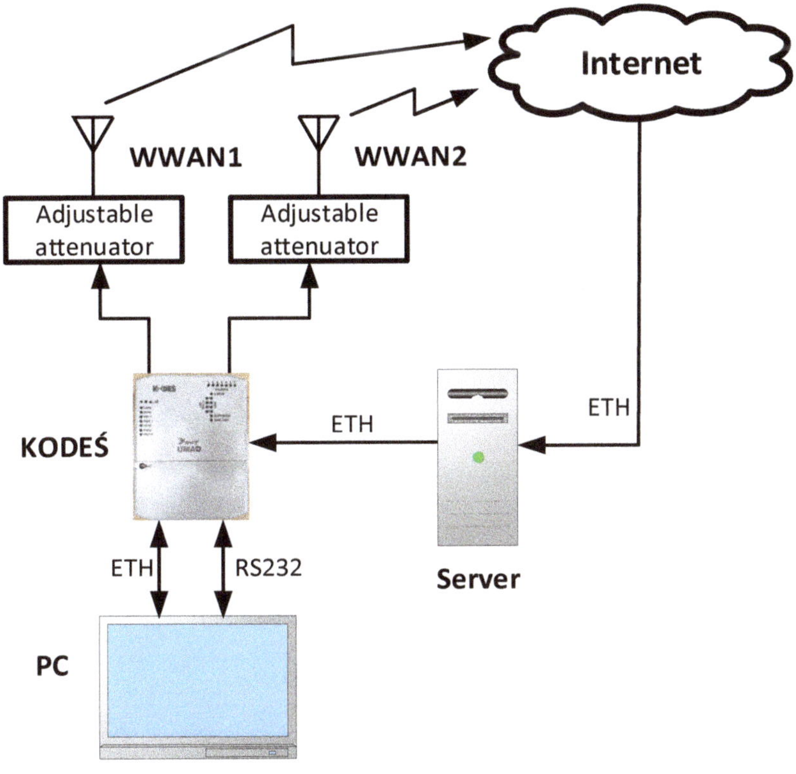

Figure 3. Laboratory stand.

4. Results for NB-IoT

NB-IoT connection in KODEŚ is realized with the utilization of the Neoway N27 radio module. It reports the following signal parameters: RSRP, RSRQ, RSSI and SINR. For NB-IoT technology, 4500 measurements were done—UDP packets were sent from Server. Each packet had 100 bits in length. Simultaneously to each transmission attempt, all reported parameters by the radio module were saved. In Figure 4 are presented all realized measurements. A delay lower than 0 means that a particular packet was not received. It can be seen that delay values strongly fluctuated; however, the mean value was evidently increasing while the transmission quality deteriorates (adjustable attenuation increased). Due to the high values of the delay, Figure 4 is difficult to analyze. The scale on ordinate axes was changed and presented in Figure 5.

In Figure 5 it can be noticed that due to increased attenuation the RSRP and RSSI parameters decreased, while from the RSRQ and SINR parameters the radio link availability could not be predicted.

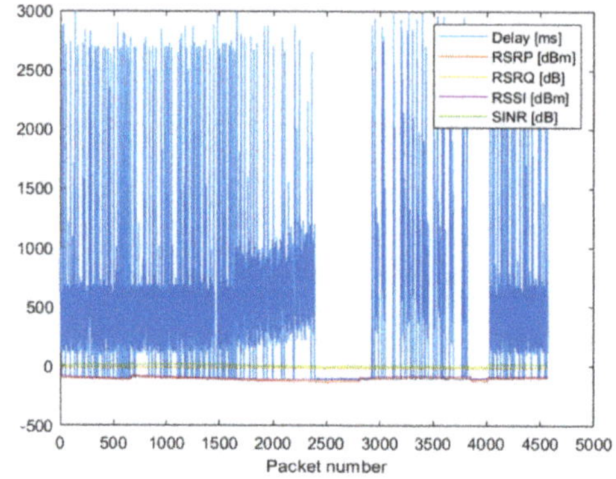

Figure 4. Measurements results of RSRP, RSRQ, RSSI, SINR and transmission delay for NB-IoT technology.

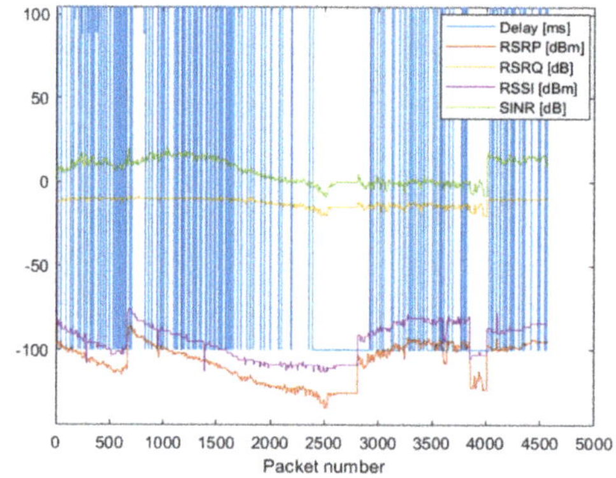

Figure 5. Enlarged plots from Figure 4.

The conducted research aimed to develop a multilink algorithm. Multilink must decide, based on the knowledge of RSRP, RSRQ, RSSI and SINR values, if the NB-IoT is suitable for transmission. From Figure 5 it can be concluded that if values of particular parameters will be lower than some threshold, multilink must decide that the link is not available. To find those thresholds for each possible configuration of quality parameters the probability of correct decision and a probability of false positive decision were calculated for different values of thresholds. The correct decision is calculated as:

$$P_{CD} = \frac{L_{CD}}{L}, \tag{1}$$

where L_{CD} is the number of correct decisions and L is the number of all made decisions. The false positive decision means that the multilink algorithm considered the link as reliable when the packet could not be received—L_{FP} is the number of such situations:

$$P_{FP} = \frac{L_{FP}}{L}. \tag{2}$$

Threshold values for each quality parameter should maximize the probability of a correct decision. However, as constructed data concentrator has two WWAN interfaces, the worst case is when the multilink algorithm will decide to send a packet via an inefficient link. Therefore, thresholds must be chosen such that the probability of a false positive will be the lowest. As this probability does not cover the mistake when the algorithm decides that the link is inefficient while transmission could be realized with success, both probabilities (P_{CD} and P_{FP}) must be analyzed simultaneously and compromised values must be chosen. Moreover, as KODEŚ has implemented more interfaces than one, the authors decided that particular link quality must be evaluated with a probability not less than 70% and the probability of a false decision should not be greater than 30%. Therefore, we looked for the highest value of P_{CD} and the lowest value of P_{FP}. It was found that the multilink algorithm should decide that the NB-IoT connection has enough quality for transmission when:

- RSRP ≥ -125 dBm and
- RSRQ ≥ -20 dB and
- RSSI ≥ -114 dB and
- SINR ≥ 0 dB.

Then the probability of a correct decision is 77.7% and the probability of a false decision does not exceed 18.4%. A probability of correct decision for different threshold values for RSRP and SINR when the threshold for RSRQ was set to -20 dB and the threshold for RSSI was set to -114 dB is presented in Figure 6 whereas the probability for false positive decision is in Figure 7. In Figure 8 the probability of correct decision for different threshold values for RSSI and RSRQ when the threshold for RSRP was set -125 dBm and threshold for SINR was set to 0 dB is presented, while a false positive probability is shown in Figure 9. The chosen values for those plots are the selected best threshold values for ML algorithm.

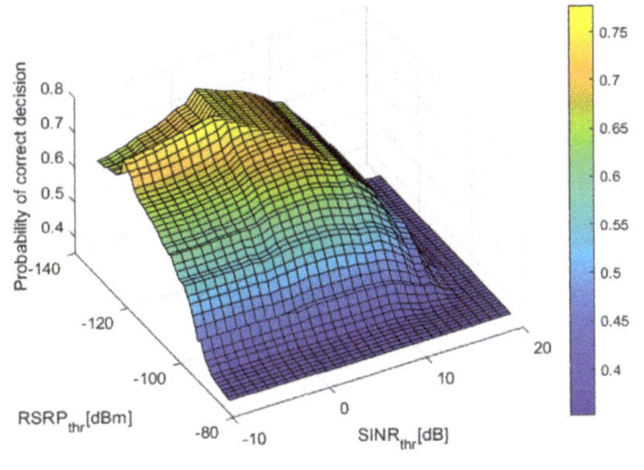

Figure 6. The probability of correct decision for different threshold values for RSRP and SINR, while the threshold for RSRQ was set to -20 dB and the threshold for RSSI, was set to -114 dB.

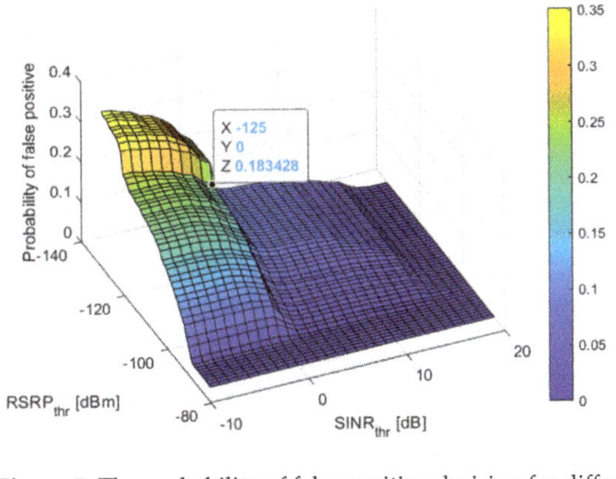

Figure 7. The probability of false positive decision for different threshold values for RSRP and SINR while the threshold for RSRQ was set to -20 dB and the threshold for RSSI was set to -114 dB.

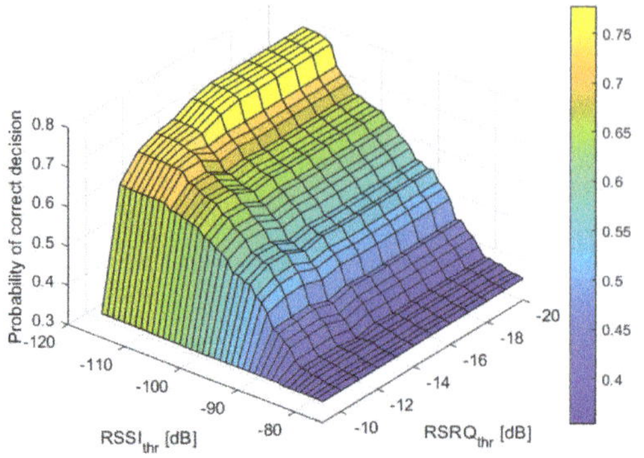

Figure 8. The probability of correct decision for different threshold values for RSSI and RSRQ while the threshold for RSRP was set to −125 dBm and the threshold for SINR was set to 0 dB.

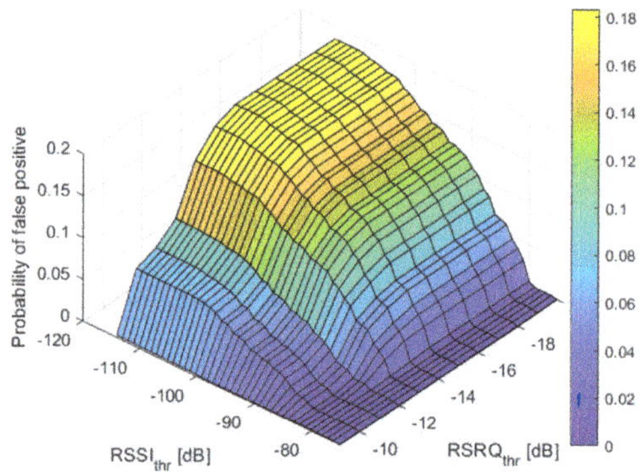

Figure 9. The probability of false positive decision for different threshold values for RSSI and RSRQ while the threshold for RSRP was set to −125 dBm and the threshold for SINR was set to 0 dB.

5. Results for LTE

In KODEŚ the LTE technology is utilized in the WWAN2 interface where the Quectel RM500Q-GL radio module is used. This module reports the following values of quality parameters: RSRP, RSRQ, SINR, RSSI, CQI and MCS. Each of those parameters was saved at the time of each realized transmission. During measurements, 5000 UDP packets of length 100 bits were sent. The measurement procedure was the same as with NB-IoT technology. In Figure 10 are presented all realized measurements. On the plot are presented values of measured delay [ms], RSRP [dBm], RSRQ [dB], RSSI [dBm], MCS, CQI and SINR [dB] for each transmitted packet. A delay lower than 0 means that a particular packet was not received. It is clearly seen that the delay increases as the signal deteriorate—the attenuation increases. Large values of delays at the end of some breaks of transmissions result from the measurements procedures where packets were sent in a predetermined period even when they could not reach the receiver. Due to the high values of the delay in Figure 10, it is difficult to analyze the relations of values of other parameters. To improve the clarity of plots, the scale on ordinate axes was changed and modified plots are presented in Figure 11. The other parameters, except for MCS, decreased with the increasing attenuation. It leads to a conclusion similar to the one in the previous section. The multilink algorithm must decide that the LTE interface is not available when values of quality parameters get lower than some thresholds. The MCS values do not follow any pattern; however, from previous research [5], we know that modulation and coding scheme has a great impact on transmission robustness. However, during measurements with the utilization of a commercial cellular network, it was impossible to influence on MCS parameter. Therefore, the search for the most suitable thresholds was conducted without and with the consideration of MCS. In Table 1 are given the number of transmission and number of lost packets for each possible MCS during measurement. It can be seen that the ratio of realized transmission to lost packets is very good for MCS 0–3 and 8–30 while the ratio is very poor for MCS 4, 5 and 6 when over 50% of packets were not received within the settled timeout. The reasons for such results should be seen in the algorithm of changes in transmission parameters adopted by the cellular operator. Unfortunately, we have no way of viewing this algorithm. Based on data in Table 1 the MCSs values for which a sufficient number of transmissions for further analysis can be selected. These are 0, 1, 3, 4, 5, 6. For transmissions realized with given modulation and coding schemes and for all realized transmissions (without differentiation on MCS), a similar search for threshold values for each quality parameter was realized as for NB-IoT in the previous section.

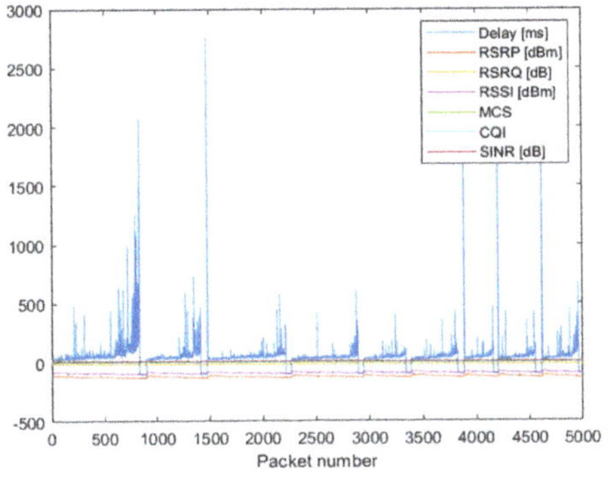

Figure 10. Measurements results of RSRP, RSRQ, RSSI, SINR, MCS, CQI and transmission delay for LTE technology.

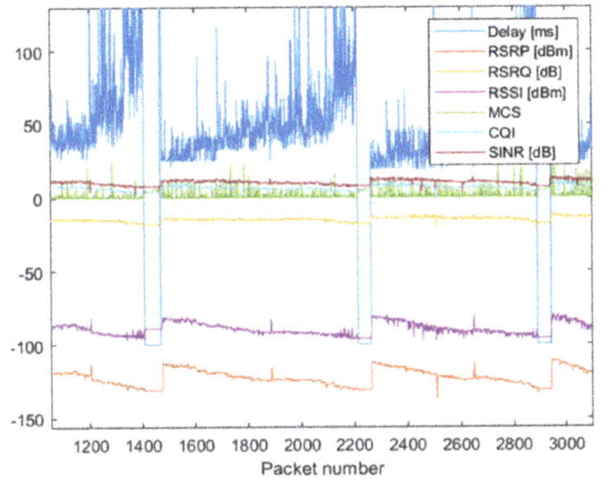

Figure 11. The enlarged plot from Figure 10.

It was found that for all realized transmissions the best threshold values for each quality parameter are:

- $RSRP_{thr} = -133$ dBm and
- $RSRQ_{thr} = -17$ dB and
- $RSSI_{thr} = -100$ dB and
- $SINR_{thr} = 0$ dB.

Table 1. The number of realized transmissions and lost packets for each MCS.

MCS	Number of Transmissions	Number of Lost Packets	MCS	Number of Transmissions	Number of Lost Packets
0	2772	13	16	8	0
1	944	5	17	1	0
2	20	7	18	9	0
3	368	4	19	7	0
4	192	138	20	12	0
5	213	107	21	6	0
6	255	212	22	0	0
7	30	2	23	6	0
8	32	0	24	3	0
9	20	0	25	3	0
10	0	0	26	5	1
11	29	0	27	0	0
12	28	0	28	2	0
13	16	0	29	2	0
14	10	0	30	0	0
15	7	0			

Then the probability of a correct decision is 96% and the probability of a false positive decision is 3%. The CQI was omitted in the multilink algorithm as it is strongly connected with the modulation scheme and MCS parameter. For each individual of selected MCS transmissions, both probabilities were calculated for selected threshold values. Results are presented in rows with white background in Table 2. For MCS 4 and 6 achieved probability values for such thresholds are very poor. Almost reach assumed limits. As the worst case appears to be for MCS = 4, so for this modulation and coding scheme has performed another search for the best threshold values. These are:

- $RSRP_{thr} = -133$ dBm and
- $RSRQ_{thr} = -17$ dB and
- $RSSI_{thr} = -95$ dB and
- $SINR_{thr} = 0$ dB.

Table 2. Probability of correct decision and positive false decision for selected threshold parameters and different MCS.

MCS	$RSRP_{thr}$ [dBm]	$RSRQ_{thr}$ [dB]	$RSSI_{thr}$ [dBm]	$SINR_{thr}$ [dB]	P_{cd} [%]	P_{FP} [%]
0–30	−133	−17	−100	0	96	3
	−133	−17	−95	0	95.7	0.4
0	−133	−17	−100	0	92.5	6.1
	−133	−17	−95	0	95.7	0.1
1	−133	−17	−100	0	98.8	0.5
	−133	−17	−95	0	96.5	0.4
3	−133	−17	−100	0	98.6	1.1
	−133	−17	−95	0	96.5	0.3
4	−133	−17	−100	0	71.4	22.9
	−133	−17	−95	0	91.1	0
5	−133	−17	−100	0	92.5	6.1
	−133	−17	−95	0	94.8	0.5
6	−133	−17	−100	0	76.5	21.6
	−133	−17	−95	0	96.9	1.2

For this new threshold values were calculated probabilities P_{CD} and P_{FP} for each MCS and all transmissions. Results are also given in Table 2; however, in rows with a grey background. For new threshold values P_{CD} is always greater than 94% and P_{FP} lower than 1.3%. Therefore, these thresholds should be implemented in ML for LTE technology.

A probability of correct decision for different threshold values for RSRP and RSRQ when the threshold for SINR was set to 0 dB and the threshold for RSSI was set to −100 dB is presented in Figure 12 whereas the probability for false positive decision is in Figure 13. In Figure 14, a probability of correct decision for different threshold values for SINR and RSSI when the threshold for RSRP was set −133 dBm and threshold for RSRQ was set to −17 dB is presented, while a false positive probability is shown in Figure 15. The chosen values for those plots are the selected best threshold values for ML algorithm.

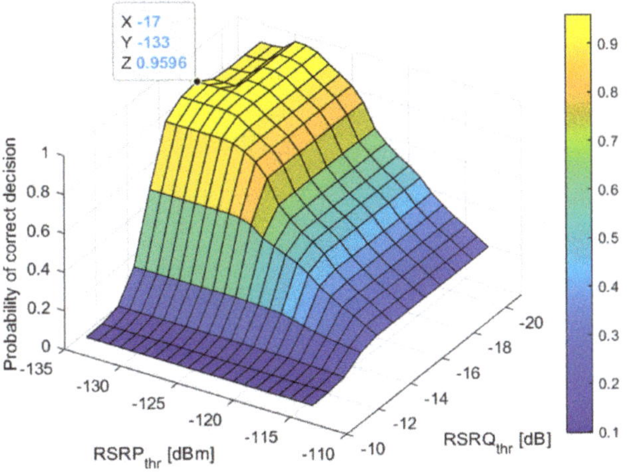

Figure 12. The probability of correct decision for different threshold values for RSRP and RSRQ while the threshold for RSSI was set to −100 dBm and the threshold for SINR was set to 0 dB.

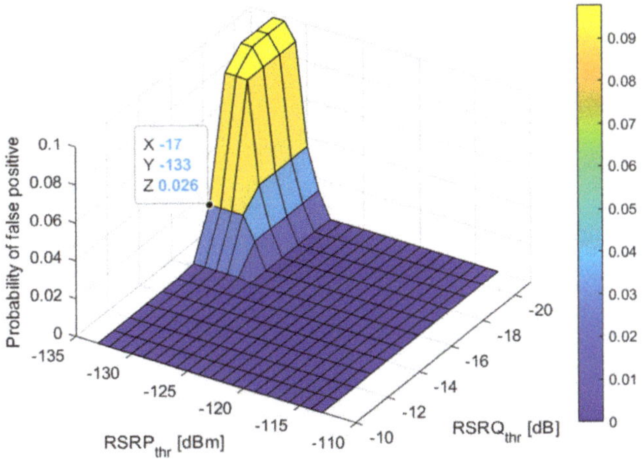

Figure 13. The probability of false positive decision for different threshold values for RSSI and RSRQ while the threshold for RSSI was set to −100 dBm and the threshold for SINR was set to 0 dB.

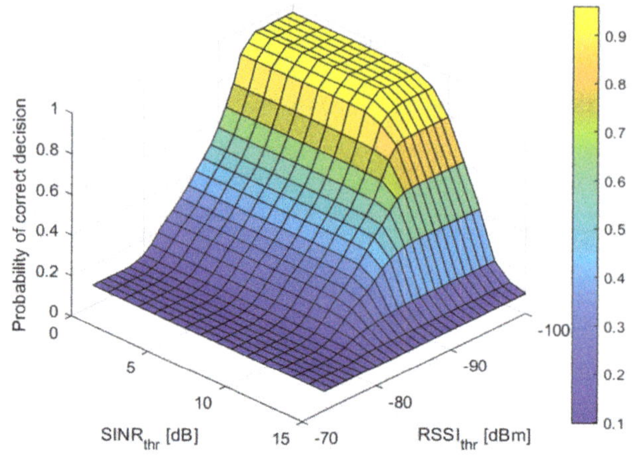

Figure 14. The probability of correct decision for different threshold values for RSSI and SINR while the threshold for RSRP was set to −133 dBm and the threshold for RSRQ was set to −17 dB.

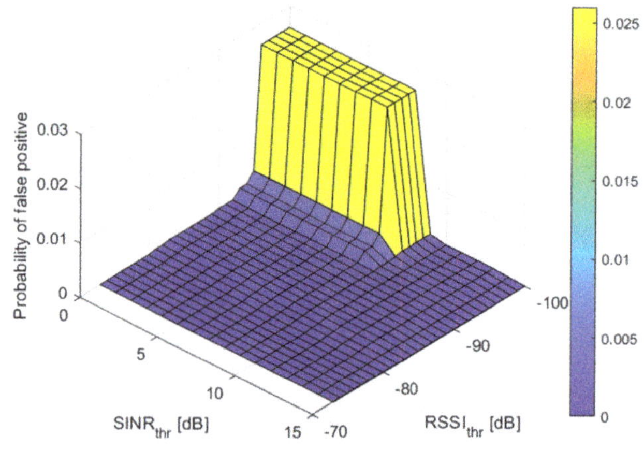

Figure 15. The probability of false positive decision for different threshold values for RSSI and SINR while the threshold for RSRP was set to −133 dBm and the threshold for RSRQ was set to −17 dB.

6. Conclusions

In the article, the authors presented a data concentrator device dedicated to an energy grid that integrates many different communication technologies. Especially, it enables the work of AMI systems at the interfaces of SCADA, HAN and IoT. As this device ought to realize edge computing and has gateway functions it must be equipped with decision algorithms. Among others, those algorithms must decide which of all radio communication interfaces is best for transmission at a particular time and for particular data. To elaborate such an algorithm a measurement campaign was performed with the utilization of a real cellular network. The novelty of these measurements results, among others, from the use of commercial radio modules instead of specialized measuring equipment. This approach enables the assessment of the credibility of the signal quality parameters reported by a given radio module. From the presented research it can be noticed that for NB-IoT technology the connection was broken very often while the module utilizing LTE technology worked very steadily. Despite this inconvenience, for both technologies, it was possible to determine the method of implementation of the ML algorithm. As all of the values of reported quality parameters were decreasing as the quality of signal deteriorated, ML must decide that a particular link is unavailable when all quality parameters get below some thresholds. Thresholds that maximize the probability of a correct decision and minimize the probability of a false positive decision are given in Table 3. It is worth seeing that, unexpectedly, the SINR parameter has the least impact on the decision making as its found threshold is equal 0 dB.

Table 3. Found thresholds for the ML algorithm.

Quality Parameter	LTE	NB-IoT
RSRP [dBm]	−133	−125
RSRQ [dB]	−17	−20
RSSI [dBm]	−95	−114
SINR [dB]	0	0

The presented results show that it is possible to assess the link state for LTE and NB-IoT technologies. Therefore, the current work aims to implement the ML algorithm in the target device and perform field measurements to make final ML validation. It was decided to use this link which enables greater throughput. Therefore, if both LTE and NB-IoT will be available the ML will choose LTE to send packet. Future extensions of the data concentrator will be related to improving its hardware implementation and to expanding the possibilities of its applications not only in electrical grid but also i.e., in harbors.

Author Contributions: Conceptualization, S.G. and M.G.; methodology, S.G., M.G., R.B. and A.C.; software, R.B., A.C.; validation, S.G., M.G., R.B. and A.C.; formal analysis, A.C.; investigation, A.C. and R.B.; data curation, R.B. and A.C.; writing—original draft preparation, A.C.; writing—review and editing, S.G., M.G. and R.B.; visualization, A.C.; supervision, S.G. and M.G.; project administration, M.G.; funding acquisition, M.G. All authors have read and agreed to the published version of the manuscript.

Funding: This work has been carried out as part of the project entitled "Power data concentrator with innovative decision functionality and gate functionality in AMI, SCADA, HAN, IoT environments", co-financed under the European Regional Development Fund within the Smart Growth Operational Program, agreement no. POIR.01.01.01-00-0651/19-00 on 09.03.2020r.

Informed Consent Statement: Not applicable.

Conflicts of Interest: The authors declare no conflict of interest.

References

1. Ghasempour, A.; Moon, T. Optimizing the Number of Collectors in Machine-to-Machine Advanced Metering Infrastructure Architecture for Internet of Things-based Smart-Grid. In Proceedings of the IEEE Green Technologies Conference, Kansas City, MO, USA, 7–8 April 2016.
2. Lawrence, A.; Iqbal, T. Development of an IoT Based Open Source SCADA System for PV System Monitoring. In Proceedings of the IEEE Canadian Conference of Electrical and Computer Engineering, Edmonton, AB, Canada, 5–8 May 2019.
3. Soniya, J.; Divya, M. A novel architecture for efficient communication in smart grid home area network. In Proceedings of the IEEE International Conference on Computational Intelligence and Computing Research, Tamil Nadu, India, 10–12 December 2015.
4. Błaszkiewicz, O.; Burczyk, R.; Czapiewska, A.; Gajewska, M.; Gajewski, S.; Marczak, A.; Miszewski, M.; Sadowski, J.; Stefański, J. Laboratory stand for automatic analysis of the transmission quality of the KODEŚ concentrator with LTE, NB-IoT and ISM interfaces. *Przegląd Telekomunikacyjny i Wiadomości Telekomunikacyjne* **2021**, *1*, 10–19.
5. Błaszkiewicz, O.; Burczyk, R.; Czapiewska, A.; Gajewska, M.; Gajewski, S.; Sadowski, J. Transmission quailty estimation in LTE basing on quality indicators measured by radio module. *XX Krajowa Konferencja Elektroniki* **2022**. [CrossRef]
6. 3GPP TS 36.214 LTE. Evolved Universal Terrestrial Radio Access (E-UTRA); Physical layer; Measurements, Release 16, 2020–07. Available online: https://www.etsi.org/deliver/etsi_ts/136200_136299/136214/16.01.00_60/ts_136214v160100p.pdf (accessed on 16 August 2022).
7. Xian, H.; Muqing, W.; Jiansong, M.; Cunyi, Z. The impact of channel environment on the RSRP and RSRQ measurement of handover performance. In Proceedings of the International Conference on Electronics, Communications and Control (ICECC), Ningbo, China, 9–11 September 2011.

8. Harja, S.L.; Hendrawan. Evaluation and optimization handover parameter based X2 in LTE network. In Proceedings of the 3rd International Conference on Wireless and Telematics (ICWT), Palembang, Indonesia, 27–28 July 2017.
9. Luan, L.; Wu, M.; Chen, Y.; He, X.; Zhang, C. Handover parameter optimization of LTE system in variational velocity environment. In Proceedings of the IET International Conference on Communication Technology and Application (ICCTA 2011), Beijing, China, 14–16 October 2011.
10. Petrut, I.; Otesteanu, M.; Balint, C.; Budura, G. HetNet handover performance analysis based on RSRP vs RSRQ triggers. In Proceedings of the 38th International Conference on Telecommunications and Signal Processing (TSP), Prague, Czech Republic, 9–11 July 2015.
11. Eyceyurt, E.; Egi, Y.; Zec, J. Machine-Learning-Based Uplink Throughput Prediction from Physical Layer Measurements. *Electronics* **2022**, *11*, 1227. [CrossRef]
12. Chang, K.; Wicaksono, R.P. Estimation of network load and downlink throughput using RF scanner data for LTE networks. In Proceedings of the International Symposium on Performance Evaluation of Computer and Telecommunication Systems (SPECTS), Seattle, WA, USA, 9–12 July 2017.
13. Raida, V.; Lerch, M.; Svoboda, P.; Rupp, M. Deriving Cell Load from RSRQ Measurements. In Proceedings of the Network Traffic Measurement and Analysis Conference (TMA), Vienna, Austria, 26–29 June 2018.
14. Chang, K.; Wicaksono, R.P.; Kunishige, S.; Takagaki, N. LTE idle mode optimization improving end user experiences. In Proceedings of the 16th International Conference on Advanced Communication Technology, Pyeongchang, Korea, 16–19 February 2014.
15. Putra, E.M.; Budiman, E.; Malewa, Y.; Cahyadi, D.; Taruk, M.; Hairah, U. 4G LTE Experience: Reference Signal Received Power, Noise Ratio and Quality. In Proceedings of the 3rd East Indonesia Conference on Computer and Information Technology (EIConCIT), Surabaya, Indonesia, 9–11 April 2021.
16. Stafecka, A.; Lizunovs, A.; Bobrovs, V. Mobile LTE network signal and Quality of Service parameter evaluation from end-user premises. In Proceedings of the 2018 Advances in Wireless and Optical Communications (RTUWO), Riga, Latvia, 15–16 November 2018.
17. Ugwuanyi, S.; Paul, G.; Irvine, J. Survey of IoT for Developing Countries: Performance Analysis of LoRaWAN and Cellular NB-IoT Networks. *Electronics* **2021**, *10*, 2224. [CrossRef]
18. Dangana, M.; Ansari, S.; Abbasi, Q.H.; Hussain, S.; Imran, M.A. Suitability of NB-IoT for Indoor Industrial Environment: A Survey and Insights. *Sensors* **2021**, *21*, 5284. [CrossRef] [PubMed]
19. Routray, S.K.; Sharmila, K.P.; Akanskha, E.; Ghosh, A.D.; Sharma, L.; Pappa, M. Narrowband IoT (NBIoT) for Smart Cities. In Proceedings of the Third International Conference on Intelligent Communication Technologies and Virtual Mobile Networks (ICICV), Tirunelveli, India, 4–6 February 2021.
20. Routray, S.K.; Gopal, D.; Pallekonda, A.; Javali, A.; Kokkirigadda, S. Measurement, Control and Monitoring in Smart Grids using NBIoT. In Proceedings of the 6th International Conference on Inventive Computation Technologies (ICICT), Coimbatore, India, 20–22 January 2021.
21. Wei, X.; Wang, Y.; Zhu, W.; Chen, J.; Dong, Y.; Guo, J. Design of a Universal Marine Energy Remote Monitoring System. In Proceedings of the 12th International Conference on Measuring Technology and Mechatronics Automation (ICMTMA), Phuket, Thailand, 28–29 February 2020.
22. Mezei, I.; Lukić, M.; Bebakov, L.; Pavković, B.; Radovanović, B. Grapevine Downy Mildew Warning System Based on NB-IoT and Energy Harvesting Technology. *Electronics* **2022**, *11*, 356. [CrossRef]
23. Ferreira, B.; Gaspar, B.; Paiva, S.; Santos, A.; Cabral, J. Coverage and Deployment Analysis of NB-IoT Technology under Various Environment Scenarios. In Proceedings of the 2nd International Conference on Societal Automation (SA), Funchal, Portugal, 26–28 May 2020.
24. Poddar, N.; Khan, Z.; Mass, J.; Srirama, S.N. Coverage Analysis of NB-IoT and Sigfox: Two Estonian University Campuses as a Case Study. In Proceedings of the International Wireless Communications and Mobile Computing (IWCMC), Limassol, Cyprus, 15–19 June 2020.

electronics

MDPI

Article

Selection of Robotic Machining Parameters with Pneumatic Feed Force Progression

Andrzej Burghardt [1,*], Magdalena Muszyńska [1], Piotr Gierlak [1], Krzysztof Kurc [1], Dariusz Szybicki [1], Artur Ornat [2] and Marek Uliasz [2]

1 Department of Applied Mechanics and Robotics, Faculty of Mechanical Engineering and Aeronautics, Rzeszow University of Technology, al. Powstancow Warszawy 8, 35-959 Rzeszow, Poland
2 Pratt and Whitney Rzeszow S.A., 35-078 Rzeszow, Poland
* Correspondence: andrzejb@prz.edu.pl; Tel.: +48-(17)-8651905

Abstract: The subject matter of the above paper presents part of the research carried out as part of the robotization of the manufacturing processes of aircraft engine components. The paper concerns robotic deburring of the V2500 diffuser's sharp edges. The diffuser is a precision casting characterized by a slight variation in the geometry of the workpiece depending on the accuracy of the casting molds and the phenomenon of shrinkage during solidification. Due to the small degree of deburring and thus low cutting forces, robotic machining with the FDB150 tool was used. This tool is characterized by compliance with adjustable force, which enables machining of workpieces with a randomly changing shape. The authors of the paper propose a procedure for carrying out work allowing for the selection of suboptimal process parameters. In the analyzed case, these parameters are the speed of movement of the characteristic point of the tool (TCP) and the tool/workpiece contact force. The proposed procedure for determining the parameters of the force and speed of movement allowed for indicating a set of parameters ensuring the performance of the product in accordance with the requirements defined in the documentation. The proposed solution is of an engineering nature and is not a classic search for the extreme of a function from the point of view of the adopted criteria (quality indicators). Its advantage is simplicity, which is very important from the point of view of an industrial application.

Keywords: robotic deburring; machining; machining parameters; pneumatic tools

Citation: Burghardt, A.; Muszyńska, M.; Gierlak, P.; Kurc, K.; Szybicki, D.; Ornat, A.; Uliasz, M. Selection of Robotic Machining Parameters with Pneumatic Feed Force Progression. *Electronics* **2022**, *11*, 3211. https://doi.org/10.3390/electronics11193211

Academic Editors: Piotr Szymak, Stanisław Hożyń and Paweł Piskur

Received: 12 September 2022
Accepted: 5 October 2022
Published: 7 October 2022

1. Introduction

For a long time, deburring was understood as the removal of burrs occurring on the edges of workpieces. Nowadays, the term has been extended to include the process of making chamfers or rounds on the edges of workpieces and the process of melting sharp edges. Today, there are about 20 methods of deburring edges and making chamfers. These methods include hand deburring with countersinks and scrapers, magneto-abrasive machining, abrasive blasting, abrasive grinding, electrochemical treatment, chemical treatment, thermal deburring, container smoothing, belt grinding, turbo abrasive machining, machining using bevelers with elastic blades, and machining with multi-purpose tools [1–3]. In terms of the degree of automation, graining methods can be divided into three categories: manual, manual-machine, and machine. The choice of the best machining method is related to the size of the burrs, the properties of the material, the size of the workpiece, the number of pieces to be made, and the degree of automation of the enterprise.

Manual deburring is used primarily in small-batch production. Burrs are removed using deburring tools and scrapers of various geometries and files. It is a process characterized by low repeatability and high time consumption.

Hand-machine deburring is often used because of the versatility and convenience of using air- or electric-powered tools. Machining tools such as abrasive belts, grinding wheels, and wire brushes are usually used.

The fastest growing group of deburring methods is machining. The popularity of these methods is associated with high productivity, automation of production, and implementation of the idea of Industry 4.0 [4–6]. In this group of methods using CNC machine tools, the most common tools used are chamfers, milling tools, ceramic, or wire brushes. In the case of machining, many of the methods used use an abrasive medium. It is used in magneto-abrasive and embossing-abrasive machining. In the case of vibratory-abrasive machining, various types of abrasive shapes are used, with the hardness depending on the type of workpiece material. In addition to the aforementioned methods, chemical, electrochemical, and thermal deburring are well suited for automation [7–11].

Robotic deburring can be classified as a machine-to-machine method and is associated with high repeatability and efficiency. One of the advantages of using robots is the versatility of the method. In addition, it is characterized by the possibility of using various types of electrically or pneumatically driven tools and machining tools usually used in manual machining. In the case of robots, their use in deburring [12–14] is usually possible when ensuring the susceptibility of the tool. Vulnerability can be understood as the ability to adapt the tool to the workpiece. Adaptation can be realized by achieving pressure on the entire tool, for example through force control systems or systems with springs. Pressure during deburring is important because the position of the surfaces or edges being machined can be variable, for example, in the case of castings. Adequate contact pressure may also be necessary due to tool wear during the process, as is the case with abrasive tools. The susceptibility solutions of whole tools were analyzed in papers [15–17]. In the paper [17], the authors devoted an entire chapter of the monograph to the analysis of methods for adapting the trajectories of industrial robots.

Grinding, polishing, or deburring are the stages of surface finishing that constitute a significant part of the production process of elements in the aviation industry, such as: elements of gears, housings, guide vanes, or blades. Often, these operations are carried out manually by skilled workers using traditional techniques. The high level of skill and dexterity required makes this process difficult to automate. Many of these difficulties result from the lack of data on the influence of the process parameters on the obtained surface roughness, although some experimental studies have been published in this respect in Refs. [18–20]. This subject is a series of papers related to automatic machines, robots, and machining centers [21,22].

In robotic systems solutions, there are several ways to adapt the robot's trajectory in real time or based on previous measurements. The adaptation of the path to changing external conditions is determined by the conditions of the process. For quick-changing processes with low tool/workpiece contact forces, pneumatic tools with variable contact forces are used [23–26]. In the processes where the contact force exceeds 10 N, robotic systems with force control are used [26–29]. An alternative solution is to generate the tool path based on the shape measurement performed with the use of laser or vision measuring systems. Learning systems are also worth paying attention to. It is a software combination of the force control system, with the software recording consecutive discrete trajectory points. The process is then divided into the learning stage and the process of proper robot movement. Such a solution, although accurate, requires much more time to be implemented.

Pneumatic polishing and deburring tools are usually classified as active or passive. Active tools use a closed-loop control system to correct the force exerted on the workpiece, regardless of the tool's trajectory [21]. Passive tools rely on the compliance of the tool itself to maintain the nominal contact force [30]. In some implementations, process parameters, such as forces and the amount of allowance removed, are measured, and the process parameters are then adjusted to obtain decent machining quality. This approach is known as a hybrid [31].

In machining applications using flexible pneumatic tools, there are two kinds of problems due to air compressibility and long system delays. The first is to determine the dependence of the amount of air pressure, and thus the feed force, on the quality of processing, e.g., surface quality or the amount of material removed. The second problem is to program the start of the process, i.e., the phase of entering the tool into the material, so that, despite the high compressibility of the air, the assumed surface quality is achieved. While the first problem is widely described in the literature, the second, although very important, has not received as much attention.

The authors of [32] present experimental studies on the operation of robotic grinding of steel dies, which allowed for the determination of the process parameters in relation to the surface quality. The angle of inclination of the grinding wheel in relation to the workpiece, feed speed, and contact force as a function of surface roughness were analyzed. The same was done in Ref. [33], which also applies to the polishing of dies, where the complex shape of the surface was additionally corrected with the use of a PC.

Integrated with the six-axis robot, the pneumatic flexure spindle is used for deburring and grinding a wide range of freeform topographies. The treatment results depending on the parameters are presented in Ref. [34]. The authors of [35] describe concerns about the generation of collision-free trajectories of the motion of a six-axis robot with a pneumatic flexible tool in the polishing process. The paper solves the problem of taking into account the dimensions of the tool in precision machining. The trajectory was corrected on the basis of the solid model of the workpiece, by taking into account the sizes of the tools used. A very unusual application is described in Ref. [36], which concerns robotization in the seafood processing industry. A robotic station was proposed to replace the hitherto difficult manual work of removing long, sharp crab spines (porcupine crab). Data from a 3D point cloud was used as input to generate the trajectory of a robot with a pneumatic tool. The digital model of the crab is obtained by scanning.

To sum up, the subject of determining the parameters of robotic machining with a rotary pneumatic tool with compliance is topical, and there are several areas that require supplementation, such as the selection of process initiation parameters.

2. Materials and Methods

The subject matter of the article presents part of the research carried out as part of the robotization of the manufacturing processes of aircraft engine components. The paper concerns the diffuser of the V2500 engine (Figure 1a). It is a two-shaft high-bypass turbofan engine. It is produced by the International Aero Engines consortium (East Hartford, CT, USA), which was established in 1983. The V2500 unit powers the Airbus A320 and McDonnell Douglas MD-90 aircraft. The engine was certified by the FAA in 1988. As part of its activities, Prat & Wittney, Rzeszów S.A., which is part of UTC, performs technological operations consisting of processing diffuser castings of the V2500 engine (Figure 1b). One of the many technological operations is edge deburring.

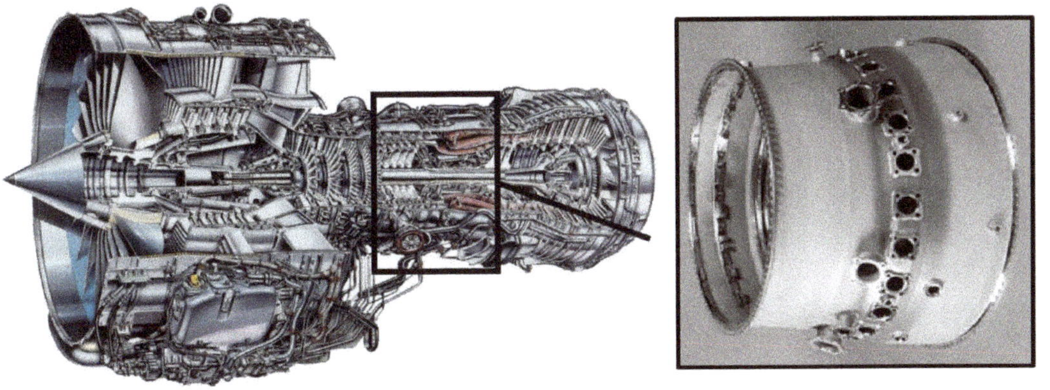

(a) (b)

Figure 1. (**a**) V2500 engine model; (**b**) turbofan engine diffuser model.

In the process of making the diffuser, there are a number of edges that require deburring, which is done manually. Manual processing is due to the fact that there are elements with a shape that varies randomly to a limited extent. It is caused by the applied precision casting technology, characterized by the variable geometry of the workpiece depending on the accuracy of the casting molds and the phenomenon of shrinkage during solidification. This generates difficulties with precisely defining the shape, which in turn results in the need to use manual finishing. This fact introduces a high risk of defective elements related to the presence of human factors (error, fatigue, and stress).

There are nine types of shapes that can be deburred in the diffuser of the V2500 engine. The triangular boss shown in Figure 2 has been selected as an example.

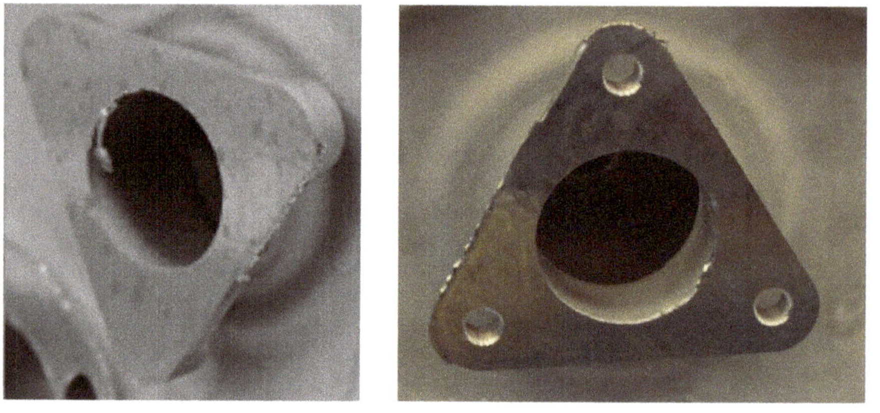

(a) (b)

Figure 2. Workpiece sections intended for deburring: (**a**) raw casting; (**b**) preprocessed casting with edges to be deburred.

Control of deburring at several selected points is carried out using optical methods. The vast majority of the quality of the operations performed is carried out visually by a quality control employee. The purpose of quality control is to eliminate workpieces with sharp edges, notches, or cuts. It is obvious that the presence of surface notches reduces the fatigue life of the workpiece. As part of the work, a robotic station for deburring edges (Figure 3a), using a pneumatic tool (Figure 3b), was proposed. The use of a pneumatic tool with progression of the feed force was dictated by the necessity to make chamfers with a small dimension of 0.1–0.5 mm.

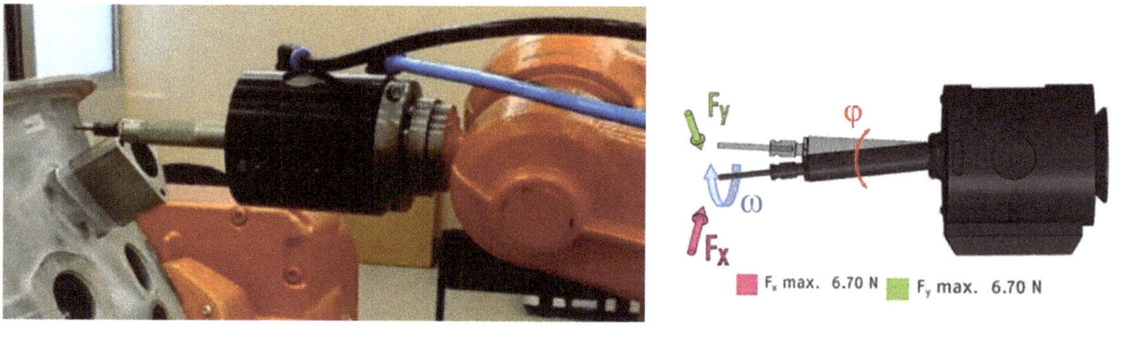

(a) (b)

Figure 3. Deburring edges with a pneumatic tool with compliance: (**a**) view of the station; (**b**) the concept of a rotary file with pneumatic progression of the feed force.

The process is performed by the ABB IRB140 robot. The robot controller communicates digitally with the pressure setting system, cooperating with the FDB150 tool. The most important parameters of the FDB150 tool are the maximum speed of 65,000 rpm and the adjustment of forces in the range from 3.1 to 6.7 N depending on the pressure. The possible range of control pressures is programmed from 1.4 to 4.1 bar.

For the deburring process carried out, the FDB150 tool acts as a pneumatic spring to dampen the vibrations that occur during the deburring process. The spindle stiffness is pneumatically regulated by means of compressed air. The dependence of the force generated on the tool on the set pressure has been included by the manufacturer in the form of a chart in the documentation. These values were checked with a force gauge and a digital pressure regulator during the static test. A PNEUMAX 171E2N.T.D.0009 pressure regulator, with a range of 9 bar and a sensitivity of 0.01 bar, was used to apply pressure during the tests. The accuracy of the applied force obtained during static tests was 0.1 N. Tests of the accuracy of the applied force during the deburring process were not conducted. Testing during the process would require a multi-axis force and moment sensor mounted on the robot or workpiece.

Initial research work carried out with the use of a pneumatic tool allowed for the drawing of conclusions that effective control of the pressure change, and thus the tool/workpiece contact force, is impossible. This is because the machining is performed on parts of a workpiece of a complex shape. Since the amount of material removed is small, the tool moves at a high speed with low contact forces. The speed of movement of the TCP is so high that the response time of the system is greater than the necessity to change the inputs due to its complex shape. The compressibility of the air causes a delayed reaction of the tool, which in turn results in the suppression of rapid changes in the value of the force set by the robot controller. This fact generates the necessity to select the parameters of the machining process, i.e., to define the strategy of the movements, air pressure values, and thus the tool/workpiece contact force, and to determine the speed of the TCP.

3. Results and Discussion

The insensitivity of the machining system to shifts of the boss due to inaccuracies in the form and the phenomenon of solidification of the casting is ensured by the tool with pneumatic progression of the feed force, which is presented in Figure 3b. The possibility of deviation of the spindle axis from the neutral position is within the range of ±5 mm, which completely ensures insensitivity to the limited inaccuracies of the machined surfaces and allows for the offset trajectory to be programmed. Assuming the maximum dispersion of the shape of the boss is at the level of ±2 mm, it is possible to design an offset trajectory. The offset trajectory was determined in relation to the nominal model and its value was adopted at the level of doff = 1.7 mm. The view of the nominal trajectory is shown in Figure 4a, while the offset trajectory is shown in Figure 4b.

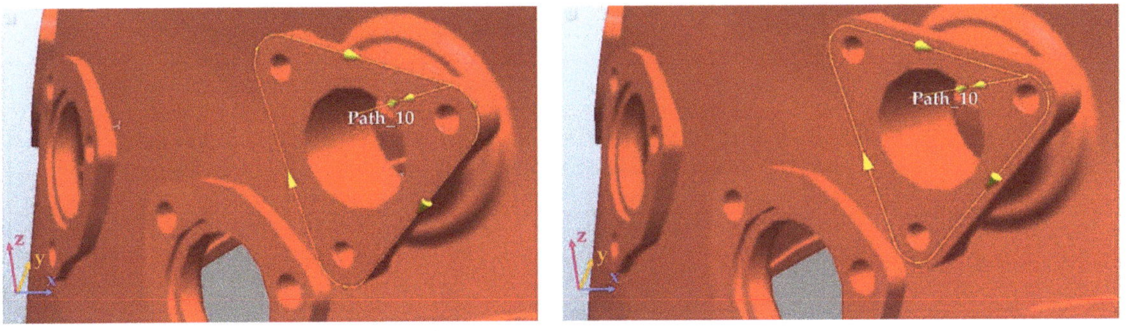

(a) (b)

Figure 4. Model of the machining process in RobotStudio: (**a**) nominal trajectory; (**b**) offset trajectory.

In the classic approach to programming the robot's movement, a fixed workpiece with its trajectory is used. In this case, it is necessary to change the angles of the joints of the robot in order to carry out the movement along complex shapes. The more complex the shape, the greater the velocity and acceleration values in the joints. This causes large inaccuracies in the implementation of the set TCP speed. In machining processes, this results in a change, for example, in the width of the obturation performed. In order to ensure the smoothness of the surface and the absence of places with a significant fluctuation in the value of the material removed, the work was carried out using dependent MultiMove movements. They consist of the simultaneous movement of the workpiece and the tool. This allows the speed of movements in the robot joints to be reduced and the TCP speed to be implemented better, since the positioner movements help to implement the trajectory. Figure 5a shows the modeling of triangular machining using the MultiMove function, while Figure 5b shows the implementation of the modeled process.

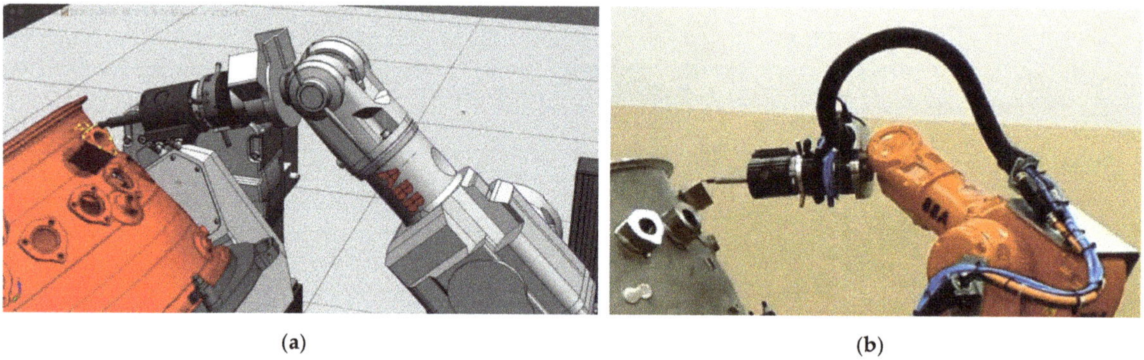

Figure 5. Model of edge deburring in dependent movements: (**a**) Robot Studio software; (**b**) real object.

In the initial period of research, two tool shapes were analyzed: end mill (HFC 0313.03.Z5 Lukas) and conical milling cutter (HFM 0307.03.Z7). In further parts of the work, the conical cutter was abandoned because there were problems with stabilizing the cutting process, and in particular with stabilizing the width of the chamfer.

3.1. Selection of Process Parameters Influencing the Width of Deburring

The development for a robotic technology of edge deburring with the use of a pneumatic rotary cutter should begin with the determination of two important parameters: the tool feed rate and the tool/workpiece contact force. The rotational speed of the tool is constant and results from the type of pneumatic tool used. After selecting the above-mentioned parameters or several of their values, the repeatability of the process should be checked and the parameters for its initiation selected.

The selection of process parameters is related to the necessity to perform a very large number of measurements. Therefore a robotic automatic measuring system was used, consisting of the IRB1600 robot, a 3D scanner, and the ATOS Professional software (Figure 6b). Control measurement points were proposed in places where the greatest inaccuracy of the embedding was expected.

The diagram in Figure 7b summarizes the values of the chamfer width spread for different robot TCP speeds: 50, 100, 250, 500, 750, and 1000 mm/s. The diagram was prepared for the forces 3.1, 4, 5, 6, and 6.7 N. Values of parameters meeting the imposed limitations resulting from the width of the chamfer being implemented are marked in green. An example of the result of the deburring process is shown in Figure 6a.

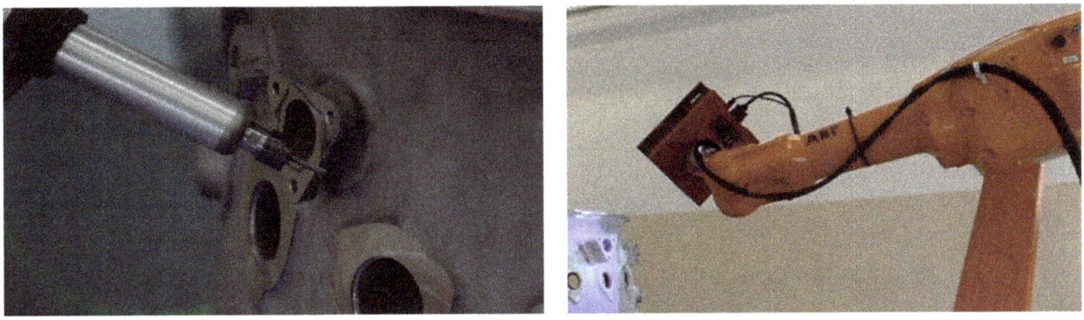

(**a**) (**b**)

Figure 6. (**a**) View of deburring process; (**b**) measurement with a 3D scanner.

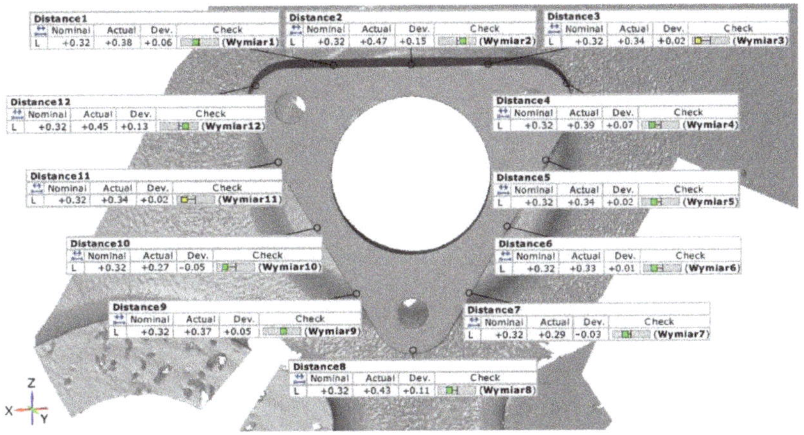

(**a**)

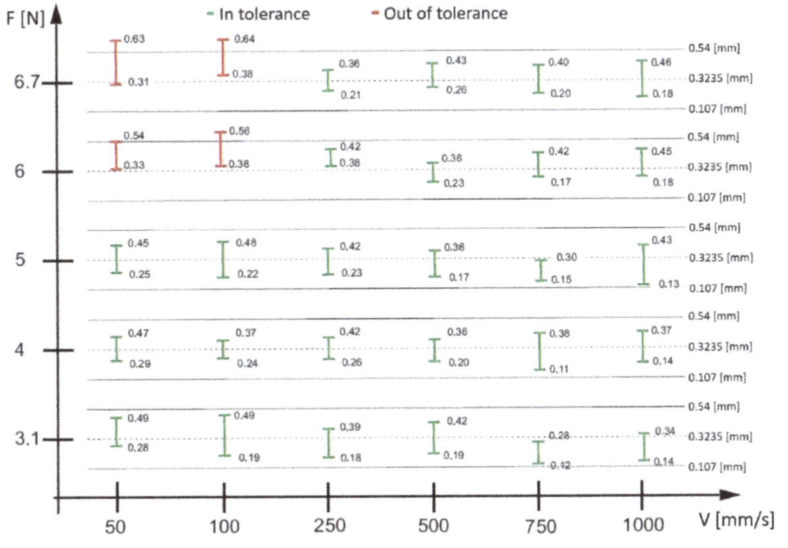

(**b**)

Figure 7. (**a**) Exemplary measurement report; (**b**) research results.

A total of 360 measurement points were used to prepare the graph presented in Figure 7b. An exemplary measurement report for one tool/workpiece contact force value and one set TCP movement speed is shown in Figure 7a.

The analysis of Figure 7b allows for the conclusion that there is a wide range of process parameters, i.e., contact forces and TCP movement speeds, which, at a given nominal tool speed, ensure edge deburring in accordance with the requirements defined in the documentation. In order to select from the set of parameters those that will ensure the best possible implementation of the process, the following optimization procedure described in [26] was proposed.

3.2. *Selection of Process Parameters at the Time of Its Initiation*

During the machining process of steel samples by the manipulator at a constant speed of movement of the tool and constant pressure of the spindle in the area of starting the machining, the process is disturbed. Air compressibility, proportional to the set force, requires time to stabilize the process. This results in excessive or insufficient material removal. It takes until the spindle operating parameters are established. The research work aims to register changes and to experimentally select the spindle pressure force or change the tool feed rate during milling, so that the machining is uniform along the entire length of the sample edge. An example of a disturbed chamfer width is presented in Figure 8.

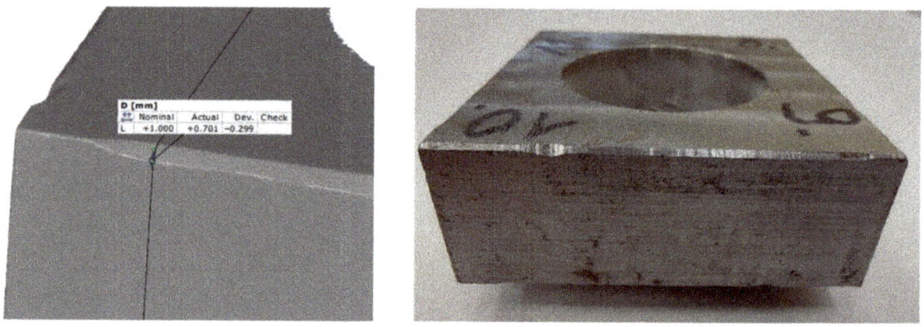

(a) (b)

Figure 8. An example of a disturbed chamfer width: (**a**) scan; (**b**) view of the sample.

A number of Inconel 718 steel samples were prepared for the research. The experiment was carried out in such a way that the robot performs a chamfer at an angle of 45°, for which a path consisting of five points was programmed. Point P10 is the start of movement. Point P20 defines the chamfer depth. Point P30 is approximately 12 mm from point P20 and defines the chamfer depth. It is at the point where the disturbance ends. Although the distance between the points P20 and P30 is not exactly 12 mm in every case, this average value was adopted based on experience. Point P40 defines the end of the milling process, and P50 is the end point outside the workpiece. The shape of the adopted path is shown in Figure 9.

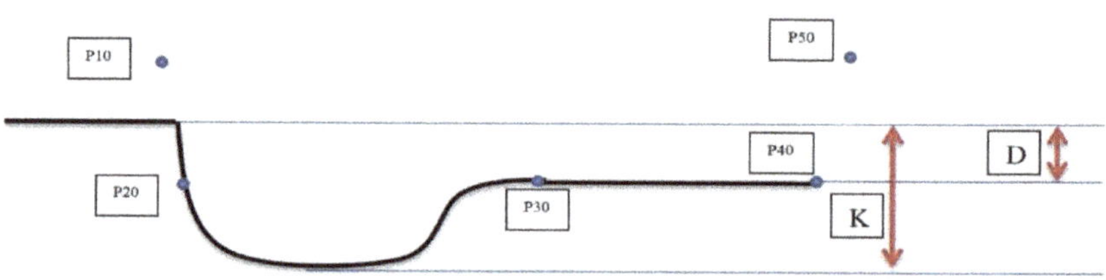

Figure 9. Characteristic points of the robot's path.

Additionally, the dimensions K and D, marked in Figure 9, were introduced, which are, respectively, the depth of machining when starting the cutting process (K), and the nominal assumed depth of material removal (D). As part of the research, parameters such as the spindle feed force between the points P20 and P30 or the change in the speed of the robot's movement are looked for so that the depth of machining along the entire length is D. In the first step, based on the diagram in Figure 7b, the values of the robot's TCP force and speed were selected, for which the suboptimal parameters for the beginning of the cutting process were selected. The selected values are summarized in Table 1.

Table 1. Selected values of process parameters.

Measurement No	Force [N]	Velocity [mm/s]
1	6.7	250
2	6.7	500
3	6.7	750
4	6.7	1000
5	6	500
6	6	750
7	6	1000
8	5	50
9	5	100
10	5	250
11	5	500
12	4	50
13	4	100
14	4	250
15	4	500
16	4	750
17	4	1000

The next step was to determine the D–K difference (Figure 9), assuming a constant speed and changing the contact forces within the range 3.1–6.7 N. The test results for the velocity values of 250, 500, 750, and 1000 mm/s are presented in Figure 10. The value of the force at a given speed for the correct implementation of the machining process initiation is obtained in the case of a zero value of the difference D–K. At the speed of 1000 mm/s, this situation does not occur. The other tests show that it would be reasonable to use force changes at the entrance at TCP speeds of 250 and 500 mm/s. In these cases, the zero value of D–K is in the half of the available force adjustment range, while at the speed of 750 mm/s, the value of D–K is equal to zero when the force is close to the maximum value. It is not possible to increase it.

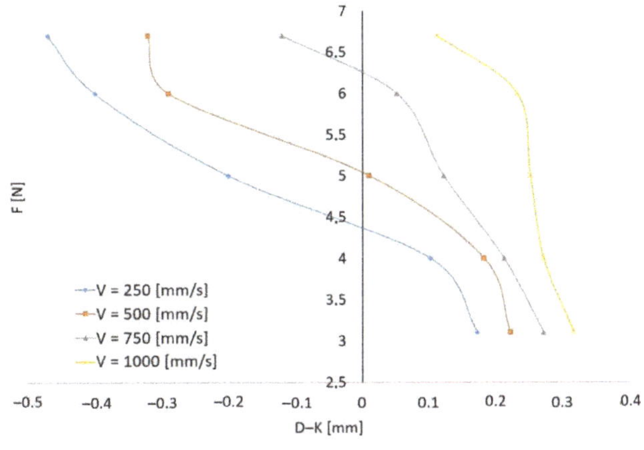

Figure 10. Graph of correctness of machining as a function of force for different values of the speed of the TCP.

For the three correct values resulting from the zero D–K difference (Figure 10), a sensitivity test of the process parameters was carried out, shown in Figure 11. The sensitivity of the process was tested for the parameter D. The tests were carried out for the entry parameters V = 250 mm/s, F = 4.5 N and V = 500 mm/s, F = 5 N.

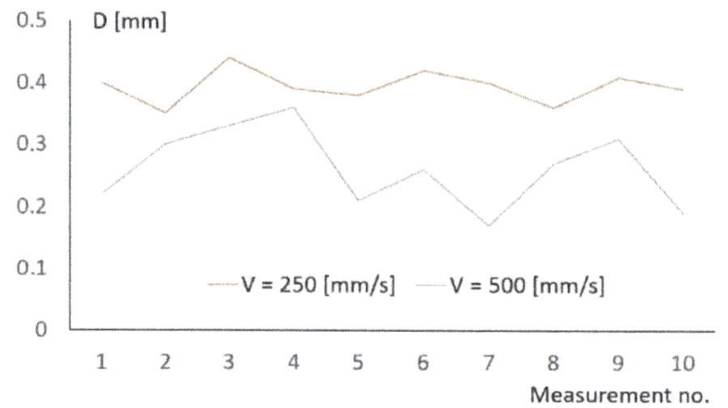

Figure 11. Examination of the stability process for two speed values.

As can be seen at low feed rates due to the longer time of pressure stabilization, and therefore also the force, the best repeatability of the process is obtained. This conclusion prompted the performance of the analysis, the results of which are presented in Figure 12. It concerned the study of the impact of changes in the speed of starting the process at a constant tool/workpiece contact force.

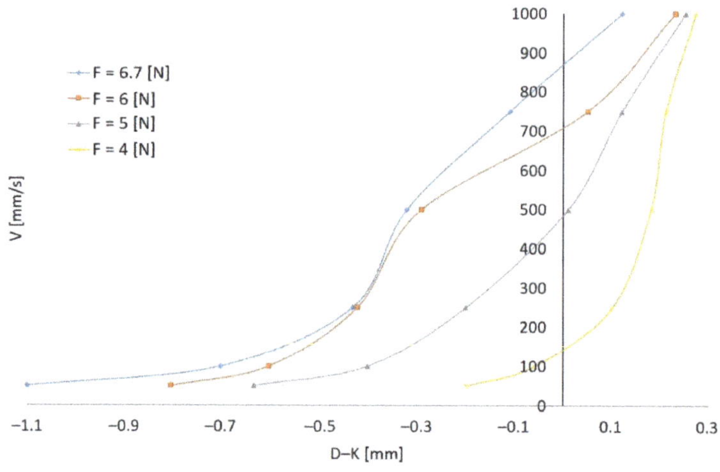

Figure 12. Graph of correctness of machining as a function of speed for different values of contact forces.

Figure 12 shows the values of the starting speed of the machining process with the assumed constant value of the contact force, and the value of the TCP movement speed was converted. That is, unlike in Figure 10, where the movement speed was constant, and the variable force was the parameter influencing the chamfer size. As can be seen, a zero value of the difference D–K for four sets of force and velocity parameters is obtained. For the value of the force of 5 N and 6 N, it can be changed as it is not the limit value for the FDB150 tool used, which, in the event of the need for correction in production conditions, gives such an opportunity. Therefore, as parameters for the correct start of the deburring process when changing the tool feed rate parameter, the values V = 750 mm/s, F = 6 N and V = 500 mm/s, F = 5 N were selected. Similarly, as above, for the obtained process parameters with the zero value of the difference D–K, the stability of the obtained solution was tested in 10 trials, the results of which are presented in Figure 13. The tests were carried out for the entry parameters V = 750 mm/s, F = 6 N and V = 500 mm/s, F = 5 N.

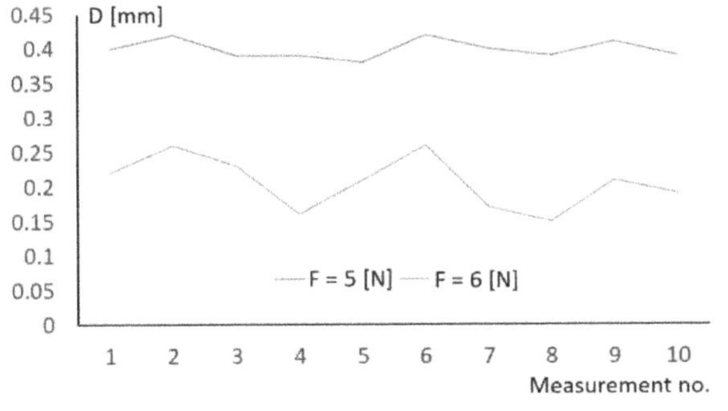

Figure 13. Examination of the stability process for different force values.

Figure 13 shows that better repeatability of cutting with a pneumatic tool is obtained for a force value of 5 N. The results of the repeatability of the process were summarized on the basis of 10 tests. The obtained results of research on the impact of changes in force and tool feed rate on the width of the deburred chamfer allow for the following conclusions to be drawn: better results of stabilization of the chamfer width at the start of the process are obtained with changes in the value of the process start speed, and there is a force value for which the chamfer accuracy spread is at the level of 10% of the nominal value.

The problem presented in the paper concerns the optimization of two parameters, force and velocity, and the objective function contains one parameter that must be satisfied, and that is the difference of geometric parameters D–K must be 0. When the relation D–K = 0 is satisfied, the second parameter considered for optimization is the stability of the process. The stability of the process is shown in Figures 11 and 13.

The tool with pneumatic force application works without an external feedback loop. The process is very fast. The forces of 4–6.7 N are very small. The value of the force was measured for given pressures in a static way. In the experiment, the set force has a large inaccuracy due to the compressibility of the air and the speed of the pressure regulator. Therefore, only four levels of force values (4, 5, 6, and 6.7 N) were digitized in the study. With such small forces, the amount of material removed is small, therefore the speed of movement of the TCP was assumed to be 50, 100, 250, 500, 750, and 1000 mm/s. Due to the failure to meet the criterion D–K = 0 for large forces of 6.7 N, the values of velocity 50 and 100 were rejected. For a force of 6 N, the values of 50, 100, and 250 were rejected. For a force of 5 N, the value of velocity was 1000. In addition, under these operating conditions, high tool wear was noted.

All possible parameters are summarized in Table 1. This is a set of data including 17 items for which the combination of parameters for which the D–K relation is equal to 0 is to be indicated. Since the amount of input data is small, it is possible to analyze all possible combinations. Therefore, the optimization methods known from the literature were not applied. Thus, seven combinations of contact force and velocity parameters for which D–K = 0 were obtained. The D–K = 0 relationship satisfies the imposed technological parameters. In the following part, for the seven pairs of parameters, one was selected for which in 10 consecutive attempts to start machining, the maximum error resulting from the difference D–K had the smallest relative value.

In summary, the dimensionality of the task is so small that the process of selecting a suboptimal solution does not require the use of complex optimization methods. At the same time, it is transparent and simple in perception, by which it has high application value.

4. Conclusions

The research results presented in the paper provided information on the accuracy with which edge deburring with the use of flexible pneumatic tools can be performed. Therefore, as the most difficult element of deburring is its initiation, a procedure was proposed that allowed for the indication of a set of process parameters that ensured the implementation of the process in accordance with the requirements defined in the documentation. The proposed solution is of an engineering nature and is not a classic search for an optimal solution. The advantage of the proposed solution is simplicity, which ensures correct implementation of deburring, which is very important from the point of view of an industrial application.

The obtained results of the research work allow for the conclusion that the best results of stabilization of the chamfer width at the start of the process are obtained with changes in the value of the process start velocity parameter. It is not advisable to look for the possibility of influencing the machining volume during the start by changing the value of the tool/workpiece contact force. This conclusion is universal.

It should be noted that the obtained results relate to Inconel 718 and the HFC 0313.03.25 tool and they are not universal. The authors propose a procedure for carrying out work allowing for the selection of suboptimal process parameters. In the analyzed case, these parameters are the speed of movement of the characteristic point and the tool/workpiece contact force.

The obtained results were used during the construction of a robotic station at Prat & Wittney S.A. in Rzeszow.

Author Contributions: Conceptualization, A.B., D.S. and K.K.; methodology, K.K., P.G. and D.S.; software, A.B., P.G. and K.K.; validation, M.M. and A.O.; formal analysis, M.M. and M.U.; investigation, A.B.; resources, K.K.; data curation, M.U.; writing—original draft preparation, A.B.; writing—review and editing, M.M. and A.B.; visualization, P.G. and K.K.; supervision, A.B.; project administration, P.G.; funding acquisition, A.B. All authors have read and agreed to the published version of the manuscript.

Funding: This research received no external funding.

Data Availability Statement: Data is contained within the article.

Conflicts of Interest: The authors declare no conflict of interest.

References

1. Cichosz, P.; Kuzinovski, M. Metody wykonywania fazek i gratowania krawędzi. Cz. 1. *Mechanik* **2011**, *84*, 553–554.
2. Cichosz, P.; Kuzinovski, M. Metody wykonywania fazek i gratowania krawędzi. Cz. 2. *Mechanik* **2011**, *84*, 674–681.
3. Kim, J.-D.; Kim, K.-D. Deburring of burrs in spring collets by abrasive flow machining. *Int. J. Adv. Manuf. Technol.* **2004**, *24*, 469–473. [CrossRef]
4. Kosmol, J.; Dyrbuś, G.; Kazimierczak, M.; Kolka, A.; Lis, K. Automatyzacja fazowania i zatępiania krawędzi uzębień kół zębatych. *Mechanik* **2015**, *88*, 66–80.
5. Caesarendra, W.; Pappachan, B.K.; Wijaya, T.; Lee, D.; Tjahjowidodo, T.; Then, D.; Manyar, O.M. An AWS Machine Learning-Based Indirect Monitoring Method for Deburring in Aerospace Industries Towards Industry 4.0. *Appl. Sci.* **2018**, *8*, 2165. [CrossRef]
6. Kianpour, P.; Gupta, D.; Krishnan, K.K.; Gopalakrishnan, B. Automated job shop scheduling with dynamic processing times and due dates using project management and industry 4.0. *J. Ind. Prod. Eng.* **2021**, *38*, 485–498. [CrossRef]
7. Matuszak, J.; Zaleski, K. Analiza Sił w Procesie Obróbki Krawędzi Szczotkami Ceramicznymi. *Sci. Lett. Rzesz. Univ. Technol.-Mech.* **2017**, *89*, 509–516. [CrossRef]
8. Graham, D. Electro-Chemical Deburring. In Proceedings of the Twentieth International Machine Tool Design and Research Conference; Tobias, S.A., Ed.; Palgrave Macmillan: London, UK, 1980; pp. 613–616. [CrossRef]
9. Ko, S.L. Measurement and effective deburring for the micro burrs in piercing operation. *Int. J. Precis. Eng. Manuf.* **2000**, *1*, 152–159.
10. He, T.; Huang, S.; Fang, S.; Deng, H.; Ding, Q.; Zhang, M. Study on the Electrochemical Deburring for the External Surface of the Microhole Caused by Mechanical Drilling Process. *Machines* **2022**, *10*, 726. [CrossRef]
11. Kadam, S.; Mitra, S. Electrochemical deburring—A comprehensive review. *Mater. Today Proc.* **2021**, *46*, 141–148. [CrossRef]

12. Onstein, I.F.; Semeniuta, O.; Bjerkeng, M. Deburring Using Robot Manipulators: A Review. In Proceedings of the 2020 3rd International Symposium on Small-scale Intelligent Manufacturing Systems (SIMS), Gjovik, Norway, 10–12 June 2020; pp. 1–7. [CrossRef]
13. Zhang, H.; Chen, H.; Xi, N.; Zhang, G.; He, J. On-Line Path Generation for Robotic Deburring of Cast Aluminum Wheels. In Proceedings of the 2006 IEEE/RSJ International Conference on Intelligent Robots and Systems, Beijing, China, 9–15 October 2006; pp. 2400–2405. [CrossRef]
14. Villagrossi, E.; Pedrocchi, N.; Beschi, M.; Tosatti, L.M. A human mimicking control strategy for robotic deburring of hard materials. *Int. J. Comput. Integr. Manuf.* **2018**, *31*, 869–880. [CrossRef]
15. Burghardt, A.; Kurc, K.; Szybicki, D. Robotic Automation of the Turbo-Propeller Engine Blade Grinding Process. *Appl. Mech. Mater.* **2016**, *817*, 206–213. [CrossRef]
16. Burghardt, A.; Szybicki, D.; Kurc, K.; Muszyńska, M. Optimization of Process Parameters of Edge Robotic Deburring with Force Control. *Int. J. Appl. Mech. Eng.* **2016**, *21*, 987–995. [CrossRef]
17. Burghardt, A.; Cieślik, J.; Flaga, S.; Kurc, K.; Minorowicz, B.; Nawrocki, M.; Pluta, J.; Stefański, F.; Szybicki, D.; Zając, M. Wybrane problemy współczesnej robotyki. In *Monografie Katedry Automatyzacji Procesów AGH w Krakowie*; Akademia Górniczo–Hutnicza w Krakowie: Krakow, Poland, 2015; pp. 110–137. Available online: https://www.researchgate.net/profile/Stanislaw-Flaga/publication/283055718_Wybrane_problemy_wspolczesnej_robotyki/links/5627fcd808ae04c2aead8203/Wybrane-problemy-wspolczesnej-robotyki.pdf (accessed on 21 October 2015).
18. Saito, K.; Miyoshi, T.; Jeong, J. A dual-axis micro-finishing tool for free-form metal surface of a mold cavity. In Proceedings of the Grinding Symposium (ASME Winter Annual Meeting), Miami Beach, FL, USA, 17–22 November 1985; pp. 229–230.
19. Clayton, P.; Elbestawi, M. Investigation of die polishing strategies using five-axis machining centre. In Proceedings of the 1996 American Society of Mechanical Engineers International Engineering Congress—Dynamics Systems and Control Division, Atlanta, GA, USA, 17–22 November 1996; Volume 58, pp. 243–250.
20. Zhao, J.; Saito, K.; Kondo, T.; Narahara, H.; Igarashi, S.; Sasaki, T.; Zhang, L. A new method of automatic polishing on curved aluminium alloy surfaces at constant pressure. *Int. J. Mach. Tools Manuf.* **1995**, *35*, 1683–1692. [CrossRef]
21. Güvenç, L.; Srinivasan, K. An overview of robot-assisted die and mold polishing with emphasis on process modeling. *J. Manuf. Syst.* **1997**, *16*, 48–58. [CrossRef]
22. Asada, H.; Sawada, Y. Design of an adaptable tool guide for grinding robots. *Robot. Comput. Manuf.* **1985**, *2*, 49–54. [CrossRef]
23. Pires, J.N.; Afonso, G.; Estrela, N. Force control experiments for industrial applications: A test case using an industrial deburring example. *Assem. Autom.* **2007**, *27*, 148–156. [CrossRef]
24. Odham, A. Successful robotic deburring is really a matter of choices. *Tool Prod. Mag.* **2007**, 14–19.
25. Shiakolas, P.; Labalo, D.; Fitzgerald, J. RobSurf: A Near Real Time OLP System for Robotic Surface Finishing. In Proceedings of the 7th Mediterranean Conference on Control and Automation (MED99), Haifa, Israel, 28–30 June 1999.
26. Burghardt, A.; Szybicki, D.; Kurc, K.; Muszyńska, M.; Mucha, J. Experimental Study of Inconel 718 Surface Treatment by Edge Robotic Deburring with Force Control. *Strength Mater.* **2017**, *49*, 594–604. [CrossRef]
27. Burghardt, A.; Szybicki, D.; Gierlak, P.; Kurc, K.; Muszyńska, M. Robotic Grinding Process of Turboprop Engine Compressor Blades with Active Selection of Contact Force. *Teh. Vjesn.* **2022**, *29*, 15–22. [CrossRef]
28. Muszyńska, M.; Burghardt, A.; Kurc, K.; Szybicki, D.; Szczęch, T. Monitoring the Parameters of the Robot-Operated Quality Control Process. *Adv. Sci. Technol. Res. J.* **2017**, *11*, 232–236. [CrossRef]
29. Burghardt, A.; Kurc, K.; Szybicki, D.; Muszyńska, M.; Nawrocki, J. Robot-operated quality control station based on the UTT method. *Open Eng.* **2017**, *7*, 37–42. [CrossRef]
30. Rasmussen, B.; Derby, S. Design and evaluation of robotic end-effectors for an automated die finishing system. In Proceedings of the Symposium on Computer-Aided Design and Manufacture of Dies and Molds (ASME Winter Annual Meeting), Anaheim, CA, USA, 8–13 November 1987; pp. 61–74.
31. Furukawa, T.; Rye, D.; Dissanayake, M.; Barratt, A. Automated polishing of an unknown three-dimensional surface. *Robot. Comput. Manuf.* **1996**, *12*, 261–270. [CrossRef]
32. Huissoon, J.; Ismail, F.; Jafari, A.; Bedi, S. Automated Polishing of Die Steel Surfaces. *Int. J. Adv. Manuf. Technol.* **2002**, *19*, 285–290. [CrossRef]
33. Ryuh, B.-S.; Park, S.M.; Pennock, G.R. An automatic tool changer and integrated software for a robotic die polishing station. *Mech. Mach. Theory* **2006**, *41*, 415–432. [CrossRef]
34. Wilbert, A.D.; Behrens, B.; Zymla, C.; Dambon, O.; Klocke, F. Robotic finishing process—An extrusion die case study. *CIRP J. Manuf. Sci. Technol.* **2015**, *11*, 45–52. [CrossRef]
35. Takeuchi, Y.; Ge, D.; Asakawa, N. Automated polishing process with a human-like dexterous robot. In Proceedings of the IEEE International Conference on Robotics and Automation, Atlanta, GA, USA, 2–6 May 1993; pp. 950–956. [CrossRef]
36. Wu, H.; Zou, T.; Burke, H.; King, S.; Burke, B. A Novel Approach for Porcupine Crab Identification and Processing Based on Point Cloud Segmentation. In Proceedings of the 20th International Conference on Advanced Robotics (ICAR), Ljubljana, Slovenia, 6–10 December 2021; pp. 1101–1108. [CrossRef]

electronics

Article

Application of a 3D Scanner in Robotic Measurement of Aviation Components

Krzysztof Kurc [1,*], Andrzej Burghardt [1], Piotr Gierlak [1], Magdalena Muszyńska [1], Dariusz Szybicki [1], Artur Ornat [2] and Marek Uliasz [2]

[1] Department of Applied Mechanics and Robotics, Faculty of Mechanical Engineering and Aeronautics, Rzeszów University of Technology, al. Powstancow Warszawy 8, 35-959 Rzeszow, Poland
[2] Pratt and Whitney Rzeszow S.A., 35-078 Rzeszow, Poland
* Correspondence: kkurc@prz.edu.pl; Tel.: +48-(17)-8651814

Abstract: The aviation industry is associated with high precision and accuracy standards of the manufactured components, and thus the need to ensure precise quality control. Measurement processes, depending on the manufactured components, take place before, during and after the processing stage. Optical scanners can be used for these measurements, the measurement results of which can be displayed on the operator panel or used to prepare a report. The innovative approach is to measure, compare the results with a pattern, send the deviations to a neural decision-making system, select the forces and send the results to a robot controller for adaptive machining. The presented proprietary solution includes a data acquisition system, a neural decision-making system and a robot that carries out the machining process via force control. The proposed solution was verified on aviation components. During the process parameter optimization stage for the diffuser and ADT gearbox, the points describing the change in width of the chamfer being performed and the blade thickness in the control sections were approximated.

Keywords: industrial robots; optical scanner; programming; neural decision-making system; communication

Citation: Kurc, K.; Burghardt, A.; Gierlak, P.; Muszyńska, M.; Szybicki, D.; Ornat, A.; Uliasz, M. Application of a 3D Scanner in Robotic Measurement of Aviation Components. *Electronics* **2022**, *11*, 3216. https://doi.org/10.3390/electronics11193216

Academic Editor: Hamid Reza Karimi

Received: 8 September 2022
Accepted: 4 October 2022
Published: 7 October 2022

1. Introduction

The production of components for the aviation industry is associated with high precision and accuracy standards in the manufacture of components, and thus the need to ensure quality control at various stages of production. Measurement processes, depending on the manufactured semi-finished products, take place before, during and after the processing stages. For control and final measurements, various measurement techniques can be used, from taking a shape imprint and displaying the deviation value on a measuring projector, non-contact, e.g., with a laser profilometer or optical scanner, and by contact with the use of a probe and Coordinate Measuring Machine (CMM).

Taking the imprint and then cutting it and displaying it on the measurement projector is very accurate, but time-consuming, and the accuracy of the measurement depends on the correctness of manual cutting operations. The manual method with the laser profilometer requires the correct positioning of the measuring head in relation to the measured detail. Conducting measurements with coordinate machines requires additional time to mount the workpiece in the measuring area of the CMM. The advantages of this solution are high versatility and accuracy, while the disadvantages are the high cost of purchasing and maintaining the CMM and the long measurement time [1,2]. The aforementioned methods require a large amount of work time, while the detail must be inspected sometimes and in several hundred places, some of them are susceptible to the so-called human factor and do not allow automatic generation of a measurement report.

For the above reasons, it was decided to design and build a stand for conducting robotic measurements using a 3D optical scanner. Communication between devices and data exchange represents a very important stage in the design and construction of robotic stations, and details the integration of machine systems and the introduction of changes in production processes aimed at increasing efficiency and introducing adaptation possibilities. Machining, e.g., grinding, are examples of processes that require common communication between many station components in a fast and continuous manner. While browsing the literature, one can find examples of studies on machining by robots [3,4]. Force control algorithms and their applications during processes have been presented in [5,6]. Communication between many robots, and surface shaping using an industrial robot and tool condition monitoring are described in works [7,8]. The authors of works [9–11] use the Profibus protocol for communication. The TCP/IP protocol is employed by the authors of [12–16]. Various programming languages can be used to create a dedicated network, with one increasingly popular example being Python [17,18], used by the authors in the application [19]. The authors of the article [20] describe the improvement in the grinding process of, for example, metal molds with pressure control using CAD/CAM data, and the improvement of the process by scanning is presented in articles [21–23]. Laser displacement sensor in the application of aero-engine blade measurement is presented by authors of the work [24]. The authors of the work [25] presented an innovative 3D measurement system for the material removed from the engine blades. The use of the GOM 3D scanner to measure the geometry of the blades in the process of inspection, servicing and repairs is presented in the works [26].

2. Measuring System and Robotic Station

The applied three-dimensional optical scanner obtains geometric data from an existing physical object. These data are used to reflect a virtual 3D model of the scanned object that can be used for a variety of applications such as reverse engineering, rapid prototyping, quality control, and cultural heritage documentation. Robotized geometry measurements are performed using the Atos Core 3D scanner (GOM: Oberkochen, Germany) and the Atos Professional software (GOM 7.04: Oberkochen, Germany). The 3D scanner is mounted on the ABB IRB 1600 manipulator, which allows for any positioning of the scanner head in front of the measuring element.

The Atos Professional software allows for recording the position and measurement orientation of the 3D scanner in space in order to automate and repeat the measurements of subsequent components. After completing part of the scan, the scanner moves and rotates to areas not recorded in the previous scan. The individual measurements are automatically processed using reference points relative to one common coordinate system, resulting in a complete 3D point cloud.

The Atos Core 3D scanner contains a fringe projector and two stereoscopic cameras, working on the triangulation principle (calculating the intersection of a plane and a ray in space). These fringes are recorded by two cameras shown in Figure 1, which creates a chamfer shift based on the sinusoidal distribution of the intensity on the camera matrices. Atos Core uses multiple chamfer shifts based on the heterodyne principle to achieve the highest sub-pixel accuracy. Based on the optical transformation equations, independent 3D coordinates are automatically calculated for each camera pixel.

The computed polygon mesh describes free-form surfaces and geometric elements. Its verification is possible by comparing the surface with the technical drawing or directly with the CAD data set. The software also enables the implementation of 3D surface analysis and 2D analysis of sections or points.

The Atos system measures the deviation between CAD data and actual 3D coordinates, providing full-field measurement data. These data, in addition to the representation of surface deviations in relation to CAD data, contain complete information about the object from which the software automatically determines other details. After the scanning stage and obtaining a 3D point cloud, it is possible to generate a report of the properties of a given scan as in Figure 2. The accuracy of the scanner depends on many factors, e.g., whether one scan is sufficient for the measurement or several scans are needed, and if so, whether or not reference points were used to connect (stick) them.

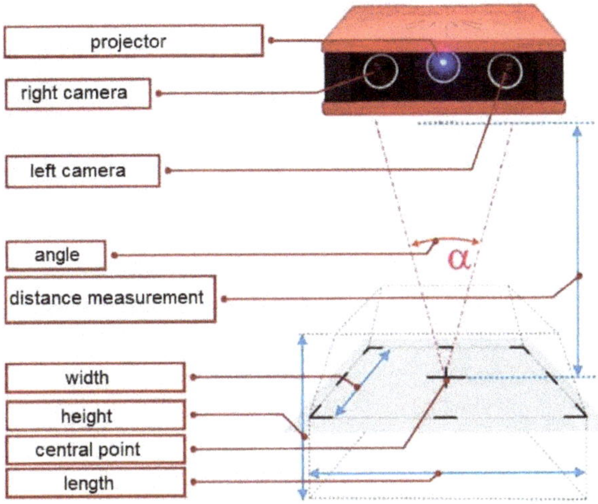

Figure 1. Atos Core—the idea of measurement.

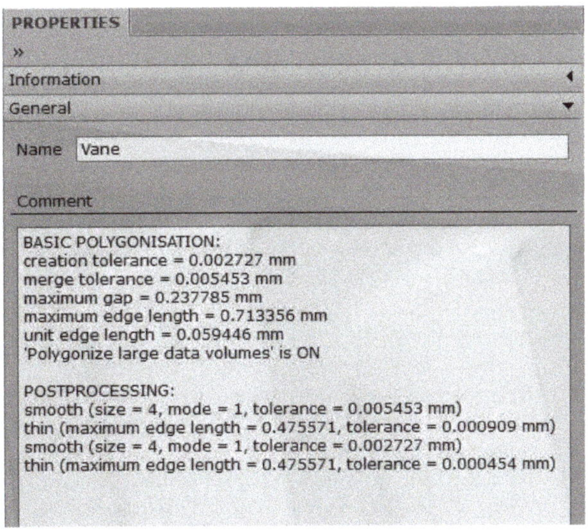

Figure 2. 3D-scanning properties.

Optical measurement systems owe their accuracy to the use of the latest solutions in the field of optoelectronics, precise image-processing technologies and mathematical algorithms, combined with unchanging precision standards and automatically performed calibration.

A novelty in the discussed issue is the robotic acquisition of a scan of the aerial part before processing, in comparison with the pattern, selection of the grinding pressure force using a neural decision-making system and sending the processed information in automatic mode to the robot controller in order to adapt the processing parameters.

The designed robotic station shown in Figure 3 for grinding aviation elements displays: the IRB 140 robot holding the detail, the electrospindle and the IRB 1600 robot with a 3D scanner. The Atos Professional software works with the Atos Core 3D scanner and communicates with the IRC5 robot controller via the TCP/IP protocol.

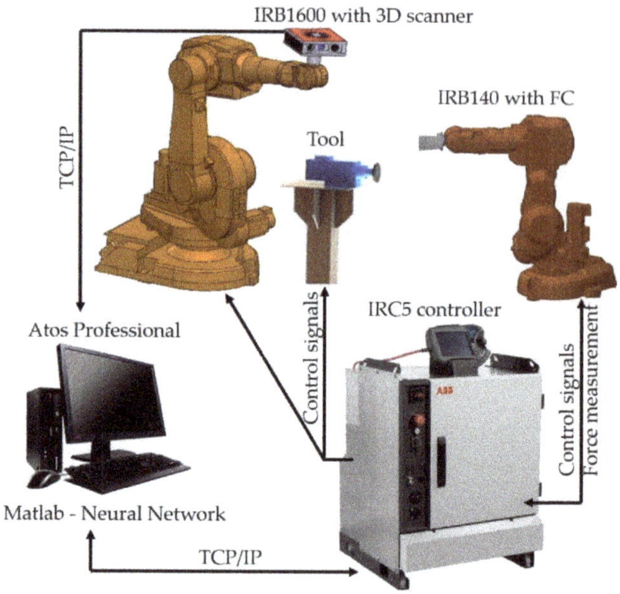

Figure 3. Schematic diagram of communication between robotic cell devices.

The RobotWare robot controller software (ABB 6.08: Zurych, Szwajcaria) has a force control option and provides control of two robots. The value of the force with which the detail is pressed against the electro-spindle tool determines the precise neural network based on information from the 3D scanner.

Atos Professional performs the tasks of the measuring system and, together with the Matlab software (2021B: Natick, MA, USA), communicates with the IRC5 robot controller by sending data via the TCP/IP protocol. The grinding tool speed is regulated by a frequency converter connected to the robot controller using DeviceNet.

The initial scenario for a robotic process is:

- Start;
- Measurement with a scanner;
- Comparison of the point cloud with the pattern;
- Data recording with deviations;
- Sending deviations to Matlab;
- Selection of the pressure force with a neural decision system;
- Sending received values to the controller;
- Automatic correction of the contact pressure for the programmed path;
- Machining;
- Control measurement by comparing with the reference (Yes/No);
- Stop and detail change.

3. Comparison of the Point Cloud with the Pattern

One of the components selected for testing is an aircraft blade. The blades are elements of gas turbine engines. Their shape makes it possible to convert the energy of the flowing medium into mechanical energy. The production of blades is carried out mainly for the aviation and energy industries. They are produced by the monocrystalline casting process, made on CNC machines, up to electrochemical drilling. Shaping the blades by electrochemical drilling is cost-effective, but the technology is sensitive to a number of factors, such as temperature, the intensity of the current flowing, the width of the gap, etc.

The uniqueness of the blade shape requires an individual approach for each piece during the grinding process.

For geometric measurement of the blades, contact and non-contact methods are used. The use of the contact method consists in the use of measuring probes to determine the geometric dimensions at selected points. The authors of [27,28] present this type of probe in various applications. There are also solutions with measuring systems placed on a robot [29]. The authors of the work [30] use coordinate measuring machines (CMM) to measure the grinding process. In non-contact measurements, laser heads [24] or optical scanners [15] are used. Optical scanners are mounted on stands [4,25] or on robot arms [31]. Contact devices are not sensitive to: dust, glare and lighting, but are slow compared to the number of measurements collected using non-contact methods.

Figure 4 shows a view of the blade after electrochemical drilling before grinding. Measurement with a 3D scanner as in Figure 5 allows one to measure the blade at each point of interest.

Figure 4. A blade before the griding operation.

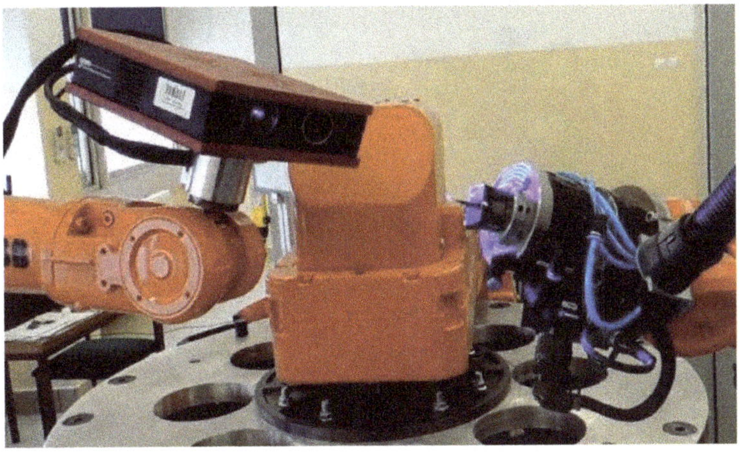

Figure 5. Measurement of blade geometry with the use of a 3D scanner.

The scanning program in the Atos Professional application (example shown in Figure 6) is an extensive hierarchical structure of dependencies between the nominal element (CAD pattern) and the current element (scan).

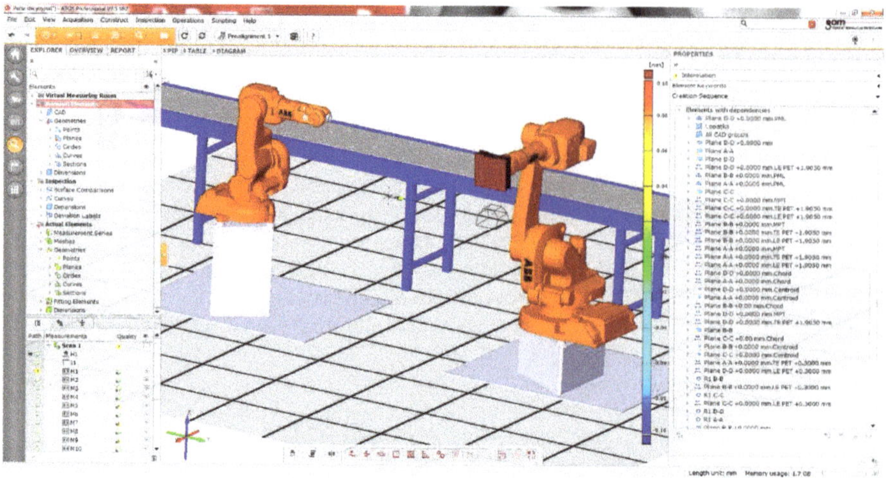

Figure 6. Virtual environment scanning program in the Atos Professional application.

In the Atos Professional software, after comparing the scan with the CAD model, allowances in 80 programmed points are determined shown in Figure 7.

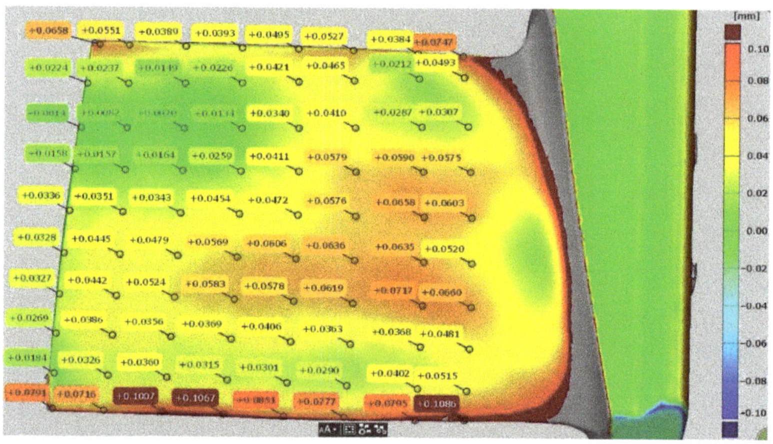

Figure 7. Determined allowances at 80 points on the blade.

Then the obtained dimensions are automatically saved in the variables defined in Python. In the process of determining the allowances, the Atos Professional Blade software add-on is used, dedicated to the measurement inspection of all types of blades.

4. Recording Data with Deviations and Sending It to Matlab

With the Atos Professional software, it is possible to save sequentially executed commands thus creating a macro. Recorded macros can be called automatically and they can contain advanced program functions shown in Figure 8. This allows a recorded sequence of operations to be performed repeatedly. By editing a recorded script, it can be adapted to other tasks or generalized. Script programming is based on modifying or combining recordings. The software in the Atos Professional environment allows users to record a script.

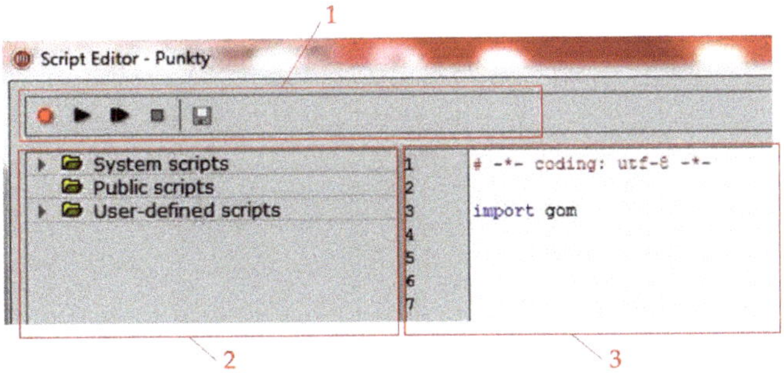

Figure 8. Creating scripts in Atos Professional.

In the top bar (Figure 8, reference 1) the editor contains buttons for recording scripts: record, start and stop. On the left side (Figure 8, reference 2) there is a catalog of preinstalled and user-defined scripts. There you can find items for creating new scripts, for renaming and removing items, just as you would with regular documents. It is also possible to export and import scripts and script archives. The scripts directory consists of three parts. The first contains preinstalled system scripts, the second is a public folder for multi-user access, and the third contains the user's private scripts. The editor is used to modify the currently selected script, in this section the block structure of the script is also displayed. In addition to the common operations (copy, paste, find and replace), the context menu of the editor also contains analog recorder buttons. The third part of the script editor (Figure 8, reference 3) is the exit area. The results obtained during the script execution are displayed there, see Figure 9.

```
import gom

MCAD_ELEMENT=gom.script.section.create_multisection_by_pa
rallel_planes (
        direction_mode='symmetric',
        name='Section 1',
        position=0.00000000e+00,
        properties=gom.Binary ('eAHtV11o...'),
        reference_plane=gom.app.system['system_plane_x'],
        section_distance=5.00000000e+01,
        separated_elements=False)
```

Figure 9. One of the recording results.

Initially, the function is selected from the gom library, which is divided into several layers. Therefore, a few additional parameters are needed to achieve the create_multisection_by_parallel_planes function. An open parenthesis at the end of a line indicates the start of a function call. The following lines are the parameters one per line. The closing brace indicates the end of the function call. Everything is assigned to the MCAD_ELEMENT variable.

The following data types are used as parameter values:

- Strings (direction_mode, name);
- Whole numbers (num_sections);
- Floating point numbers (position and section_distance);
- Boolean values (separated_elements).

For research purposes, a script was developed that implements the process of measuring, determining deviations and sending the obtained values. It is possible to change the script and the developer can adapt it to their needs.

For the correct implementation of grinding, the amount of allowance to be removed in the given areas must be defined. The 3D scan (point cloud) obtained as a result of the measurement was compared with the reference CAD model and the surpluses at 80 points were determined using geometric operations in the places shown in Figure 10.

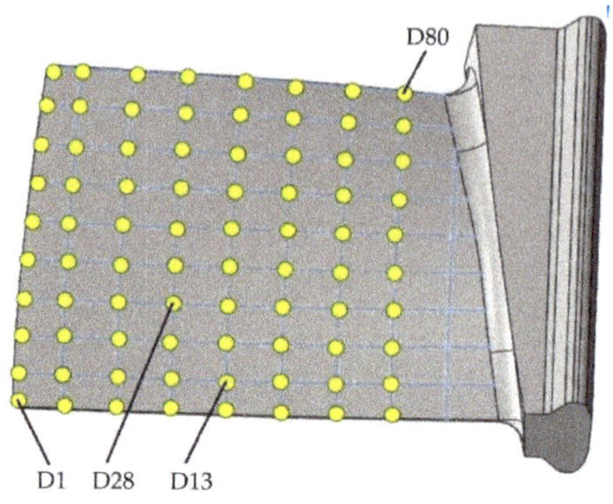

Figure 10. The back of the shoulder blade with a grid of points.

In this way, the procedures performed for 80 points D1, D2, ... D80, which were registered in the script, were prepared shown in Figure 11.

```
import gom
import os
import socket

#D1
MCAD_ELEMENT=gom.script.inspection.create_distance_by_2_points (

MCAD_ELEMENT=gom.script.inspection.inspect_dimension (
#D2
MCAD_ELEMENT=gom.script.inspection.create_distance_by_2_points (

MCAD_ELEMENT=gom.script.inspection.inspect_dimension (
#D3
MCAD_ELEMENT=gom.script.inspection.create_distance_by_2_points (

MCAD_ELEMENT=gom.script.inspection.inspect_dimension (
                        .
                        .
                        .
#D80
MCAD_ELEMENT=gom.script.inspection.create_distance_by_2_points (

MCAD_ELEMENT=gom.script.inspection.inspect_dimension (
```

Figure 11. Script with recorded allowances.

The recorded script starts with the lines: gom, os, socket, which are responsible for including the Python libraries. The gom library is responsible for cooperation with Atos Professional software and the user does not have access to it, but can be used in the recording option. The os library refers to the options of the currently used operating system. Next, the script is the results of activities on the point cloud obtained from measurements with a 3D scanner. After saving the allowances, the data are sent from the Atos Professional client to the Matlab server in order to select the pressure forces with a neural decision-making system [32].

5. The Selection of the Pressure Force with a Neural Decision-Making System

The purpose of the developed system is to select the force in contact of the blade with the electro-spindle tool during grinding. The pressure is carried out by an industrial robot using the Force Control add-on. The grinding pressure force is generally presented as:

$$F_n = f(Q, x_C, y_C, z_C) \tag{1}$$

In order to avoid the problem of modeling the relationship between the force and the allowance Q at the points x_C, y_C, z_C on the blade, an artificial neural network was used, which was taught the relationship (1) with the following assumptions:

- constant rotational speed of the electro-spindle tool n_t = 4500 rpm;
- constant feed speed (blade speed relative to the tool) v_r = 0.02 m/s;
- variable pressure force of the blade against the tool in the range of 2–10 N.

After the assumptions, two variables were obtained which determine the pressure force of the blade against the tool:

$$F_n = f(Q,\ n) \tag{2}$$

where Q—allowance, n—number of a point on the D1, D2, … D80 blade instead of the x_C, y_C, z_C coordinates.

Using Matlab software and "Neural Network Toolbox" libraries, a neural controller of the blade grinding process was realized. In order to approximate the tool pressure force from the material allowance, a feed-forward neural network was used, learned in accordance with the error back propagation algorithm [33]. The research was conducted and based on them, a network with two hidden layers with nine neurons in each of these layers was selected. In the hidden layers, sigmoid unipolar neuron activation functions were used, and in the output layer, linear activation functions were used. The weights of the neural network were learned according to an algorithm based on the Levenberg–Marquardt optimization method.

The data for training the artificial neural network were obtained during measurement experiments. In 80 points located on the surface of the ridge of the blade, geometrical measurements were made in order to determine the size of the material allowance shown in Figure 7. On this basis, the dependence of the thickness of the collected material layer on the tool pressure force was determined. During the tests, the applied values of the pressure forces were: 2, 3, 4, 5, 6, 7, 8, 10 N. The discussed tests were carried out randomly on 30 blades in the manner presented in Table 1.

Table 1. The structure of learning neural networks.

		Pressure [N]												
		2				3				…	10			
Point no.	1	Thickness of collected allowances				Thickness of collected allowances				…	Thickness of collected allowances			
	2													
	…													
	80													
		Blade 1	Blade 2	…	Blade 30	Blade 1	Blade 2	…	Blade 30	…	Blade 1	Blade 2	…	Blade 30

The task of the neural process controller is to generate at the selected D1, D2, … D80 measuring points on the blade (Figure 10) the values of the pressure forces against the tool, if the value of the allowance at these points is known. The structure of the input and output layers of the neural network results from the structure of the measurement data. The neural network structure has 80 inputs and 80 outputs as in Figure 12, which is as many as there are measurement points and a visible threshold value—b.

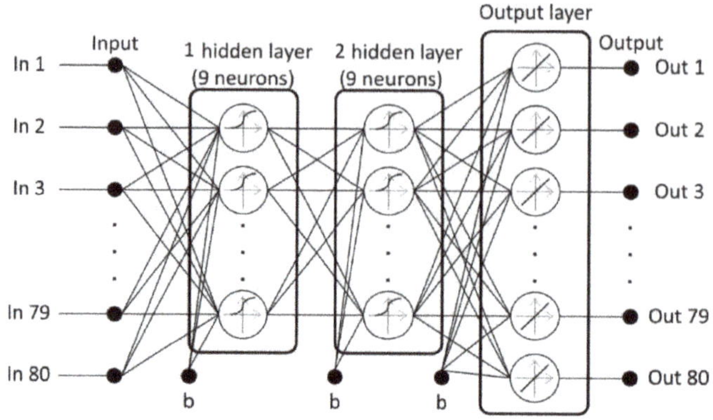

Figure 12. The structure of a neural network.

The system is built on the basis of an artificial neural network, which trains the relationship between the pressure force and the material allowance to be removed at individual points of the blade D1, D2, ... D80. After the neural decision-making system determines 80 values of the forces for individual points, they are sent to the robot's IRC5 controller in accordance with the communication parameters as in Figure 13.

```
#Communication parametrs
HOST = '192.168.1.105' #server adress
PORT = 9000
BUFFER_SIZE = 1024

#Saving distance D1 as variable
dyst1 = gom.app.project.inspection[ 'D1' ].get ( 'scalar_value' )
dyst_string = "{ }" .format (dyst1)
print (dyst_string)

#TCP/IP communication
DATA1 = dyst_string.encode ( 'UTF-8' ) #Data prepared for sending
s = socket.socket(socket.AF_INET, socket.SOCK_STREAM)
s.connect ( ( HOST, PORT) )
print ( "Connected. . . " )
s.send (DATA1)
data = s.recv (BUFFER_SIZE) #Received data
print ( "Received: { } " . format (data.decode ( 'UTF-8' ) ) )
s.close ( )
print ( "Connection closed. " )
```

Figure 13. Definition of communication parameters.

Communication parameters must be defined: server address, port and buffer needed to receive data. Then, the determined force values are taken and saved to a variable. It has to be converted to byte format because it is sent over the TCP/IP protocol. After establishing communication with the robot controller, the data determined by the neural network are sent. After the data have been transferred, the connection is closed. The transferred data are processed and the automatic change in pressure force of the workpiece to the grinding wheel is achieved.

6. Automatic Correction of the Pressure Force for the Programmed Path

The robotic process of grinding or polishing is performed with the use of belt grinders [4,15,34,35] or rotary tools [36,37]. A novelty is the development of a robotic grinding process that will take into account the individual geometrical parameters of the detail (e.g., blades) on the basis of automated measurements obtained with a 3D optical scanner and determination of processing parameters with a neural decision system. A solution was chosen in which the robot with the force control option holds the blade and presses it against the tool. A rotary grinding tool was used as the machining tool. The non-contact measurement of the blade geometry is performed with the use of a 3D scanner. To control the process, a superior system generating the set pressure force of the blade against the tool was used.

Figure 14 shows graphically the values of the pressure forces generated by the neural process controller at individual measuring points, which in the next process become points belonging to the programmed robot path.

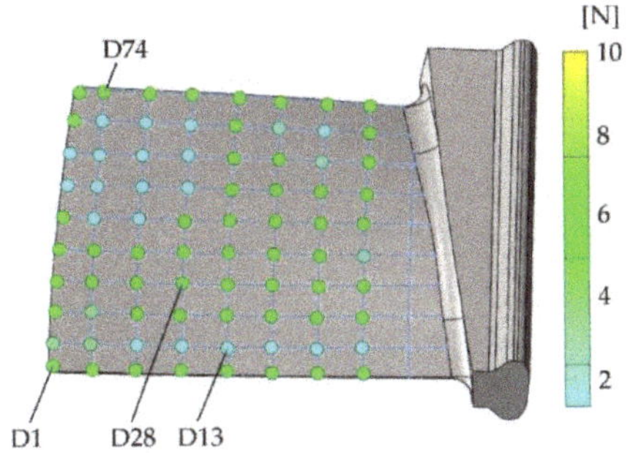

Figure 14. Graphical presentation of the pressure forces at individual points on the blade.

Depending on the allowance at the measuring points on the blade, a pressure force was generated. In the case of blade processing, adaptation of the robot's trajectory to the changing shape is required. This can be ensured by one of ABB's additions, namely the Force Control package and the FC Pressure option used by the authors. After the tests, a solution was proposed in which a stationary tool was used (electrospindle with a grinding wheel), and the object reference system (WorkObject) associated with the robot arm (blade in the gripper) was movable. The advantage of this solution is the reduction in disturbance level, because when the electrospindle is located on the robot's arm, the force value is recorded with disturbances resulting from dynamic phenomena.

A view of the robot application with a blade, set motion trajectory and directions of forces in RobotStudio are presented in Figure 15. The use of the approach with force control allows for the adaptation of the trajectory (change of the pressure force at selected points) to the shape of the part, which varies to a certain extent.

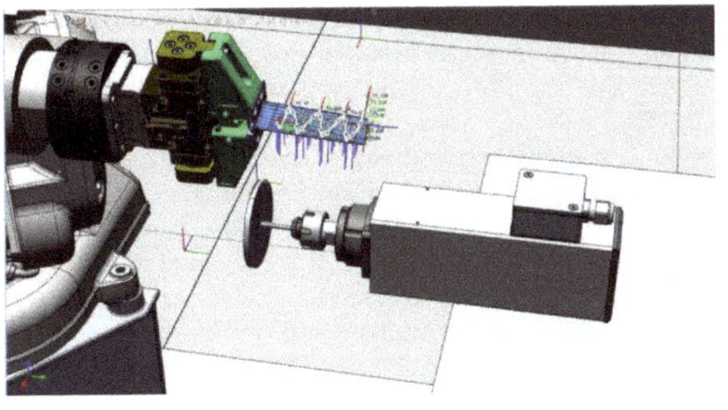

Figure 15. Application view in RobotStudio.

7. Machining

Tests were conducted on the designed robotic station with a 3D measuring system, a neural process controller and a robot with force control. The developed algorithm of the robotic machining process is shown in Figure 16.

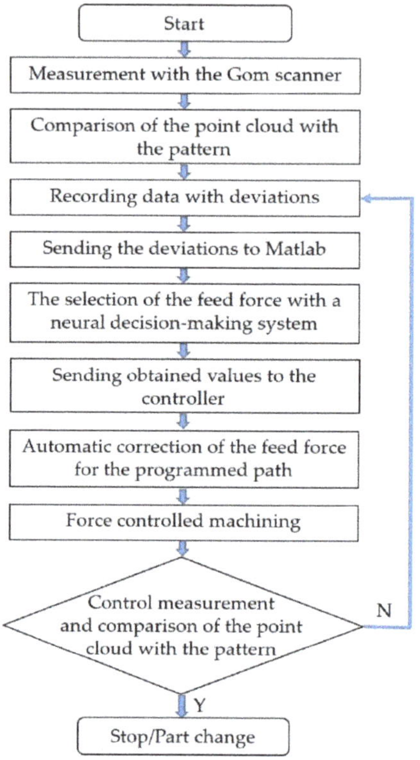

Figure 16. Information flow algorithm in robotic processing in correlation with a 3D scanner and a neural decision-making system.

The process begins by measuring the blade geometry and comparing it with a standard in Atos Professional to determine the deviation grid at 80 points. The saved deviations are sent to the Matlab software for the selection of the pressure forces by the neural decision system for each point. The obtained pressure force values are sent to the robot controller and automatically adapted to the programmed path at selected points. In the next stage, the machining process begins (robot grinding with the option of force control) along the programmed path. After the grinding process, the geometry is measured again and the allowances are determined. When the measured allowances are within the tolerated limits, the grinding is completed and the blade is deposited and the next one is taken. When the measured allowances are outside the tolerated limits, the blade grinding process is repeated until the assumed dimensions are achieved. The entire process of processing the blades was tested on the designed station and the visual results were obtained as in Figure 17, and the measurement of the allowance as in Figure 18.

When comparing the CAD model of the blade with the measurements obtained with a 3D scanner, the tolerance at the measurement points should not exceed $\pm$ 0.05 mm. The conducted verification tests confirm the correctness of the adopted objectives regarding measurements, determining the value of pressure forces from the measured allowances and machining in a robotic process.

Figure 17. View of the blade after grinding.

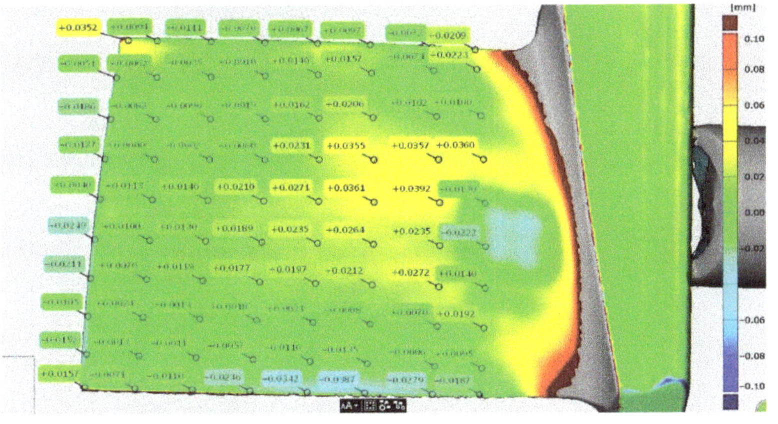

Figure 18. The measured allowances after grinding.

Figure 19 shows the results of the measurement after machining for the blade, namely its thickness in cross-sections (tolerance ± 0.2 mm) for the control allowing the blade to be installed in the engine.

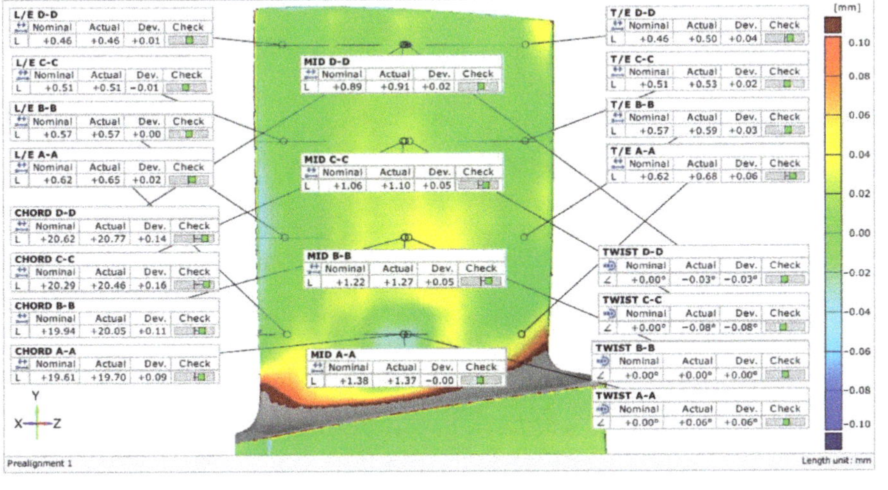

Figure 19. Results of the measurement of the blade in control sections.

Other aviation components subjected to the developed process include deburring the edges of the ADT housing with measurement of the radius repeatability shown in Figure 20, and deburring on the bosses with the measurement of the obtained chamfer shown in Figure 21.

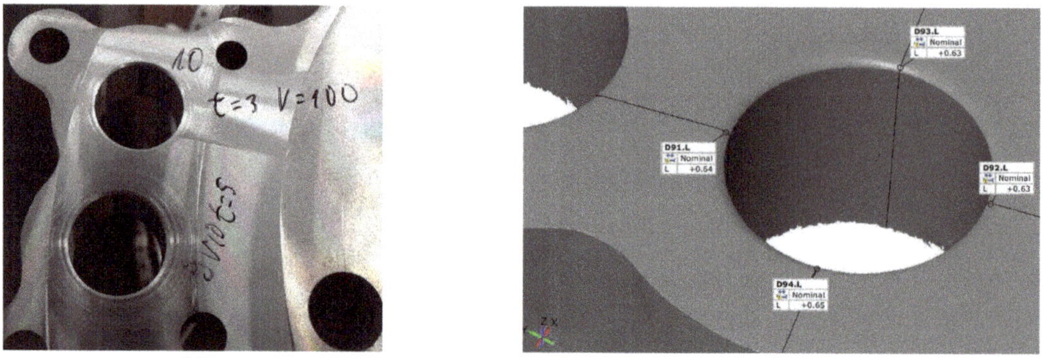

(a) (b)

Figure 20. Deburred ADT gearbox housing: (**a**) block section; (**b**) measuring the repeatability of the radius.

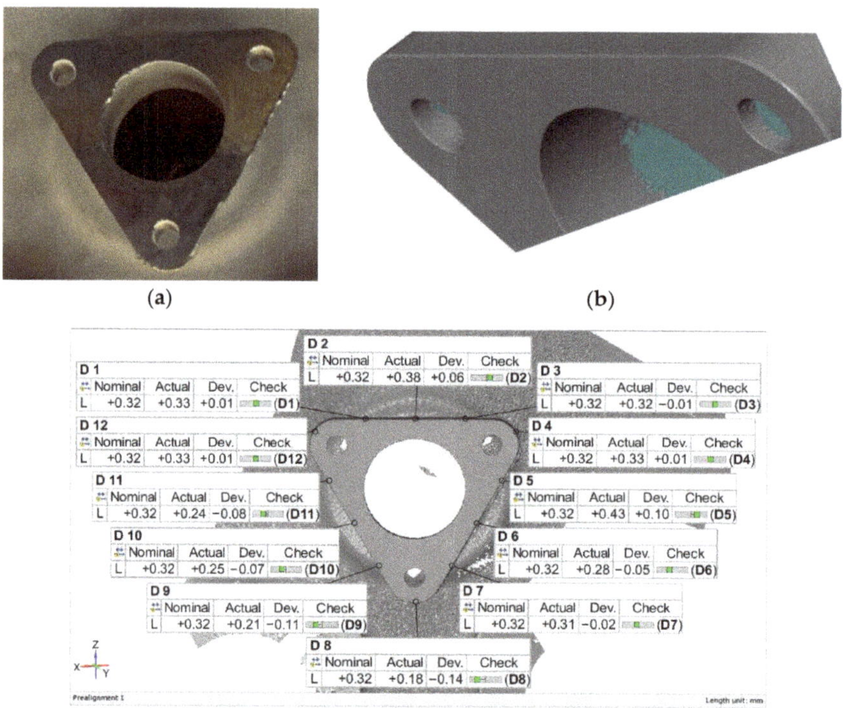

(a) (b) (c)

Figure 21. Deburred boss: (**a**) before the process; (**b**) after the process; (**c**) measurement of the obtained chamfer.

The presented approach allows for adaptive machining by determining the pressure force of the blade against the tool (grinding wheel) depending on the measured allowance. This avoids a situation in which the pressure force is constant over the entire surface of the blade, which can cause a large allowance in some areas of the blade or damage it in the case of large losses.

8. Conclusions

The paper presents the proprietary application of a 3D scanner and a method of communication containing a script sent via TCP/IP protocol in robotic processes of aviation components. The innovative approach is based on conducting a measurement, saving the results, sending their size (allowances), comparing them with the pattern to the decision system and then to a robot controller for adaptive processing by a robot with force control. The proposed solution was verified on three objects: in the selection of parameters during blade grinding, in the process of deburring bosses and the edges in the ADT housing. A set of experimental data was employed in the research, by means of which the relationship between the location of a point on the trajectory of the blade, the pressure force and the thickness of the collected material layer was determined. An artificial neural network was used to determine the variable pressure force of the blade by the robot against the tool (grinding wheel) on the basis of the measured allowances. The presented solution was implemented for the needs of and in cooperation with Pratt & Whitney Rzeszów S.A.

Author Contributions: Conceptualization, D.S. P.G. and K.K.; methodology, K.K. P.G. and D.S.; software, A.B. P.G. and K.K.; validation, M.M. and A.O.; formal analysis, M.M. and M.U.; investigation, A.B.; resources, K.K.; data curation, M.U.; writing—original draft preparation, P.G.; writing—review and editing, M.M. and K.K.; visualization, P.G. and K.K.; supervision, A.B.; project administration, P.G.; funding acquisition, A.B. All authors have read and agreed to the published version of the manuscript.

Funding: This research received no external funding.

Institutional Review Board Statement: Not applicable.

Informed Consent Statement: Not applicable.

Data Availability Statement: Data are contained within the article.

Conflicts of Interest: The authors declare no conflict of interest.

References

1. Bomba, G.; Ornat, A.; Gierlak, P. Geometric Measurements on a CNC Machining Device as an Element of Closed Door Technology. *Sensors* **2021**, *21*, 4852. [CrossRef]
2. Bomba, G.; Gierlak, P. Dimensional Control of Aircraft Transmission Bodies Using CNC Machines and Neuro-Fuzzy Systems. *Appl. Sci.* **2019**, *9*, 4094. [CrossRef]
3. Łygas, K.; Danilczuk, W. Stanowisko do aplikacji pick'n'place z wykorzystaniem systemu wizyjnego i autorskiego interfejsu komunikacji. *Autobusy: Tech. Eksploat. Syst. Transp.* **2017**, *18*, 913–916.
4. Song, Y.; Liang, W.; Yang, Y. A method for grinding removal control of a robot belt grinding system. *J. Intell. Manuf.* **2012**, *23*, 1903–1913. [CrossRef]
5. Jiang, Y.F. The Application of the TCP/IP on the Robot System. *Appl. Mech. Mater.* **2012**, *159*, 351–354. [CrossRef]
6. Kinder, K. Event-driven programming with Twisted and Python. *Linux J.* **2005**, 131.
7. Hui-Ping, L.; Dai-min, C.; Miao, Y. Communication of multi-robot system on the TCP/IP. In Proceedings of the 2011 International Conference on Mechatronic Science, Electric Engineering and Computer (MEC), Jilin, China, 19–22 August 2011; IEEE: Piscataway, NJ, USA, 2011. [CrossRef]
8. Latos, H.; Mikolajczyk, T. Surface shaping with industrial robot. In Proceedings of the 1st International Conference "Optimization of the Robots and Manipulators" OPTIROB-2006, Predeal, Romania, 26–28 May 2006; University "POLITEHNICA" of Bucharest, Faculty IMST, Department MSP: Bucharest, Romania, 2006; pp. 265–269.
9. Sempere, V.M.; Silvestre, J. Multimedia applications in industrial networks: Integration of image processing in profibus. *IEEE Trans. Ind. Electron.* **2003**, *50*, 440–448. [CrossRef]
10. Valera, A.; Salt, J.; Casanova, V.; Ferrus, S. Control of industrial robot with a fieldbus. In Proceedings of the 1999 7th IEEE International Conference on Emerging Technologies and Factory Automation. Proceedings ETFA'99 (Cat. No. 99TH8467), Barcelona, Spain, 18–21 October 1999; IEEE: Piscataway, NJ, USA. [CrossRef]
11. Zolkiewski, S.; Pioskowik, D. Robot control and online programming by human gestures using a Kinect motion sensor. In *New Perspectives in Information Systems and Technologies, Volume 1. Advances in Intelligent Systems and Computing*; Rocha, Á., Correia, A., Tan, F., Stroetmann, K., Eds.; Springer: Cham, Switzerland, 2014; Volume 275. [CrossRef]

12. Burghardt, A.; Kurc, K.; Szybicki, D.; Muszyńska, M.; Nawrocki, J. Software for the robot-operated inspection station for engine guide vanes taking into consideration the geometric variability of parts. *Teh. Vjesn.* **2017**, *24* (Suppl. 2), 349–353. [CrossRef]
13. Burghardt, A.; Kurc, K.; Szybicki, D.; Muszyńska, M.; Nawrocki, J. Robot-operated quality control station based on the UTT method. *Open Eng.* **2017**, *7*, 37–42. [CrossRef]
14. Burghardt, A.; Kurc, K.; Szybicki, D.; Muszyńska, M.; Szczęch, T. Monitoring the parameters of the robot-operated quality control process. *Adv. Sci. Technol. Res. J.* **2017**, *11*, 232–236. [CrossRef]
15. Li, W.L.; Xie, H.; Zhang, G.; Yan, S.J.; Yin, Z.P. 3-D shape matching of a blade surface in robotic grinding applications. *IEEE/ASME Trans. Mechatron.* **2016**, *21*, 2294–2306. [CrossRef]
16. Piotrowski, A. An Analysis of the use of the Python Language in Robot Applications. *Appl. Comput. Sci.* **2016**, *12*, 5–21.
17. Lutz, M. *Programming Python*; O'Reilly: Sebastopol, CA, USA, 1996.
18. Rhodes, B.; Goerzen, J.; Beaulne, A.; Membrey, P. *Foundations of Python Network Programming*; Apress: New York, NY, USA, 2014.
19. Python Scripting for GOM Applications. Available online: https://www.gom.com/en/services/gom-training/gom-scripting (accessed on 20 June 2022).
20. Mizugaki, Y.; Sakamoto, M.; Kamijo, K.; Taniguchi, N. Development of metal-mold polishing robot system with contact pressure control using CAD/CAM data. *CIRP Ann.* **1990**, *39*, 523–526. [CrossRef]
21. Mikołajczyk, T. Robot application to surface finish machining. *J. Pol. CIMAC* **2010**, *5*, 107–112.
22. Tam, H.Y.; Lui, O.C.H.; Mok, A.C. Robotic polishing of free-form surfaces using scanning paths. *J. Mater. Processing Technol.* **1999**, *95*, 191–200. [CrossRef]
23. Zhao, Y.; Zhao, J.; Zhang, L.; Qi, L. Development of a robotic 3D scanning system for reverse engineering of freeform part. In Proceedings of the 2008 International Conference on Advanced Computer Theory and Engineering, Phuket, Thailand, 20–22 December 2008; IEEE: Piscataway, NJ, USA. [CrossRef]
24. Sun, B.; Li, B. Laser displacement sensor in the application of aero-engine blade measurement. *IEEE Sens. J.* **2015**, *16*, 1377–1384. [CrossRef]
25. Qi, L.; Gan, Z.; Yun, C.; Tang, Q. A novel method for Aero engine blade removed-material measurement based on the robotic 3D scanning system. In Proceedings of the 2010 International Conference on Computer, Mechatronics, Control and Electronic Engineering, Changchun, China, 24–26 August 2010; IEEE: Piscataway, NJ, USA. [CrossRef]
26. Yilmaz, O.; Gindy, N.; Gao, J. A repair and overhaul methodology for aeroengine components. *Robot. Comput.-Integr. Manuf.* **2010**, *26*, 190–201. [CrossRef]
27. Szybicki, D.; Burghardt, A.; Kurc, K.; Pietruś, P. Calibration and verification of an original module measuring turbojet engine blades geometric parameters. *Arch. Mech. Eng.* **2019**, *66*, 97–109. [CrossRef]
28. Szybicki, D.; Burghardt, A.; Kurc, K.; Gierlak, P. Device for Contact Measurement of Turbine Blade Geometry in Robotic Grinding Process. *Sensors* **2020**, *20*, 7053. [CrossRef]
29. Huang, J.; Ma, Z.; Hu, Y.; Wang, Y.; Yu, S. Measurement of aero-engine BLISK using industry robot. In Proceedings of the 2008 International Conference on Computer and Electrical Engineering, Phuket, Thailand, 20–22 December 2008; IEEE: Piscataway, NJ, USA. [CrossRef]
30. Fan-Jun, M.; Xun, L.; Zhi-Tong, C.; Xiao-Wen, W. Study on the cantilever grinding process of aero-engine blade. *Proc. Inst. Mech. Eng. Part B J. Eng. Manuf.* **2014**, *228*, 1393–1400. [CrossRef]
31. Ornat, A.; Uliasz, M.; Bomba, G.; Burghardt, A.; Kurc, K.; Szybicki, D. Robotised Geometric Inspection of Thin-Walled Aerospace Casings. *Sensors* **2022**, *22*, 3457. [CrossRef]
32. Kurc, K.; Burghardt, A.; Szybicki, D.; Gierlak, P.; Łabuński, W.; Muszyńska, M.; Giergiel, J. Robotic machining in correlation with a 3D scanner. *Mech. Mech. Eng.* **2020**, *24*, 36–41. [CrossRef]
33. Hagan, M.T.; Demuth, H.B.; Beale, M.H. *Neural Network Design*; PWS Publishing: Boston, MA, USA, 1996.
34. Yixu, S.; Hongbo, L.; Zehong, Y. An adaptive modeling method for a robot belt grinding process. *IEEE/ASME Trans. Mechatron.* **2011**, *17*, 309–317. [CrossRef]
35. Zhsao, P.; Shi, Y. Composite adaptive control of belt polishing force for aero-engine blade. *Chin. J. Mech. Eng.* **2013**, *26*, 988–996. [CrossRef]
36. Xiao, G.; Huang, Y.; Yin, J. An integrated polishing method for compressor blade surfaces. *Int. J. Adv. Manuf. Technol.* **2017**, *88*, 1723–1733. [CrossRef]
37. Burghardt, A.; Szybicki, D.; Gierlak, P.; Kurc, K.; Muszyńska, M. Robotic Grinding Process of Turboprop Engine Compressor Blades with Active Selection of Contact Force. *Teh. Vjesn.* **2022**, *29*, 15–22. [CrossRef]

electronics

MDPI

Article

Development of a Dedicated Application for Robots to Communicate with a Laser Tracker

Dariusz Szybicki *, Paweł Obal, Paweł Penar, Krzysztof Kurc, Magdalena Muszyńska and Andrzej Burghardt

Department of Applied Mechanics and Robotics, Faculty of Mechanical Engineering and Aeronautics, Rzeszow University of Technology, 35-959 Rzeszow, Poland
* Correspondence: dszybicki@prz.edu.pl; Tel.: +48-(17)-8651843

Abstract: The paper presents the concept of operation and methods of using laser trackers in robotics. So far, a small amount of research on software for sharing and exchanging data with trackers has been done. As a result of the identified demand, a proprietary application for communication between the laser tracker and robots, as well as other software, was developed. The developed solution is based on the software development kit (SDK) provided by Leica and the Python language. The structure and functioning of the developed software were described in detail. The software meets the goals set at the beginning of the design process regarding online communication with the tracker and using the universal, popular TCP/IP standard. The functioning of the developed software was shown in the paper in a few examples related to manipulating robots and mobile robots. The capabilities of the developed software were described, as well as the planned work on its development.

Keywords: robot; laser tracker; design of robotic system; mobile robot; measurement in robotics

Citation: Szybicki, D.; Obal, P.; Penar, P.; Kurc, K.; Muszyńska, M.; Burghardt, A. Development of a Dedicated Application for Robots to Communicate with a Laser Tracker. *Electronics* **2022**, *11*, 3405. https://doi.org/10.3390/electronics11203405

Academic Editor: Piotr Szymak

Received: 15 September 2022
Accepted: 19 October 2022
Published: 20 October 2022

1. Introduction

According to the 2022 IRF (International Federation of Robotics) report, based on 2021 numbers, the still dominant industries in terms of robot applications are electrotechnical/electronics (137,000 robots were installed in production), automotive (119,000 units) and engineering ones (64,000 units). Based on the 2020 IRF report about the use of industrial robots [1], the use of robots in machining, deburring, and grinding is growing significantly. Due to the aforementioned facts, one of the challenges of robotics is to increase the accuracy and repeatability of industrial robots. This will make it possible to perform advanced machining or assembly tasks. In certain areas, it will be possible to replace costly CNC machines. Methods to improve the accuracy and repeatability parameters of robots are numerous. Improving stiffness, modifying control systems, or using external systems for measuring and path correction. Using laser trackers is one way to accurately measure TCP orientation position and online correction for industrial robots. A laser tracker is a tool that enables various types of geometric measurements in three-dimensional space. It can be used for automated position control as well as, e.g., for 3D scanning of objects. Its most important elements are the laser interferometer and the absolute rangefinder. Due to their accuracy, robotics trackers are used in testing the accuracy and repeatability of manipulating robots and for their calibration [2–4]. Laser trackers in robotic measurements are used for a variety of applications. In the paper [5], the authors present an innovative methodology for measuring the susceptibility of articulated serial robots, and the laser tracker is used to measure the system's response.

In paper [6], the laser tracker was used to measure the deflection of the robot end effector during comparative tests and optimization of the robot position using static and dynamic stiffness models for various milling scenarios.

The authors of the papers [7,8] ponder how to increase the accuracy of industrial robots with the help of the Leica Absolute Tracker AT960 (Hexagon, Stockholm, Sweden). They propose new methods of calibrating robots with tools in their workplace. These methods improve positioning accuracy by compensating for the identified parameters. The accuracy of the robots, along with the reduction in calibration time, are key factors in the success of robotic production systems.

In paper [9], the authors discuss assembly operations in the aviation industry, which are time-consuming and require high accuracy. They emphasize that robotic assembly is a good solution that increases productivity, but they point out that the poor accuracy of industrial robots limits their use. They propose to improve it by adding an accurate online 3D positioning system, which consists of the KEYENCE LJ-V7200 vision system (Keyence, Osaka, Japan) and the Leica AT-960 + T-Mac TMC-30B (Hexagon, Stockholm, Sweden) tracking system.

Paper [10] investigated the ability of a laser tracker to measure the relative position and orientation between two mobile Stewart platforms simulating the movement of ships at sea. These ships are exposed to disturbance from waves and have cranes equipped with active compensation systems on board, which keep the cargo at a certain height from the seabed.

In paper [11], a laser tracker was used to improve the accuracy of cable-driven parallel robots. Inaccuracies are caused by deviations in cable lengths caused by elongation, elasticity, or creep.

In paper [12], the authors focus on modeling, measuring, and identifying the change in the kinematic chain of serial articulated industrial robots based on thermomechanical deformations caused by self-heating caused by drives. The assessment of the change in the positioning accuracy of the ABB IRB 1600 (ABB Ltd., Zürich, Switzerland) robot was carried out using a Leica AT960 laser tracker and a FLIR SC640 (FLIR, Wilsonville, OR, USA) thermal imaging camera.

The authors of the work [13] used a laser tracker to improve the absolute accuracy of the ABB IRB 1600 industrial robot. They developed an advanced calibration model that significantly reduced position errors. Similar work was carried out by the authors of [14], who focused on building calibration systems easily adaptable to each robot type. A computer system was built to develop and implement a calibration system. Subsequently, experimental trials were carried out using the IRB2000 robot, which resulted in a large accuracy improvement. Similar works on calibration and improvement of accuracy were carried out by researchers [15–18]. The use of trackers in robotics, however, often goes beyond accuracy testing and calibration. The subject of real-time tracker compensation is discussed in [19,20].

In paper [19], the authors described the idea of compensation for production systems by using external metrological systems to compensate machine tools and robots in real-time. On the other hand, in the article [20], they used direct feedback from the position and orientation of the end effector with the Leica laser tracker to develop an algorithm that allowed for moving along paths at a speed of 100 mm/s with an RMS of only 0.11 mm.

The compensation for errors in production processes in the aviation industry is the subject of article [21]. The authors of that paper show a method of maintaining high accuracy of robot manipulation by continuously tracking the position and orientation of the mounted tool while minimizing errors. The study of parameters of industrial robots with the use of trackers during contact operations is the subject of [22,23]. In paper [22], researchers evaluate the performance of robots during deburring, grinding, and cutting operations in terms of quasi-static path accuracy and repeatability. In article [23], the authors investigate the modal properties of industrial robots during the milling process using a tracker. The applications of laser trackers in robotics mentioned so far concern manipulating industrial robots. In the segment of mobile robots, laser trackers are most often used for calibration or as a position verification tool. Paper [24] presents a method of autonomous loading, transporting, and unloading large objects using a non-holonomic mobile manipulator. The Faro Vantage S6 laser tracker was used to calibrate and verify the operation of the target vision system.

A tracker for calibrating a six-legged walking robot with an integrated parallel manipulator was used in [25]. The authors of this work, after applying the proposed tracker method for validation, identified the kinematic parameters of the entire robot, and the accuracy of the movement of each leg and manipulator was significantly improved. In [26], a laser tracker was used to track the trajectory of a mobile robot based on linkage suspension. Using a tracker allowed for the trajectory of the physical prototype of a mobile robot to be registered and analyzed.

The subject of applications related to recording and sharing data from trackers is discussed by the authors less often than using them to measure accuracy and repeatability. Applications are most often created by manufacturers of individual trackers and are used for metrology. Proprietary software for data sharing and communication with laser trackers is often created. An example of such a solution was shown by the authors of [27]. They developed a real-time control system for the position and orientation of industrial robots using a laser tracker. Via the developed real-time interface, position data is acquired in millisecond cycles and is used to calculate the current errors of the robot path. The EtherCAT bus and a dedicated software solution were used for data transmission between the Leica tracker and the robot controller. Similar research work was presented in articles [28,29]. In these works, the authors show the software they developed for communication with a laser tracker, along with examples of applications. The paper [29] shows the most advanced example of the functioning of an application working behind a PC and communicating with a robot via KUKA RSI (robot sensor interface). The software runs on a Windows PC and supports communication with the laser tracker via Ethernet using the FARO SDK for the laser tracker.

As for other applications of laser trackers, in the paper [30], a laser scanner, alongside a manipulator and a 3D scanner, was used as part of a system for determining the position of small assembly points on large-scale components. The authors indicate that the developed method of a large-scale 3D measurement reduces the maximum and average calibration error of the measurement system by about 55% at a measurement radius of 7 m. An analogous use of trackers is presented in the paper [31]. The paper describes the use of trackers for particle accelerators. Due to their mobility and accuracy, they are used to determine the position of magnets, which is a critical part of its construction and requires tolerances of millimeters at distances of tens of meters. The paper [32] describes the use of multiple trackers forming a trilateration network to improve the accuracy of measuring the flatness of the surface of a 5 m diameter ring. The effect of the positioning of the trackers, their number in the network, and differences in the height of the measurement head on the accuracy of po-measurement were presented. In the case of measuring flatness with two trackers placed in the center of the ring, whose height difference is 1 m, the measurement accuracy is ± 6 µm.

Laser trackers are used in the construction of devices that require very high-precision manufacturing. Papers [33] discuss the use of a tracker for the precise positioning of particle accelerator systems. To increase precision, several trackers can be combined into a single system, as shown in the example of positioning a high-energy photon source High Energy Photon Source (HEPS) system [34].

Laser trackers can be used not only to validate the accuracy of the manipulators themselves. They are also used to determine the accuracy of various calibration instruments interfacing with industrial robots, such as the case described in the paper [35], where the authors checked the accuracy of a robot's TCP automatic self-calibration instrument.

Another interesting use of a laser tracker is the work [36] where the authors proposed using the tracker to measure the position of the drill stem of a drilling rig. By measuring the ground part, the authors could estimate the shape of the hole drilled by the rig.

The authors of article [37] developed a method of using a laser tracker to program paths of industrial robots. While working on this method, it turned out that it would be useful to develop a universal application that makes tracker data available to various types of robots. In the solution presented there, the data with the positions measured by the tracker was saved on a PC and then loaded from this computer by the robot controller. The presented solution was not universal and was cumbersome to use. In connection with the problem described, work was started on the development of an own application that provides data from the tracker. Software for manipulating robots, mobile robots, and data transferring to popular software used, among others, in robotics, namely the Matlab-Simulink package, was designed and developed. The idea behind the tracker and the developed software will be discussed later. Examples of its applications will also be presented.

2. Characteristics of the Tracker and the Method of Recording and Sharing the Measured Parameters

Measuring with a laser tracker is often considered a coordinate measurement method. It allows for precise measurements of the position of a selected point in three directions simultaneously. These devices are used in geodesy, scanning large-size objects in 3D, and automated control in industry. When using a laser tracker (Figure 1), it is necessary to install a mirror reflecting the laser beam, the so-called sphere-mounted retroreflector (SMR).

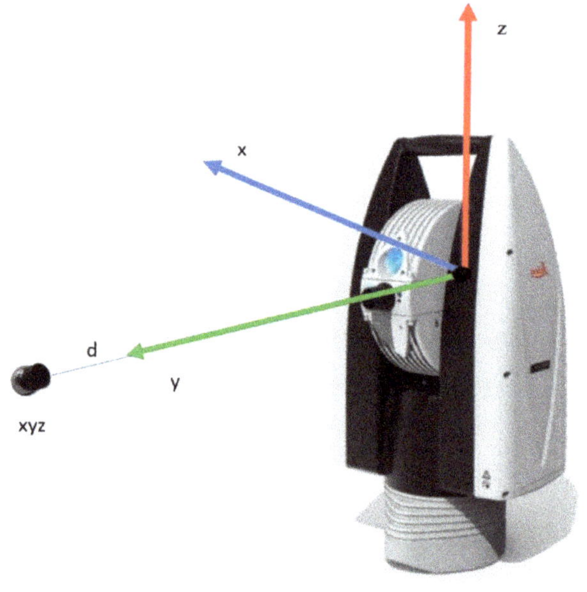

Figure 1. Measuring with a laser tracker.

Research work on the own software was carried out with the use of the Leica AT960 absolute laser tracker. It is a very accurate device that specifies its accuracy as a maximum permissible error (MPE). The maximum permissible error (MPE) is defined in the ISO 10360-10: 2016 standard as the highest value of the measurement error allowed in the specification for a given measurement. Table 1 shows the tracker specification according to ISO 10360-10: 2016, Annex E.

Table 1. The maximum permissible error of the tracker.

Laser Tracker Subsystem	Symbol	Maximum Permissible Error (MPE)
Interferometer (IFM)	eIFM	±0.4 μm + 0.3 μm/m
Absolute laser measurement (ADM)	eADM	±10 μm
Parameter R0 (R0)	eR0	±3 μm
Transverse	eT	±15 μm + 6 μm/m

The described tracker and its equipment are part of the industrial robotics laboratory of the Department of Applied Mechanics, Rzeszow University of Technology. The tracker and robots located in the laboratory are used to design and program stations dedicated to aviation industry needs. The tracker was used in the station's design to control the outlet guide vanes (OGV) described in [38]. Tracker measurements were useful when designing a module for geometrical measurements of jet engine blades described in [39]. Geometric measurements taken with a tracker turned out to be useful during the development of digital twins for the purposes of training with the use of virtual reality; this subject was described in [40]. In the mentioned works, the tracker served only as a metrological tool for measuring the geometry of a robotic station or workpieces. The aforementioned work related to programming robots using a tracker caused the need to expand the tracker's capabilities and develop its software.

The tracker was delivered with two types of software offered by the manufacturer. The basic tool for operating the Leica tracker is the Leica Tracker Pilot program (Figure 2a). The program is a graphical interface between the user and the tracker controller. It allows the user to enter all settings into the tracker, save measurements as text files, and perform maintenance and calibration. After establishing a connection with the tracker, the program displays status information, sensor states, and measurement values of the currently tracked object.

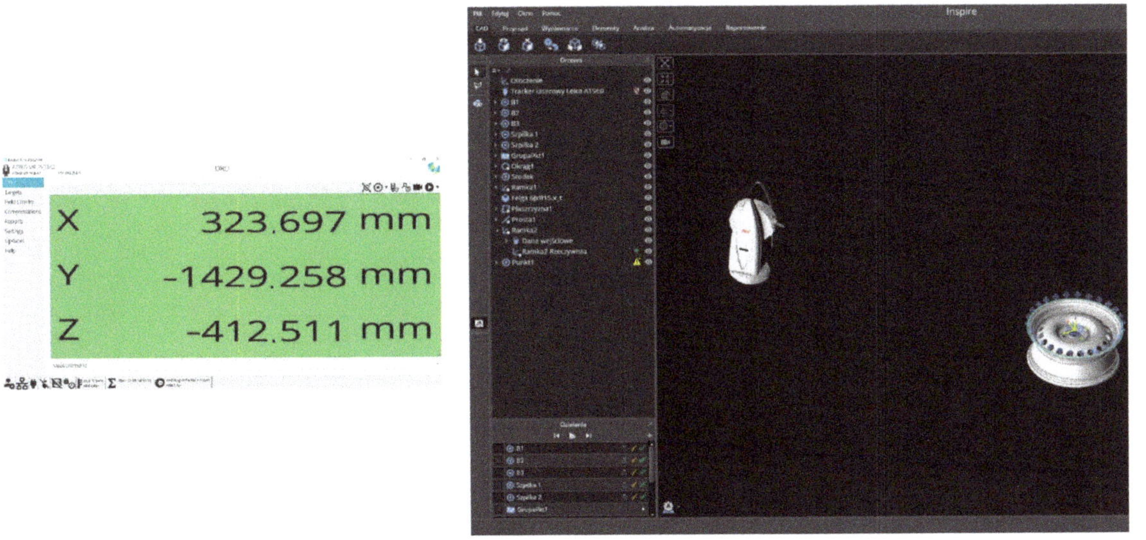

(a) (b)

Figure 2. Available tracker software: (**a**) Tracker Pilot software window; (**b**) Inspire software window.

The triggering of measurements depends on the selection of one of the 7 measurement profiles:

- single quick measurement;
- single standard measurement;
- single precise measurement;
- continuous measurement triggered at a constant distance;
- continuous measurement triggered at a specified time;
- measurement triggered by a touch probe;
- measurement triggered by an external signal.

The software allows the user to save the collected measurement data to a CSV file. In addition, it also has several tools for checking the measuring accuracy of the tracker head, reflectors, and measuring probes. Sudden temperature changes, as well as various mechanical influences, may adversely affect the accuracy of the device. Therefore, in addition to checking the accuracy of the Tracker Pilot, it is possible to perform a compensation procedure, which allows for maintaining high accuracy of measurements. The software does not have any tools for the visualization and processing of collected data.

The second program provided by the tracker manufacturer, called Inspire (Figure 2b), a metrology program, is used for this purpose. The software allows the user to create measurement procedures, automating the process of not only data acquisition but also the processing and generation of measurement reports on their basis. The software automatically connects to the measurement devices and supports all 7 measurement trigger profiles for the tracker. The great advantage of this software is the ability to define the dimensions of the shape of the stands for measuring reflectors. Then the software automatically converts the position of the measured point to the point of contact of the stand with the measured surface. The bases can come in various shapes depending on their purpose. The simplest ones are used to measure flat surfaces. More complex shapes make it easier to measure edges or characteristic points of curved shapes. The software can independently recognize the measured shape based on the arrangement of measurement points and adjust the settings for the detected shape. It also allows the user to import CAD files and has tools to define the relationship between the measurements and the shape pattern. In general, the software is very user-friendly, with a large set of advanced features that are intuitive to use. However, it has its limitations. For T-Mac probe measurements, only positions. For robot-related measurements, recording both position and orientation is very important.

Tracker Pilot software records positions and orientations but does not have any tools to process this data. A tool designed for measurements used in robotics is the RoboDyn software, which allows the user to determine the base frame of reference, the tool center point (TCP), based on measurements from the tracker. The software can also measure the accuracy of an industrial robot and perform calibration and compensation of its parameters. In addition, the tracker can work with many other types of software, such as PolyWorks®, Metrolog X4, DM Works, Silma/X4-iRobot, or RoboDK.

All of the software types mentioned above have their advantages and disadvantages. Unfortunately, none of them offers convenient and fast online communication with various types of robots and with the Matlab-Simulink package.

Thus, several possibilities for downloading measurement data from the tracker were considered:

- reading data from a CSV file;
- using the RTFP-EC accessory to communicate with the tracker via the EtherCat protocol;
- using the tools available in the SDK to communicate with the tracker via TCP/IP protocol.

Taking measurements and saving the data to a CSV file is possible in both programs supplied with the tracker, which were described earlier. This approach is convenient for a small number of individual measurements. Exporting a large series of measurements, especially continuously triggered measurements, is cumbersome because both programs we tested do not have formatting capabilities for data export. In addition, such an approach precludes the use of data for real device control. A solution designed for such applications is a tracker add-on called RTFP-EC that allows data to be sent over EtherCat protocol at a frequency of 1 kHz. This is an Ethernet-based communication standard. This protocol allows for achieving high speeds of information exchange and can be used for communication in real-time systems [41]. According to studies reported in papers [42–45], EtherCat is one of the most efficient industrial communication standards available on the market. But its implementation costs are lower than Profinet or Powerlink [41]. Many platforms used in robotics, such as the RaspberryPi, support EtherCat, particularly for controlling drives. The Matlab/Simulink environment also has a set of tools for communication over EtherCat. Unfortunately, ABB's robots do not support this communication standard. Therefore, the authors finally decided to use a simple TCP/IP connection, which will allow ABB's robot controller and other devices and applications that support an Ethernet connection to receive measurement data. Such a connection transfers data at 50 packets per second [46]. This means that for serial measurements with a high trigger frequency, data is buffered and sent every 20 ms, but without maintaining the timing regime. Thus, it is not suitable for real-time applications. On the other hand, it can be used for robotic metrology processes, registration of robot trajectories, and correction of low-dynamic motion trajectories.

The requirements that were set for the development of the new software included:

- online communication with the laser tracker;
- use of a popular, universal communication standard;
- possibility to choose the type of tracker measurement;
- possibility of triggering the measurement with a tracker at a defined time or distance;
- the possibility of triggering the measurement by a touch probe or an external signal;
- use of a popular object-oriented programming language that provides access to many libraries in the form of packages.

Analyzing the available possibilities, it turned out that two leading manufacturers of laser trackers, namely Leica and Faro, offer their own software development kits (SDK). Having a Leica tracker, it was decided to use the SDK offered by the manufacturer based on the NET programming platform. Still, it is also available in other programming languages (e.g., Python). The API, i.e., the application programming interface provided by the SDK, has the structure of an object tree. This approach means that the Tracker object is represented by subobjects containing properties, events, and methods. Their feature is the availability of events informing about changes in the tracker's status. The further part of the paper describes the developed application based on the available SDK and using the Python language.

3. The Application of the Developed Software in Robotics

The developed application was named LeicaConnector.py. It is a TCP server running on a PC under Windows. Devices implementing the TCP/IP interface can connect to it. The server processes the defined TCP frames and performs functions on the tracker based on the API (Figure 3).

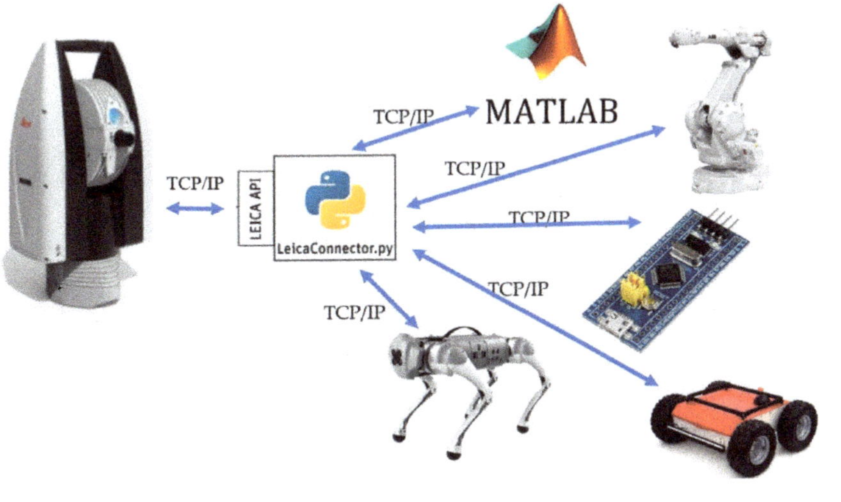

Figure 3. The concept of an intermediary program between the tracker and the end device.

The implementation of such an intermediary in communication with a Leica device allows for:

- Leica device support by microcontrollers with TCP support, industrial robot controllers, mobile robots, or engineering packages, e.g., Matlab-Simulink;
- easy program development by defining subsequent commands available to the client;
- the ability to send the position and orientation (in the case of T-MAC) read by the Leica device to multiple devices;
- easy integration of measurements from the laser tracker with data from other devices.

The choice of the Python language, in the context of the design requirements set, is determined by its versatility and universality of use, especially in applications related to the Internet of Things and data engineering [47]. Due to the available SDK, there was an opportunity to use, for example, the C# language. However, given that the software runs on commands, an intuitive direction for its expansion is to integrate further measurement devices, i.e., a 2D scanner or another Motion Capture type position determination system. This approach is another point justifying the use of Python, as due to its popularity, many manufacturers are developing their APIs in it. The described approach of software development ultimately leads to the creation of a universal tool for the acquisition and integration of measurement data. Its processing is the domain of data engineering. In this field, Python is the language of first choice due to the multitude of libraries for data processing and visualization. This is another argument for choosing Python.

The LeicaConnector.py program is based on a configuration file in which, among others, the settings of the graphic interface are saved, the client command codes are assigned to the functions they perform, and the network settings are defined, i.e.,

[TCP]
Ip = 192.168.123.11
Port = 50007
[TRACKER]
Ip = 192.168.123.31

The [TCP] section defines the IP number and port of the server being created, to which the clients connect. The [TRACKER] section defines the default Leica device IP or the simulator name, as shown in the section of the configuration file (the line marked in green is a comment). The LeicaConnector.py interface comprises six sections, identified as A–F in Figure 4.

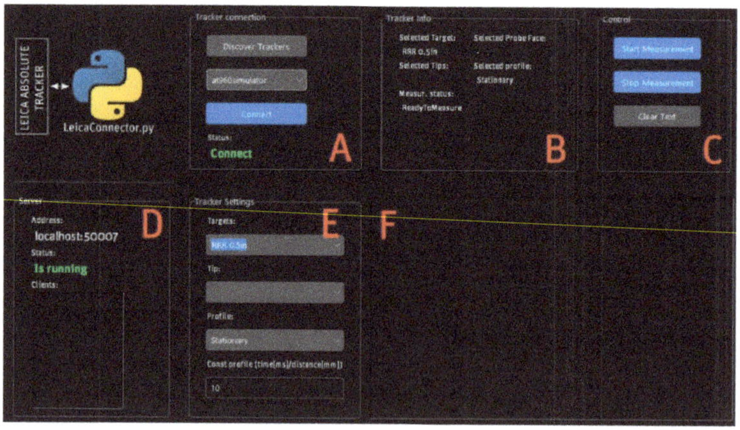

Figure 4. Program interface.

In section A, there is a tracker checkbox. The list of available devices can be updated by selecting the Discover Tracker button. Connection with the selected device takes place by selecting the Connect button, and the correctness of the connection is indicated in the Status field. Section B contains basic information about the configuration of the Tracker, related to the selected target and its configuration (ProbeFace and Tips), the measurement status, and the selected measurement profile. This can be selected by the client by sending the appropriate command in the form of a data frame with the appropriate code. The program allows the user to select several profiles:

- Stationary Profile–default profile;
- Continuous Time Profile;
- Continuous Distance Profile;
- Touch Trigger Profile.

In the case of the continuous time and constant distance profile, the configuration frame has a parameter that defines, respectively, the time interval between measurements or the change of the distance for which the measurement will be triggered. Although the default, triggering or ending the measurement is triggered by the client by sending a frame with a defined code. In section C, buttons allow the user to perform these actions from the GUI level. The Clear Text button clears the window in section F, where program logs are placed. Section D contains information about the configuration of the TCP server being created and the list of clients that are connected to it. The client who connected first has the right to issue commands. Subsequent customers only receive the measurement data. Section E is used to select a target and configure it if more than one tip is defined in the tracker controller. In this section, the user can select a measurement profile. This is equivalent to executing the appropriate command by the client. In the case of a continuous time profile and a constant distance, the set interval or distance value results from the value entered in the Const Profile field. Section F is a window where the logs of the program are displayed. LeicaConnector.py requires the following:

- Python version 3.8.10;
- Tkinter library;
- Python.NET package.

In the case of communication with a client in the local network, the IP number of the host should be specified in the IP parameter. Two types of frames are exchanged between the LeicaConnector.py proxy and the client using a TCP/IP connection.

The first type of frame consists of four fields (Figure 5):

- Command—a field in the signed int (4B) format; the field specifies the frame type through the code it contains: command frame, confirmation/error frame, or measurement frame;
- param 1—a field with data in the signed int (4B) format, the interpretation of this parameter depends on the code contained in the command field;
- param 2—a field with data in the signed int (4B) format, the interpretation of this parameter depends on the code contained in the command field;
- param 3—a field with data in the signed int (4B) format, the interpretation of this parameter depends on the code contained in the command field.

Command	param 1	param 2	param 3
signed int (4 bytes)	signed int (4 bytes)	signed int (4 bytes)	signed int (4 bytes)

Figure 5. Basic LeicaConector proxy frame.

A special frame of the first type is the frame with code 210 (Figure 6), the parameters of which param1-param3 correspond to the coordinates of the measuring point (target). Their values (returned by the tracker), due to conversions to integers, are scaled by the constant value of k.

210	xk	yk	zk
signed int (4 bytes)	signed int (4 bytes)	signed int (4 bytes)	signed int (4 bytes)

Figure 6. Measurement data frame.

When the Laser tracker works with T-MAC probes, the orientation of the T-MAC probe is returned in the form of a quaternion in addition to the position of the point in the tracker's coordinate system. In this case, the measuring frame (Figure 7) has code 211, and its fields are not only positions (x, y, z fields) but also values of the quaternion coefficients (a, b, c, d) scaled by the constant *m*.

211	xk	yk	zk	am	bm	cm	dm
signed int (4 bytes)	signed int (4 bytes)	signed int (4 bytes)	signed int (4 bytes)	signed int (4 bytes)	signed int (4 bytes)	signed int (4 bytes)	signed int (4 bytes)

Figure 7. Extended measurement data frame.

The assumed values of the constants are, respectively, k = 1000 and m = 10,000. Changing them requires modifying the GainPoz and GainRot constants in the configuration file. The frame codes are three-digit numbers, where:

- 1xy—command codes;
- 2xy—confirmation codes corresponding to the command code. The exception is the frame with the code 211 (position and orientation data) for the command 210;
- 30y, 3xy—general error codes and command error codes, where x = 1, 2..9, y = 0, 1..9.

The following command codes and the associated confirmation or error codes were adopted in the developed application:

- 110—measurement start command, runs the *startMeasurement* function;
- 112—measurement end command, triggers the *stopMeasurement* function;
- 115—command sets the stationary measurement mode;
- 116—the command sets the measurement profile of the continuous time type;
- 117—the command sets the measurement profile of the fixed distance type;
- 118—the command sets the contact type measurement profile.

The example shows how LeicaConnector.py communicates with the client when using a stationary measurement profile. It is assumed that the used retroreflector is visible to the tracker.

1. *Client: Connecting to the LeicaConnector.py server*
2. *Client: Send the frame: [0,0,0,110]*
3. *LeicaConnector.py: Frame sending: [210,1,000,000,1,523,234,5,432,984] //other customer activities*
4. *Client: Send the frame: [110,0,0,0]*
5. *LeicaConnector.py: Send the frame: [210,1,000,000,1,523,234,5,432,984]*

In the example shown, the client sends a frame with code 110, which corresponds to triggering the measurement in the stationary mode (the default). In response, LeicaConnector.py sends a frame with the code 210, the next fields of which are the coordinates of the measuring point on the x, y, and z axes. Their values are multiplied by 1000. Hence: x = 1000, y = 1523.234, z = 5432.984. Then (step 4), the client sends another frame with the code 110, requesting another measurement in stationary mode. This one is passed in step 5.

The developed application allows for handling possible communication errors, handling various methods of triggering measurements, and measuring in stationary mode with the T-MAC touch probe. The following part shows the application of the developed software in robotics.

4. Examples of Applications of the Developed Software

The software presented in this paper was originally developed to communicate with ABB's industrial robots. While working on the aforementioned paper [37], an idea emerged to provide data transmission from a laser tracker. That article concerned using a laser tracker to program the paths of industrial robots, and there is no need to quote the results here. It is important that after developing the software, the communication diagram of the industrial robot laser tracker was modified (Figure 8).

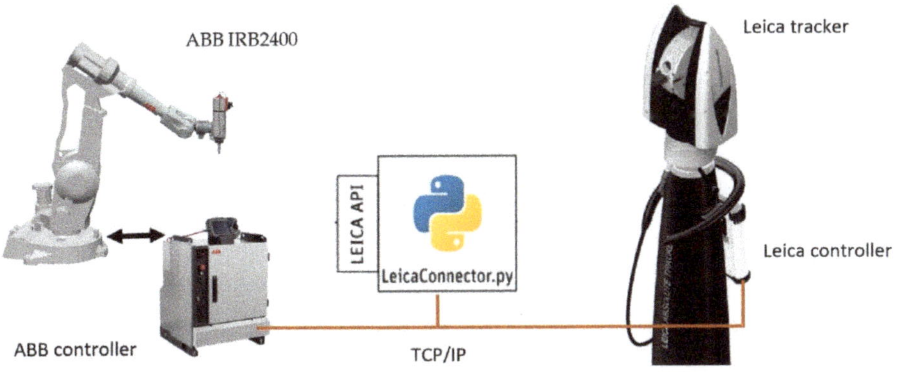

Figure 8. Communication diagram between Laser Tracker and Robot ABB.

The introduced modification allows the user to avoid the need to read text files with data. In addition, data transmission is faster, and the algorithms introduced in the future to the application will allow for the automatic determination and translation of robot coordinate systems.

In the further part of the paper, an example of the use of a laser tracker to determine the position of mobile robots is presented in the example of two constructions: a 4-wheel mobile robot with mecanum wheels and a 4-leg mobile GO1 robot.

The Department of Applied Mechanics and Robotics has a 4-wheeled mobile Panther robot manufactured by Husarion, equipped with mecanum wheels. The Raspberry Pi 4 microcomputer with the ROS system, connected via the CAN bus with the STM microcontroller implementing the lower control layer, is responsible for high-level control. Moreover, the robot allows hardware access to motors and encoders and implements the lower control layer using the rapid design environment based on the dSpace signal processor. This approach allows for the implementation of research work on traffic control algorithms or identification of the mathematical model.

In the described example, the Leica laser tracker was used to determine the position of the mobile robot in relation to the tracker coordinate system during the movement (Figure 9), which was carried out in the so-called joystick control mode. This feature is ROS-enabled. Successive positions of the mobile robot, understood as positions related to its design of the retroreflector, were saved in the Matlab software using the LeicaConector.py program (Figure 10a). In this case, the laser tracker determined the retroreflector positions in a time-related trigger mode with an interval of 0.1 s.

In mobile robotics, it is important to measure the position on the XY plane. During the drive of a 4-wheel mobile robot with mecanum wheels, measurement samples of the reflector's position were recorded. For the assumed time interval, the first 130 samples correspond to 13 s of the motion of the mobile robot. On this basis, the path of the robot's characteristic point was determined (Figure 11), related to the retroreflector in the tracker system.

The experiment in which the laser tracker determined the position of the robot while moving was carried out in a rectangular corridor. There is a tracker at one of its ends. The second one was the initial position of the mobile robot, as shown in Figure 11. The length of the radius vector

$$\left|\vec{r}\right| = \sqrt{x^2 + y^2} \tag{1}$$

of the robot's characteristic point for its initial position in the tracker coordinate system is approx. 11 m and for the end position its value is approx. 3.3 m. Moreover, it can be seen from Figures 10b and 11 that the robot's movement was towards the tracker. The path determined by the tracker indicates that the angle between the corridor axis and the OX axis of the tracker system is determined by the relationship

$$\theta = \arccos\left(\frac{x_0}{\left|\vec{r}\right|}\right) \tag{2}$$

where x_0 is the coordinate of the robot's initial position determined in the first measurement; it is approx. -46.5°, which is marked in the diagram in Figure 10b.

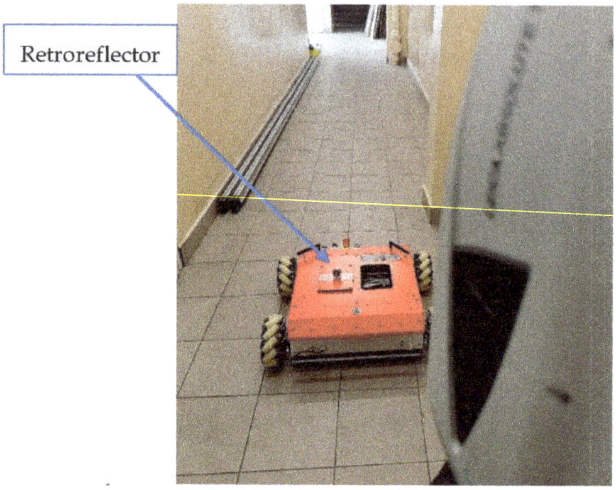

Figure 9. Positioning a 4-wheel mobile robot with mecanum wheels using a Leica laser tracker.

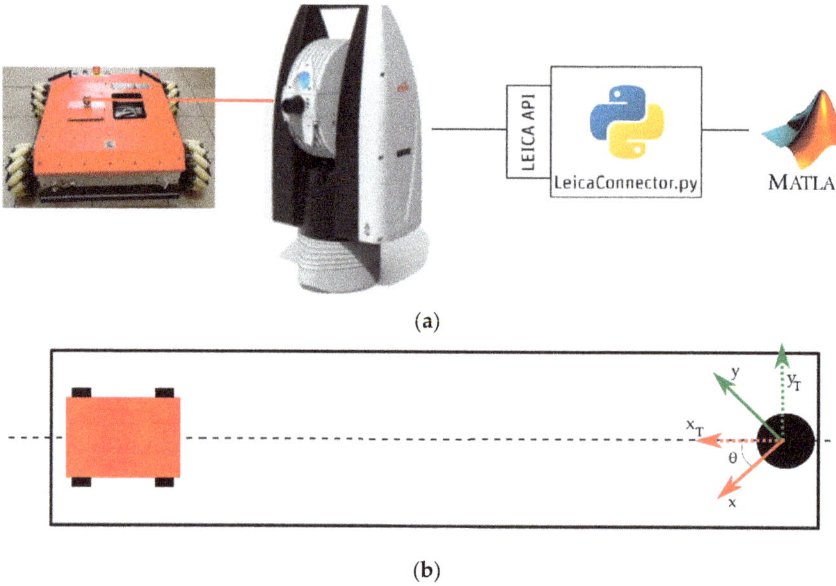

Figure 10. Test stand diagrams: (**a**) cooperation of a laser tracker and Matlab software in determining the position of a 4-wheel mobile robot with mecanum wheels; (**b**) scheme of a measurement experiment.

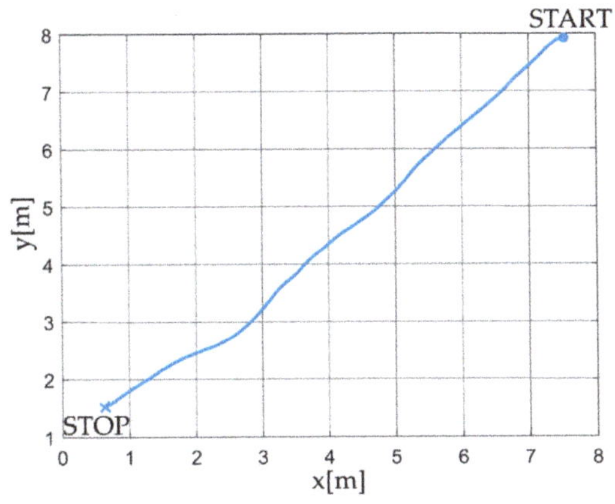

Figure 11. The path of the robot characteristic point related to the retroreflector.

Applying a homogeneous transform describing a rotation on a plane having the form

$$\begin{bmatrix} x_T \\ y_T \\ 1 \end{bmatrix} = \begin{bmatrix} \cos\theta & \sin\theta & 0 \\ -\sin\theta & \cos\theta & 0 \\ 0 & 0 & 1 \end{bmatrix} \begin{bmatrix} x \\ y \\ 1 \end{bmatrix} \tag{3}$$

it is possible to determine the path of motion of MRK in the $x_T y_T$ coordinate system related to the tracker, whose axis OX is parallel to the corridor axis (Figure 10b). The resulting trajectory is shown in Figure 12.

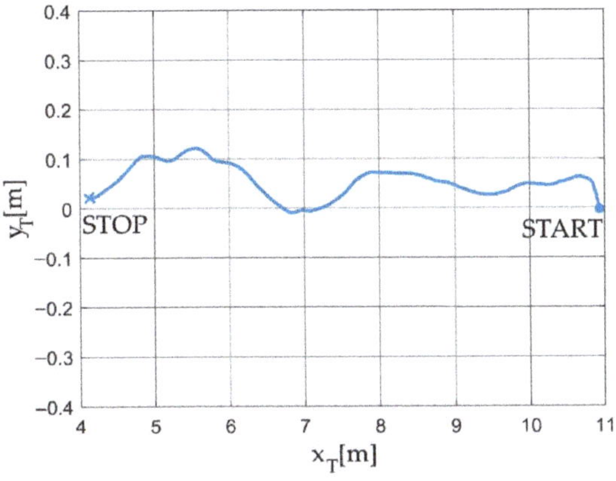

Figure 12. The path of the robot characteristic point related to the retroreflector in the transformed coordinate system.

Figure 12 shows that the 4-wheeled mobile robot with mecanum wheels changed its position in relation to x_T during the movement. Changes in the position along the y_T axis result from imprecise control and skidding of the mecanum wheels.

The example discussed here shows the possibility of using a laser tracker to determine the position of a mobile robot in relation to the adopted coordinate system. Moreover, the known time interval between measurement triggers allows for the determination of the remaining kinematic parameters of the robot. Using the finite difference method, the value of the velocity vector of the characteristic point of the robot with which the retroreflector was associated in the $x_T y_T$ system was determined. Its course is shown in Figure 13. The waveform shows that the robot movement started before the 2nd second. Velocity values remain negative because its projection on the axis is opposite to the axis direction. The observed changes in the velocity value during the movement are caused by the use of the joystick control mode.

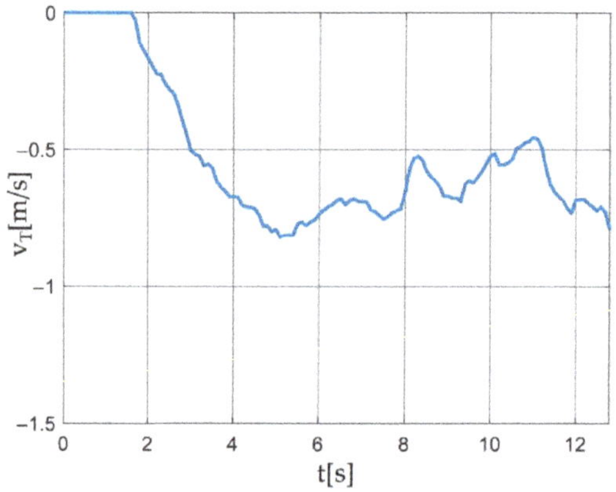

Figure 13. Robot characteristic point speed related to the retroreflector in the transformed coordinate system.

The UNITREE 4-legged GO1 robot, shown in Figure 14, is the second structure, the path of which was determined using a laser tracker.

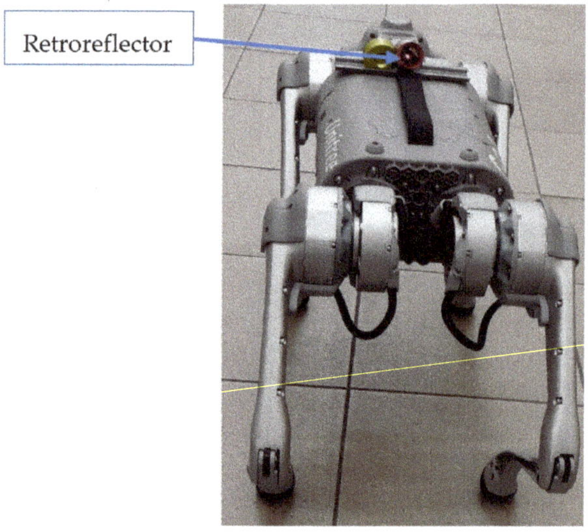

Figure 14. UNITREE 4-legged GO1 robot.

The robot weighs 12 kg, and its maximum speed is 4.7 m/s. The design is supported by the Rasberry pi 4 microcomputer and NVIDIA Nano chips, which are associated with five cameras. The construction includes a joystick that allows the user to control the robot and perform predefined movements or select modes, e.g., climbing/descending stairs. From a programming point of view, it is important that Unitree provides a ROS-based API. Like the 4-wheeled mobile robot, the GO1 robot moved along the corridor during the tests (Figure 15). However, in this case, an obstacle with a height of about 77 mm was added. Its presence is to test the ascending/descending stairs mode.

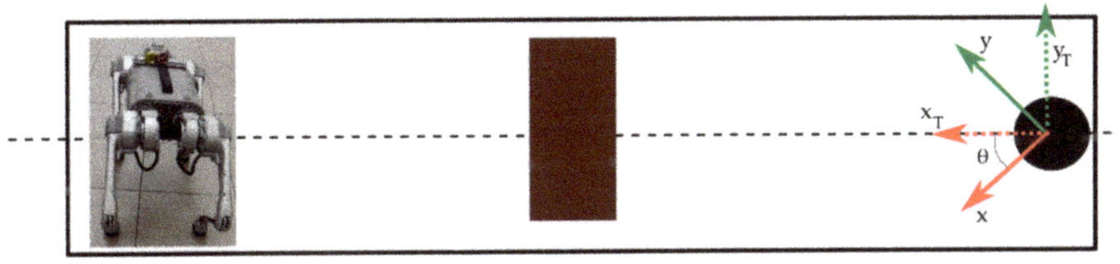

Figure 15. Scheme of a measurement experiment for the GO1 robot.

Proceeding analogously to the previous example, the path of the GO1 robot controlled by a joystick on the xy plane was determined, as shown in Figure 16a. Figure 16b shows the path in the transformed coordinate system. In this case, the angle $\theta \approx 41.5°$.

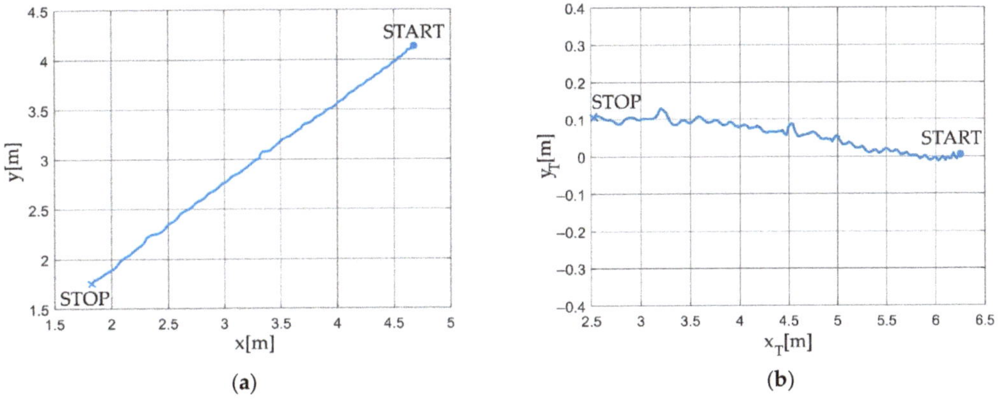

Figure 16. The path of the characteristic point of the GO1 robot related to the retroreflector in (**a**) the original coordinate system; (**b**) transformed coordinate system.

The fact that the GO1 robot overcame an obstacle placed in the test environment can be seen in the z-coordinate (in the system related to the tracker) of the GO1 robot characteristic point. Its course is shown in Figure 17a. Figure 17b shows the velocity of the characteristic point of the GO1 robot. The presented example shows the legitimacy of using the developed application and the laser tracker to record the trajectory of a walking robot.

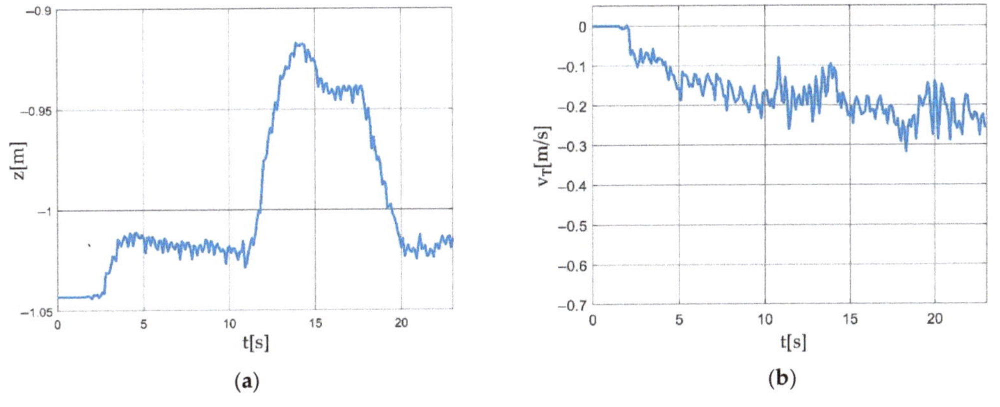

Figure 17. Graphs of (**a**) the course of the coordinate value on the OZ axis; (**b**) the velocity of the characteristic point of the GO1 robot with which the retroreflector is associated.

5. Discussion

The presented article is primarily of an engineering nature. The solution review performed was based on our own need for some software to share laser tracker data for robots. Different methods of performing a literature and solution review can be adopted. Here, the process was based on searching for applications useful in industrial and mobile robot research. According to the authors, the available commercial solutions and those described in the articles proved unsatisfactory and lacked versatility.

The adopted criteria the developed software was to comply with were based on the preference for using the Matlab package, ABB industrial robots, and Raspberry pi microcontroller-based mobile solutions. It can be assumed that, for example, in Scilab software, Kuka robots, and STM family microcontrollers, the criteria could be slightly different. Due to the need to develop a totally new application, the Python language and TCP/IP protocol were chosen. The choice made was based on the advantages of the solutions presented and experience in programming in Python and implementing projects using the TCP/IP protocol. Competing solutions based on, for example, the C# language and the EtherCAT standard are possible, but the advantages of the selected solutions proved decisive for the authors.

The developed and characterized software is in the development stage. It meets the adopted criteria and performs the main functions. However, it requires adding functions to expand its capabilities. There is a lack of visualization of measured positions, the ability to import CAD files, and the recalculation of positions with respect to a coordinate system other than the tracker has not yet been implemented. A particularly useful feature will be to recalculate the coordinates of measured points relative to the base system of an industrial robot. This function will appear in version 2.0 of the software and allow for the measurement and direct recording of points useful for programming robot paths or correcting workpiece positions. One of the novelties presented in the article is the selection and identification of tracker applications in robotics. Among the many possible applications of trackers, the authors have limited the examples to this field. In addition, for applications in robotics, the authors formulated specific criteria that the software that shares data from the tracker should comply with. Performance review, along with the advantages of programming languages and communication standards possible to use in such a project, can also be considered a novelty. The software itself is also new, with the characteristics of its construction and operation. After adding the above-mentioned necessary functions, it will be made available to Leica and all interested parties, along with the source code for its use and development. The authors hope that the developed software will be used in solutions where online robot-tracker laser communication from Leica is necessary.

6. Conclusions

The paper presents the concept of operation and methods of using laser trackers in robotics. The analysis shows that trackers can be successfully used to track the movement of mobile robots. They allow for precise measurements of traffic parameters. An interesting and intensively developed topic is the real-time correction of paths of manipulation robots. The conducted analysis indicates that the applications of trackers are described very widely; however, there is no characteristic of work on the development of software for sharing and exchanging data with trackers. A real-time data exchange tracker is a device that supports TCP/IP communication allowing for building advanced, very precise solutions in various fields. The shortcomings in the tools available on the market and our own demand resulted in developing our own software. The developed solution is based on the SDK provided by Leica and the Python language. The structure and functioning of the developed application have been described in detail. The software meets the goals set at the beginning of the design process regarding online communication with the tracker and the use of the universal popular TCP/IP standard, thanks to which it enables data transfer to most devices available on the market; especially industrial PLCs and robot controllers from various manufacturers. Python, a general-purpose, high-level programming language with an extensive library package, allows for adding new functionalities to the application. As part of further work, it is planned to expand the program with the automatic creation of coordinate systems and positions measured in relation to them. For applications in manipulative robotics, adding options related to determining the TCP of the robot with the use of a tracker, base system, and calibration of selected robots are planned. In the discussed application, it is planned to add functions related to the correction of TCP online position based on data from the tracker. The real-time correction will be performed based on the position of the robot controller and the position of the tracker. For the correction, it is also planned to use other data sent from the robot in the TCP/IP standard, namely parameters related to the contact force or the process being carried out.

Author Contributions: Conceptualization, D.S., P.O. and P.P.; methodology, K.K., M.M. and A.B.; software, P.P. and P.O.; validation, M.M. and D.S.; formal analysis, K.K. and A.B.; investigation, P.O.; resources, P.P.; data curation, P.O.; writing—original draft preparation, D.S.; writing—review and editing, M.M. and K.K.; visualization, P.P. and K.K.; supervision, A.B.; project administration, D.S.; funding acquisition, A.B. All authors have read and agreed to the published version of the manuscript.

Funding: This research received no external funding.

Data Availability Statement: Data is contained within the article.

Conflicts of Interest: The authors declare no conflict of interest.

References

1. Available online: https://ifr.org/post/the-rise-of-robots-in-machining (accessed on 15 October 2022).
2. Schröer, K.; Bernhardt, R.; Albright, S.; Wörn, H.; Kyle, S.; van Albada, D.; Smyth, J.; Meyer, R. Calibration applied to quality control in robot production. *Control. Eng. Pract.* **1995**, *3*, 575–580. [CrossRef]
3. Wang, Z.; Mastrogiacomo, L.; Franceschini, F.; Maropoulos, P. Experimental comparison of dynamic tracking performance of iGPS and laser tracker. *Int. J. Adv. Manuf. Technol.* **2011**, *56*, 205–213. [CrossRef]
4. Santolaria, J.; Conte, J.; Ginés, M. Laser tracker-based kinematic parameter calibration of industrial robots by improved CPA method and active retroreflector. *Int. J. Adv. Manuf. Technol.* **2013**, *66*, 2087–2106. [CrossRef]
5. Theissen, N.A.; Laspas, T.; Archenti, A. Closed-force-loop elastostatic calibration of serial articulated robots. *Robot. Comput. Integr. Manuf.* **2019**, *57*, 86–91. [CrossRef]
6. Cvitanic, T.; Nguyen, V.; Melkote, S.N. Pose optimization in robotic machining using static and dynamic stiffness models. *Robot. Comput. Integr. Manuf.* **2020**, *66*, 101992. [CrossRef]
7. Novák, P.; Stoszek, Š.; Vyskočil, J. Calibrating industrial robots with absolute position tracking system. In Proceedings of the 2020-25th IEEE International Conference on Emerging Technologies and Factory Automation (ETFA), Vienna, Austria, 8–11 September 2020. [CrossRef]
8. Mei, B.; Xie, F.; Liu, X.J.; Li, H. Calibration of a 6-DOF industrial robot considering the actual mechanical structures and CNC system. In Proceedings of the 2017 2nd International Conference on Robotics and Automation Engineering (ICRAE), Shanghai, China, 29–31 December 2017. [CrossRef]
9. Fernandez, S.R.; Olabi, A.; Gibaru, O. On-line Accurate 3D Positioning Solution for Robotic Large-Scale Assembly Using a Vision System and a 6Dof Tracking Unit. In Proceedings of the 2018 IEEE 3rd Advanced Information Technology, Electronic and Automation Control Conference (IAEAC), Chongqing, China, 12–14 October 2018. [CrossRef]
10. Sanden, S.; Pawlus, W.; Hovland, G. Real-time 6-DOF Vessel-to-Vessel Motion Compensation using laser tracker. In Proceedings of the OCEANS 2017-Aberdeen, Aberdeen, UK, 19–22 June 2017. [CrossRef]
11. Martin, C.; Fabritius, M.; Stoll, J.T.; Pott, A. A laser-based direct cable length measurement sensor for CDPRs. *Robotics* **2021**, *10*, 60. [CrossRef]
12. Theissen, N.A.; Mohammed, A.; Archenti, A. Articulated industrial robots: An approach to thermal compensation based on joint power consumption. *Laser Metrol. Mach. Perform.* **2019**, *13*, 81–90.
13. Nubiola, A.; Bonev, I.A. Absolute Calibration of an ABB IRB 1600 Robot Using a Laser Tracker. *Robot. Comput. -Integr. Manuf.* **2013**, *29*, 236–245. [CrossRef]
14. Ginani, L.S.; Motta, J.M.S.T. Theoretical and Practical Aspects of Robot Calibration with Experimental Verification. *J. Braz. Soc. Mech. Sci. Eng.* **2011**, *33*, 15–21. [CrossRef]
15. Sun, T.; Zhai, Y.; Song, Y.; Zhang, J. Kinematic Calibration of a 3-DoF Rotational Parallel Manipulator Using Laser Tracker. *Robot. Comput. -Integr. Manuf.* **2016**, *41*, 78–91. [CrossRef]
16. Lightcap, C.; Hamner, S.; Schmitz, T.; Banks, S. Improved Positioning Accuracy of the PA10-6CE Robot with Geometric and Flexibility Calibration. *IEEE Trans. Robot* **2008**, *24*, 452–456. [CrossRef]

17. Bai, Y.; Zhuang, H.; Roth, Z.S. Experiment Study of PUMA Robot Calibration Using a Laser Tracking System. In Proceedings of the 2003 IEEE International Workshop on Soft Computing in Industrial Applications, 2003. SMCia/03., Binghamton, NY, USA, 25 June 2003; pp. 139–144. [CrossRef]
18. Newman, W.S.; Birkhimer, C.E.; Horning, R.J.; Wilkey, A.T. Calibration of a Motoman P8 Robot Based on Laser Tracking. In Proceedings of the Proceedings 2000 ICRA. Millennium Conference. IEEE International Conference on Robotics and Automation. Symposia Proceedings (Cat. No.00CH37065), San Francisco, CA, USA, 24–28 April 2000; Volume 4, pp. 3597–3602. [CrossRef]
19. Wang, Z.; Maropoulos, P.G. Real-Time Laser Tracker Compensation of a 3-Axis Positioning System—Dynamic Accuracy Characterization. *Int. J. Adv. Manuf. Technol.* **2016**, *84*, 1413–1420. [CrossRef]
20. Droll, S. Real Time Path Correction of Industrial Robots with Direct End-Effector Feedback from a Laser Tracker. *SAE Int. J. Aerosp.* **2014**, *7*, 222–228. [CrossRef]
21. Khawli, T.A.; Anwar, M.; Sunda-Meya, A.; Islam, S. A Calibration Method for Laser Guided Robotic Manipulation for Industrial Automation. In Proceedings of the IECON 2018-44th Annual Conference of the IEEE Industrial Electronics Society, Washington, DC, USA, 21–23 October 2018; pp. 2489–2495. [CrossRef]
22. Gonzalez, M.; Hosseini, M.A.; Theissen, M.N.A.; Archenti, M.A. Quasi-Static Loaded Circular Testing of Serial Articulated Industrial Manipulators. In *ISR 2020; 52th International Symposium on Robotics*; VDE: Online, 2020; pp. 1–6.
23. Nguyen, V.; Melkote, S.N. Identification of Industrial Robot Frequency Response Function for Robotic Milling Using Operational Modal Analysis. *Procedia Manuf.* **2020**, *48*, 154–158. [CrossRef]
24. Recker, T.; Heilemann, F.; Raatz, A. Handling of Large and Heavy Objects Using a Single Mobile Manipulator in Combination with a Roller Board. *Procedia CIRP* **2021**, *97*, 21–26. [CrossRef]
25. Liu, J.; Chen, Z.; Gao, F. Kinematic Calibration of a Six-Legged Walking Machine Tool. *Chin. J. Mech. Eng.* **2022**, *35*, 1–17. [CrossRef]
26. Chen, Y.; Wang, D.; Zhong, H.; Zhu, Y.; Wang, C. Kinematics Analysis and Simulation of Mobile Robot Based on Linkage Suspension. In Proceedings of the 7th International Workshop on Advanced Computational Intelligence and Intelligent Informatics (IWACIII2021), Beijing, China, 31 October–3 November 2021; pp. 1–6.
27. Moeller, C.; Schmidt, H.C.; Koch, P.; Boehlmann, C.; Kothe, S.; Wollnack, J.; Hintze, W. Real Time Pose Control of an Industrial Robotic System for Machining of Large Scale Components in Aerospace Industry Using Laser Tracker System. *SAE Int. J. Aerosp.* **2017**, *10*, 100–108. [CrossRef]
28. Wang, Z.; Maropolous, P.G. Real-Time Error Compensation of a Three-Axis Machine Tool Using a Laser Tracker. *Int. J. Adv. Manuf. Technol.* **2013**, *69*, 919–933. [CrossRef]
29. Wang, Z.; Zhang, R.; Keogh, P. Real-Time Laser Tracker Compensation of Robotic Drilling and Machining. *Adv. Manuf. Mach. Process.* **2020**, *4*, 79. [CrossRef]
30. Zhou, Z.; Liu, W.; Wang, Y.; Yu, B.; Cheng, X.; Yue, Y.; Zhang, J. A Combined Calibration Method of a Mobile Robotic Measurement System for Large-Sized Components. *Measurement* **2022**, *189*, 110543. [CrossRef]
31. Sahu, R.K. A Review on Application of Laser Tracker in Precision Positioning Metrology of Particle Accelerators. *Precis. Eng.* **2021**, *71*, 232–249. [CrossRef]
32. Yang, D.; Zou, J. Precision Analysis of Flatness Measurement Using Laser Tracker. *Int. J. Precis. Eng. Manuf.* **2022**, *23*, 721–732. [CrossRef]
33. Leão, R.J.; Baldo, C.R.; Da Costa Reis, M.L.C.; Trabanco, J.L.A. Engineering survey planning for the alignment of a particle accelerator: Part II. Design of a reference network and measurement strategy. *Meas. Sci. Technol.* **2018**, *29*, 034007. [CrossRef]
34. Lu, S.; Liang, J.; Men, L.; Dong, L.; Wang, T.; Fan, B.; Wang, X.; He, Z.; Yan, L.; Han, Y.; et al. A laser tracker-based multilateration method for pre-alignment of High Energy Photon Source storage ring. *Nucl. Inst. Methods Phys. Res.* **2022**, 167533. [CrossRef]
35. Gaudreault, M.; Joubair, A.; Bonev, I. Self-calibration of an industrial robot using a novel affordable 3D measuring device. *Sensors* **2018**, *18*, 3380. [CrossRef] [PubMed]
36. Lu, X.; Jiang, T. Working pose measurement and quality evaluation of rotary drilling rig based on laser tracker. *Optik* **2019**, *187*, 311–317. [CrossRef]
37. Szybicki, D.; Obal, P.; Kurc, K.; Gierlak, P. Programming of Industrial Robots Using a Laser Tracker. *Sensors* **2022**, *22*, 6464. [CrossRef]
38. Burghardt, A.; Kurc, K.; Szybicki, D.; Muszyńska, M.; Nawrocki, J. Software for the robot-operated inspection station for engine guide vanes taking into consideration the geometric variability of parts. *Teh. Vjesn.* **2017**, *24* (Suppl. S2), 349–353. [CrossRef]
39. Szybicki, D.; Burghardt, A.; Kurc, K.; Pietruś, P. Calibration and verification of an original module measuring turbojet engine blades geometric parameters. *Arch. Mech. Eng.* **2019**, *66*, 97–109. [CrossRef]
40. Muszyńska, M.; Szybicki, D.; Gierlak, P.; Kurc, K.; Burghardt, A.; Uliasz, M. *Application of Virtual Reality in the Training of Operators and Servicing of Robotic Stations*; Camarinha-Matos, L., Afsarmanesh, H., Antonelli, D., Eds.; (Collaborative Networks and Digital Transformation. PRO-VE 2019. IFIP Advances in Information and Communication Technology, volume 568); Springer: Cham, Switzerland, 2019. [CrossRef]
41. Langlois, K.; Van der Hoeven, T.; Cianca, D.R.; Verstraten, T.; Bacek, T.; Convens, B.; Vanderborght, B. Ethercat tutorial: An introduction for real-time hardware communication on windows. *IEEE Robot. Autom. Mag.* **2018**, *25*, 22–122. [CrossRef]

42. Robert, J.; Georges, J.P.; Rondeau, É.; Divoux, T. Minimum cycle time analysis of ethernet-based real-time protocols. *Int. J. Comput. Commun. Control.* **2012**, *7*, 743–757. Available online: http://dx.doi.org/10.15837/ijccc.2012.4.1372 (accessed on 5 July 2012). [CrossRef]
43. Wu, X.; Xie, L.; Lim, F. Network delay analysis of EtherCAT and PROFINET IRT protocols. In Proceedings of the IECON 2014-40th Annual Conference of the IEEE Industrial Electronics Society, Dallas, TX, USA, 29 October–1 November 2014; pp. 2597–2603. [CrossRef]
44. Orfanus, D.; Indergaard, R.; Prytz, G.; Wien, T. EtherCAT-based platform for distributed control in high-performance industrial applications. In Proceedings of the 2013 IEEE 18th Conference on Emerging Technologies & Factory Automation (ETFA), Cagliari, Italy, 10–13 September 2013; pp. 1–8. [CrossRef]
45. Schlesinger, R.; Springer, A.; Sauter, T. New approach for improvements and comparison of high performance real-time ethernet networks. In Proceedings of the 2014 IEEE Emerging Technology and Factory Automation (ETFA), Barcelona, Spain, 16–19 September 2014; pp. 1–4. [CrossRef]
46. Available online: https://support.hexagonmi.com/s/article/Leica-Absolute-Tracker-AT930-AT960-User-Manual-v1-2-0en-1528374732805 (accessed on 15 October 2022).
47. Cass, S. The top programming languages: Our latest rankings put Python on top-again-[Careers]. *IEEE Spectr.* **2020**, *57*, 22. [CrossRef]

electronics

Article

TCP Parameters Monitoring of Robotic Stations

Andrzej Burghardt [1,*], Dariusz Szybicki [1], Piotr Gierlak [1], Krzysztof Kurc [1], Magdalena Muszyńska [1], Artur Ornat [2] and Marek Uliasz [2]

[1] Department of Applied Mechanics and Robotics, Faculty of Mechanical Engineering and Aeronautics, Rzeszow University of Technology, al. Powstancow Warszawy 8, 35-959 Rzeszow, Poland
[2] Pratt and Whitney Rzeszow S.A., 35-078 Rzeszow, Poland
* Correspondence: andrzejb@prz.edu.pl; Tel.: +48-(17)-865-18-14

Abstract: The impulse for writing the paper is the observation of the works related to the implementation of robotization of processes such as machining, glue application, welding and painting. The abovementioned processes, in addition to the correct implementation of the trajectory, require the definition of various parameters (e.g., speed) in the robot's software. In the trajectories where the reconfiguration of the robot arms is observed, there are significant errors in the implementation of the defined speed. Robotic technology suppliers, in the event of speed disturbances, manually increase the defined speed value or experimentally select other parameters. It is a cumbersome process, and the lack of information about the process parameters makes it time-consuming and inaccurate. In this paper, one representative process is selected, namely machining performed with various tools by ABB robots. In order for the robotic process to be controlled, it is necessary to compare the defined path with the speed profile. Then, the speed parameters can be controlled and corrected. The approach proposed in the paper allows for improving the quality of implemented robotic processes. It presents the available IT tools for station monitoring and how to use them. The advantages of the proposed solutions and their limitations are shown in the examples of implementation of robotic stations in the industry.

Keywords: signal analyzer; tune master; test signal viewer; robot monitoring

Citation: Burghardt, A.; Szybicki, D.; Gierlak, P.; Kurc, K.; Muszyńska, M.; Ornat, A.; Uliasz, M. TCP Parameters Monitoring of Robotic Stations. *Electronics* **2022**, *11*, 3415. https://doi.org/10.3390/electronics11203415

Academic Editors: Piotr Szymak, Stanisław Hożyń and Paweł Piskur

Received: 30 August 2022
Revised: 3415
Accepted: 18 October 2022
Published: 21 October 2022

1. Introduction

The motivation of the article is related to previous experiences in the robotization of processes such as machining, gluing, welding, and painting, where it is very important to ensure a defined speed. The currently encountered robots are characterized by the implementation of a path with a very high repeatability in the absence of ensuring a defined speed of movement. This also results from the provisions of the PN-ISO 9283 standard. The occurrence of significant reconfiguration of the robot arm causes large errors in the implementation of the defined speed. Robotic station integrators in the event of speed disturbances increase the defined speed value based on experience. The process of selecting these parameters is time-consuming and requires experience and dedicated tools.

In order to improve the control of a defined robotic process, it is necessary to control the defined path and observe the given speed profile. Then, we can control and correct the speed parameters. All parameters for evaluating the robot path for ABB robots are available in RobotStudio. Other parameters, such as speed, acceleration, contact forces, energy consumption, etc., can be recorded using additional software. The Signal Analyzer option embedded in RobotStudio enables the registration of the robot's kinematic parameters, displacements, velocities and accelerations, both joint parameters and parameters of the TCP (in accordance with the ISO 10218-1: 2011 tool center point—a point defined for a given application, taking into account the coordinate system of the mechanical interface), point motion in various reference systems, and recording of energy consumption, while all the abovementioned data can be saved over a wide time horizon at the station and with a remote method using network communication.

Other external programs for monitoring the operation of robotic stations are Test Signal Viewer, which is designed to measure mainly signals related to the operation of the force control add-on; it gives the option of measuring the force vector in different directions of a defined reference system.

The latest of these is the external Tune Master program, which, in addition to a friendly interface, allows for recording the parameters listed in the previous programs, and additionally allows for recording the parameters of regulators and damping, both for the robot's drive units and for added external systems.

The tools listed above are implemented depending on the problem. The study considered the use of the Tune Master add-on for the selection of the process of robotic machining of the ADT gear body, and the Test Signal Viewer program for monitoring the forces in the process of machining the V2500 engine diffuser.

The Signal Analyzer software was used to control the accelerations affecting the Keyence laser scanner in the processes of controlling the execution of rays and deburring the sharp edges of the aircraft gearbox housing, and in measuring the path correction points during robotic processing of cast iron castings.

The approach proposed in the article enables improving the quality of the robotic processes carried out. It presents the available IT tools for monitoring the parameters of robotic stations and how to use them. On the examples of the implementation of robotic stations in the industry, it shows the advantages of the proposed solutions and their limitation.

2. State of Knowledge

Almost every leading manufacturer of robots on the market offers tools that allow for recording the working parameters of robots. For example, Kuka offers the KUKA.ProcessScreen software for the visualization, monitoring, and continuous documentation of various types of robotic production processes in real time. It enables data management and documentation, and includes simple tools for analysis. Another tool offered by KUKA AG is the KUKA Connect analytical platform. It allows access to data using a web browser (Google Chrome, Safari, Microsoft Edge) or a mobile application. In this way, databases on the operation of robotic systems can be built and accessed in real time. KUKA iiQoT is the software based on the cloud, in which, thanks to the use of IIoT (Industrial Internet of Things), it is possible to view all interesting information in real time, for both one and many robots.

Another robot manufacturer, Fanuc, offers universal MT-LINK software for monitoring the parameters of not only industrial robots, but also CNC machine tools and other mechanical units.

There are a number of programs that enable the collection and analysis of data from machines and devices in the industry, including robots. These are, for example, SCADA Asix, DASYLab, or the aforementioned MT-LINKs. It is worth paying attention to solutions dedicated only to robots, such as the IXON application, supporting such robot brands as ABB, Aubo, Automata (Eva robot), Comau, Doosan, Epson, FANUC, Hanwha, Kawasaki, KUKA, Mitsubishi, Nachi, Omron, OTC Daihen, Panasonic, Sawyer, Stäubli, Universal Robots (UR+), Yaskawa, and KEBA.

This paper describes the possibilities and the way of using the ABB software for recording parameters of robotic processes.

The most readily available and frequently used data acquisition software is an add-on that is an integral part of RobotStudio called Signal Analyzer.

Its use is very wide, e.g., in article [1] a method of optimizing the energy consumption of robotic stations is presented. The problem is formulated as reducing energy consumption by finding the optimal execution time and sequence of the robot's movements. The result shows a 12% reduction in energy consumption without increasing the cycle time of the stations. A similar problem was discussed in [2], where the change of the acceleration ramp values and the speed of movement of the TCP point were used while maintaining a constant time. An alternative method of minimizing energy consumption can be found in [3], which proposes an algorithm that creates a smooth, collision-free path, taking into account energy consumption. The method of planning the probabilistic road map (PRM) path and the optimization algorithm based on the Dijkstra algorithm were used for this.

Another example is the planning of paths and modeling of robots used in foundries [4] or the optimization of their movement in industrial processes [5]. Other industrial applications include monitoring of laser cutting stations [6].

New information technologies such as virtual reality are used for training [7]; the possibility of virtual generation of complex traffic paths [8] requires knowledge of minimal digital recording of the movement of robotic stations. Obtaining such data is possible thanks to the aforementioned Signal Analyzer program, and this type of application is included in [9].

Together with the development of the addition of force control in ABB robots, the Test Signal Viewer software was launched on the market, dedicated as a solution for monitoring, controlling, and recording the signals of controllers with the IRC5 system [10,11]. The software was mainly designed to record the force values from sensors manufactured by ATI Industrial Automation. Therefore, the main applications of this software are related to processes where forces must be controlled. The main applications are force monitoring in the processes of machining [12] or polishing [13–15]. Robotic joining technologies in which forces play an important role also make extensive use of the capabilities of the Test Signal Viewer software. This applies, for example, to resistance welding (RSW) [16], or the recently very popular friction welding (FSW) [17,18]. It is also worth mentioning that the speed of data recorded by the software allows for vibration and noise analysis [19].

Other examples include assembly operations [20] or manufacturing technologies using robotic plastic processing, i.e., roboforming [21]. The use of additional force control enables the robot to be programmed by guiding the effector by hand [12].

A very original solution for the implementation of a device for energy recovery and storage on a working robot can be found in [22], in which the discussed software was used for analyses.

A newer tool for signal analysis is the Tune Master software consisting of four modules. The first one, called Robot, is designed for the analysis and selection of the dynamic coefficient responsible for damping the oscillations, which significantly improves the implementation of the planned path. The second one, called Additional Axis, has the functionality to display and modify configuration parameters for each mechanical unit in the system. For external mechanical units, it is possible to estimate the friction coefficient to identify parameters in the statistical model using the least squares method. We have the possibility to analyze quantities such as the total moment of inertia, total Coulomb friction, total viscous stick for the motor, gear and arm, and the angle between the center of rotation and the center of gravity. In addition, we have the ability to define the parameters of the PID controller, and define the values of deceleration and acceleration ramps.

The last two modules are Servo Gun and Log Signals, which are used to work with signals generated by mechanical units. Thanks to their application, it is possible to analyze and control the condition of a given machine. They allow for creating reports on the process and showing the influence of external factors on the process.

In the following subsections, the paper presents solutions using Log Signals to monitor and record quantities such as velocity and linear and angular acceleration of robot elements, values of forces, and moments, both driving and resulting from contact, e.g., a tool workpiece. Unfortunately, only one article [23] concerning the accuracy and noise immunity of the IMU sensor installed on a robot was found in the literature on the Tune Master Log Signals software application available to the authors.

Summing up, each of the robot manufacturers offers tools for analyzing and recording signals related to the robot's operation. In the case of their insufficient functionality, universal commercial tools can be used. In our work so far, during the implementation of robotic stations equipped with ABB robots, three dedicated tools have been used, and they are presented in the following subsections.

3. Application of Diagnostic Software in Industrial Examples

During many years of work with applications implemented with the use of ABB robots with the IRC5 controller, in practice many tasks would be very time-consuming or impossible to implement. Processes related to force control, minimization of energy consumption, and machining require at a certain stage of implementation information about the process parameters and the possibility of their monitoring and registration. This allows for understanding the phenomena, drawing conclusions and correctly implementing the defined task. Below are presented examples of using RobotStudio Signal Analyzer, Test Signal Viewer, and Tune Masters for the implementation of robotic industrial solutions.

3.1. *RobotStudio Signal Analyzer*

The production of parts or components of aircraft engines from the point of view of the plant's order portfolio is characterized by stability over a wide time horizon. Long-term orders for a strictly defined number of parts, on the one hand, allow for long-term production planning and provide a certain financial stability; on the other hand, they do not allow for product development or increasing the number of manufactured components in order to increase profits. The only possibility of increasing the profits from production is to carry out the production process with the least amount of human work and the lowest energy consumption. It is an impulse for optimization efforts that directly translates into the economic result of the enterprise. One of the elements of cost reduction is the optimization of electricity consumption, which we implement for robotic stations using the RobotStudio software add-on called Signal Analyzer.

The process of minimizing energy consumption is presented with the example of a real robotic station used to control the quality of outlet guide vane (OGV) castings. As part of the work to minimize energy consumption, the process parameters were recorded and the changes in the speed of the TCP point and acceleration changes were searched for, so as to minimize energy consumption while maintaining a constant process time. The robotic station this paper concerns is the equipment of the Research and Development Laboratory for Aerospace Materials (Figure 1).

The selection of process parameters allowing for the reduction of total energy consumption by the robotic system was based on the dependence on electricity. The value of the torque produced by the robot's drives is related to the description of the dynamics of the object and depends on acceleration and speed as well as resistance to motion. Summing up, changes in speed and acceleration of movement of the robot's TCP point affect the drive torque in the robot's joints and thus the drive currents, which translates into energy demand [2].

Because the entire program for controlling the thickness of the OGV airfoils of the robotic station includes over 3000 movement instructions, and the execution time is approx. 15 min, the method of selecting accelerations and speeds is illustrated by a simplified real example. Let us assume that the robot's TCP is moved from P1 to P2 (these are the TCP robot trajectory points). Figure 2a shows the point motion parameters such as speed and acceleration, power consumed by the system, and total energy. By changing the motion parameters of the point, namely increasing the speed and decreasing the acceleration value, we can reduce the energy consumption of the system (Figure 2b).

It should be noted that the time needed to travel from P1 to P2 is constant because we assume that the working cycle time of the robotic station is not extended, which is very important in our considerations.

The change of the robot's TCP point motion parameters, its monitoring and registration were carried out using RobotStudio software and the Signal Analyzer add-on. It is a tool that allows for recording a series of defined values at a speed of 40 times per second. The values of recorded parameters can be observed online in the scaled graphs or saved as text files or *.xls files. An example of the application window is presented in Figure 3.

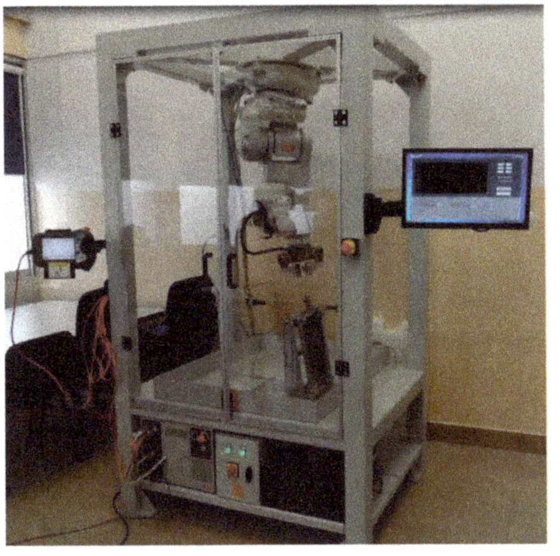

Figure 1. Robotic station for controlling the thickness of OGV airfoils.

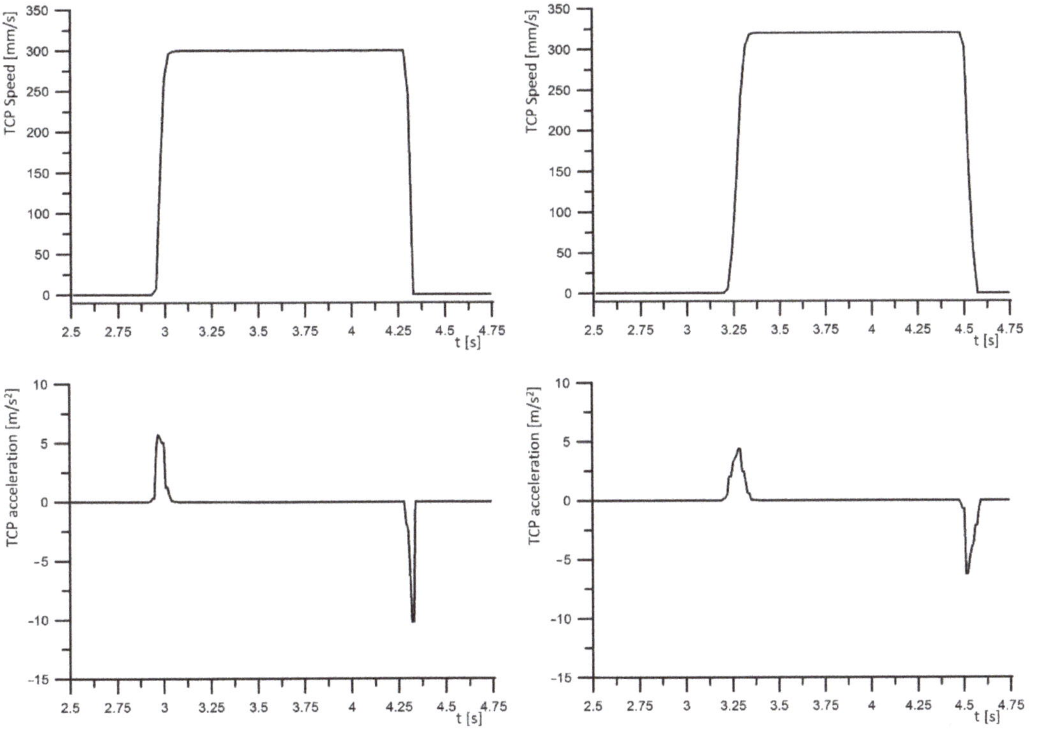

Figure 2. *Cont.*

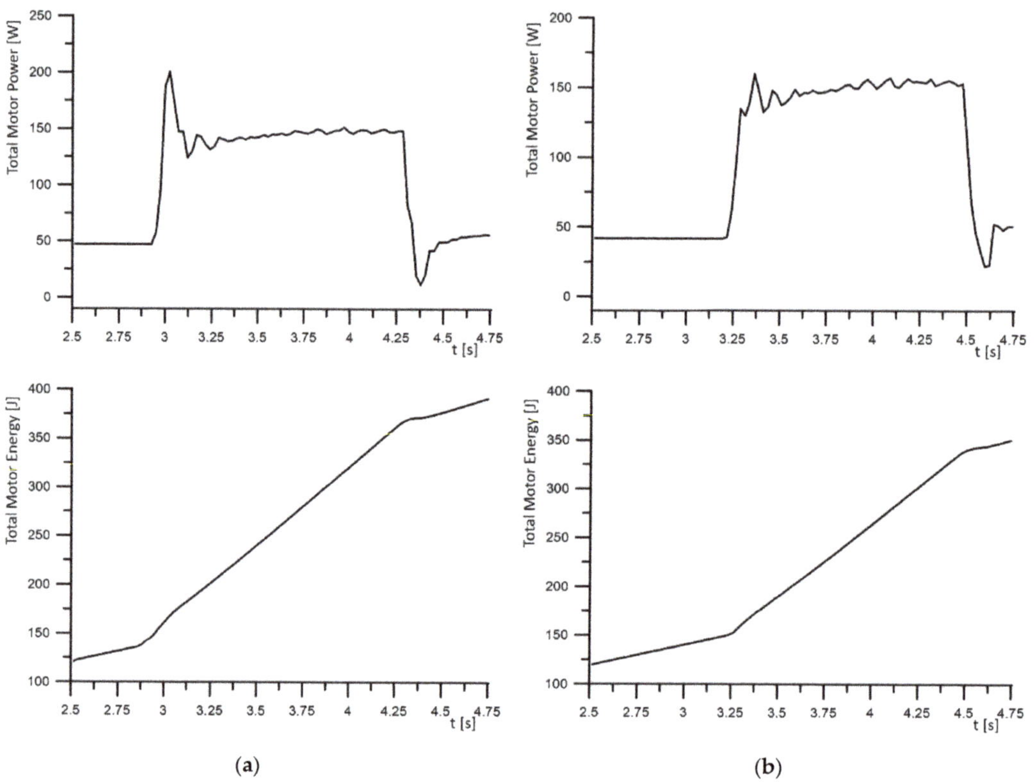

Figure 2. Parameters of the movement of the TCP point and work parameters of the robotic station: (**a**) speed, acceleration, power consumed by the system, and total energy; (**b**) changing the motion parameters of the point, increasing the speed, and decreasing the acceleration value reduce the energy.

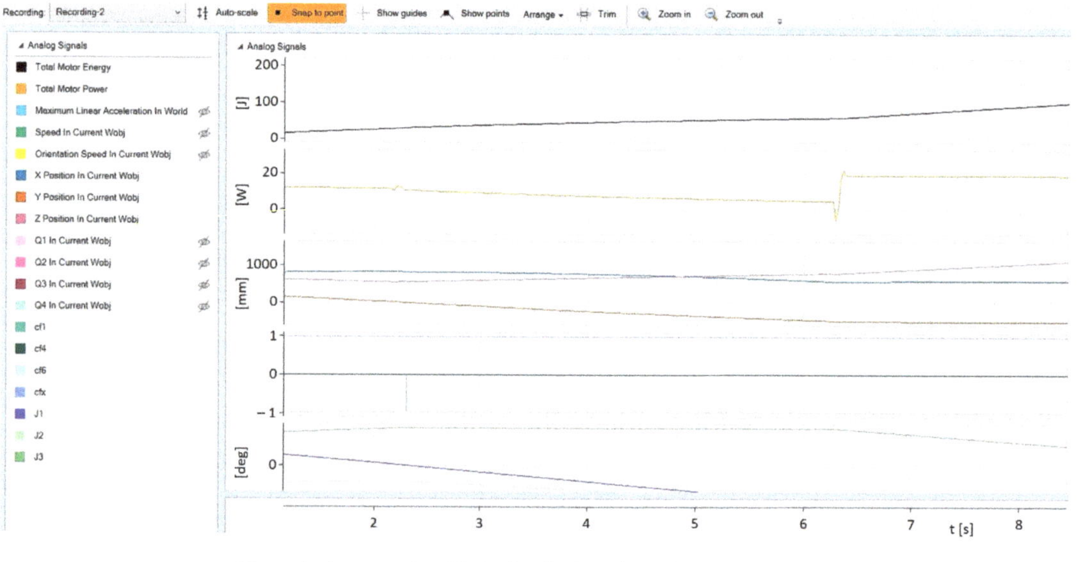

Figure 3. An example of RobotStudio Signal Analyzer application window.

Defining accelerations and decelerations in the RAPID language is possible by defining the percentage value of acceleration and the percentage value of the increase of acceleration. The method of changing the accelerations set by the AccSet function is presented in Figure 4. Figure 4a shows nominal acceleration values, at a maximum for a given station structure. Percentage change of their values is possible, e.g., by using the AccSet function with parameters. The first parameter corresponds to the limitation of the nominal acceleration value as a percentage. In Figure 4b, for example, the acceleration value is reduced to 30% of the maximum acceleration value. The second parameter corresponds to the value of the speed of acceleration increase. In the example in Figure 4c, the acceleration gain is limited to 30% of the standard value.

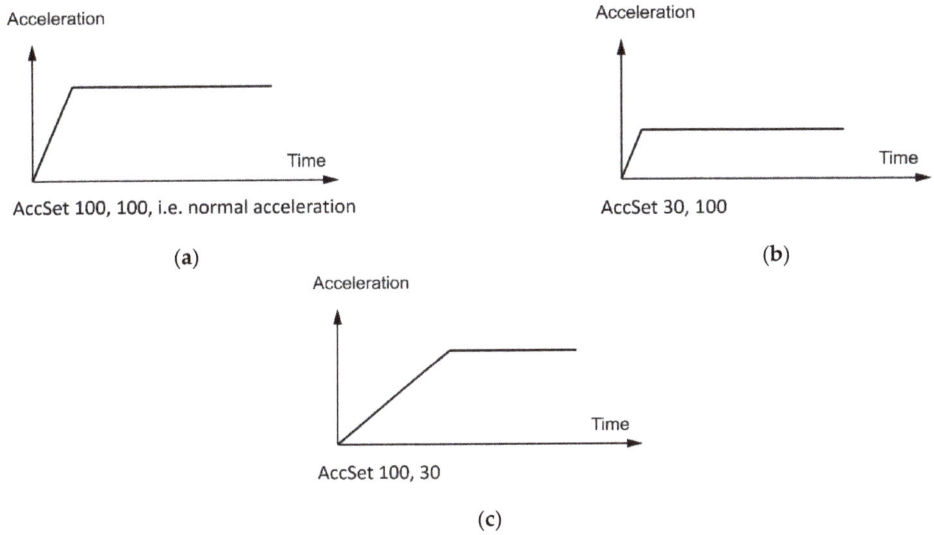

Figure 4. Methods of defining accelerations: (**a**) nominal acceleration values; (**b**) the acceleration value is reduced to 30% of the maximum acceleration value; (**c**) the acceleration gain is limited to 30% of the standard value.

In the work on reducing energy consumption in programmed movement paths, the speed was increased by software, and the acceleration was limited to a constant time.

In the solution proposed in article [2], the choice of parameters was made empirically. The invariable limitation adopted was the path execution time. Acceleration was minimized, and speed was increased until the process was no longer feasible. The most common limitation here was the minimum acceleration value of the robot.

Figure 2 shows a simple example of how it influenced the energy consumption. The same was done for all traffic stages in the real station. The conducted work allowed for reducing the value of consumed energy by 15% while maintaining the process time below 15 min. The proposed method of signal analysis and parameter changes allowed for a 15% reduction in energy consumption by a real robotic station while maintaining the same process execution time. It is worth paying attention to the economic aspect of the cited work, which over a wide time horizon brings significant financial benefits.

3.2. Test Signal Viewer

The Test Signal Viewer tool allows for selecting the parameters of force control systems. The use of the Force Control (FC) package allows controlling the force exerted by the tool on the workpiece. The use of such a solution enables the robot to interact with the environment by controlling the process using the strategy of FC Pressure or FC SpeedChange. The FC Pressure option enables polishing or grinding of elements with variable geometry (e.g., castings) while maintaining a constant pressure of the tool on the surface of the workpiece at a constant defined process speed. The idea of this solution is presented in Figure 5a. The use of this function allows the material to be processed at a constant speed and control of the force perpendicular to the processed surface. The tool travel path follows the curvature of the surface.

The FC SpeedChange function (Figure 5b) allows the casting to be machined or the machining allowances to be removed at a speed depending on the existing drag forces. The speed of movement of the robot's TCP point is reduced by the control system when the forces occurring during machining exceed the assumed values. This avoids damage to the workpiece or tool from excessive stress and heat.

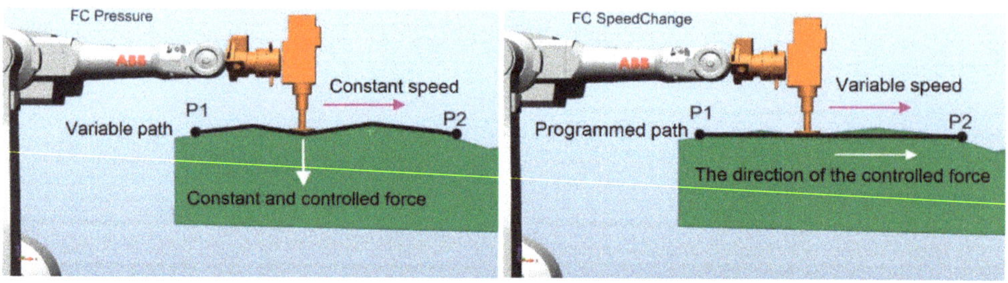

(a) (b)

Figure 5. (**a**) Principle of operation of the FC Pressure function; (**b**) principle of operation of the FC SpeedChange function.

The use of the FCPressure add-on requires the determination of a number of parameters, such as Threshold %, Force Change Rate N/s, Damping-D %, the maximum waiting time for contact (Timeout s), the percentage of the set force with which the robot starts the process (Zero Contact Force %), and the level of filtration for the low-pass filter (Noise level Hz). Regarding parameters, the rate of speed change and the percentage of the set force with which the robot starts the process are correlated with the shape of the trajectory and are selected on the basis of research practice to date. The parameters of the coefficient of force growth and the damping were selected on the basis of the experiment, so that the beginning of the process was characterized by the lack of a cavity in the performed deburring and the resulting treated surface was smooth. The selection of these parameters is related to the type of material being processed as well as the type of tool used and the cutting parameters. Adopting the parameters of the force control system allows for starting testing work aimed at determining the process parameters such as the speed of the TCP point and the value of the contact force between the tool (cutter) and the workpiece. The software tool that allows for monitoring and analyzing the values of sensor forces in different reference systems at different speeds is the Test Signal Viewer software. This software was tested on the machining of parts having a randomly changing shape (the solidification process of the detail introduces a random change of geometry to some extent). This treatment is related to the processing and determination of suboptimal values of the damping factor and the force increase factor. In the research work, the forces and speed of movement of the TCP point were searched for, ensuring the execution of edge deburring within the limits defined by the technological requirements imposed by the PWA 360 standard by Pratt and Whitney. For the tests, a V2500 engine compressor casting made of Inconell 718 was used as a demonstrator. One of the casting elements was selected, namely the so-called triangle-shaped nabe with rounded tips, and the tool used was a multi-flute conical cutter. Photos of the research work are presented in Figure 6.

(**a**) (**b**)

Figure 6. Photos from the research work: (**a**) diffuser view; (**b**) debured naba.

Figure 7a shows the path followed by the robot's TCP point and the system of the adopted reference system in the force control process (Figure 7b).

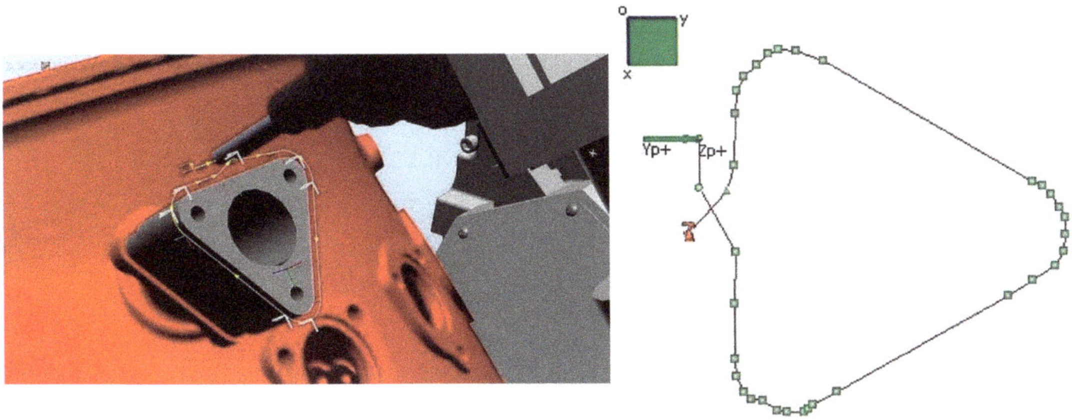

(**a**) (**b**)

Figure 7. (**a**) View of a trajectory built using RobotStudio; (**b**) a trajectory built using a centered touch panel.

The problem of loss of force control was encountered during the research work (Figure 8a). The manufacturer states that the tool works from the value of the controlled force at the level of 1 N in the Z axis.

For the workpiece, when executing the trajectory along the path (force control in the Z axis of the tool) at the last third turn (Figure 8a), the force control stops working properly. The disturbance of the force control package occurred at the set force of 2 N, and the gradual increase of the feed force minimizes this effect [24]. Figure 8a shows the effect of the decrease in the value of the forces between the tool and the workpiece while moving along a triangular path and Figure 8b shows what the correct trajectory should look like.

It was noticed that the correctness of the performed machining depends on the shape of the surface, and above all, the speed of the TCP point and the contact force. For the defined motion path and the set speed, it was necessary to define the minimum parameters of using the FC Pressure software feature. As a result of the research work, internal documentation was created, specifying for which tools and materials what minimum debris values can be obtained with the use of a robot with the force control software. The entire document was developed as a result of testing and monitoring parameters using the Test Signal Viewer software.

The paper [14] includes optimization of parameters in terms of the width of the deburring performed and the quality of the surface. As an indicator of quality, the sum of squares of lapses relative to the nominal value was proposed, which informs how much error was made in the execution of the deburring. To assess the rate of the shape change (surface quality), an indicator was introduced in the form of the maximum value of the derivative of the function describing the shape change.

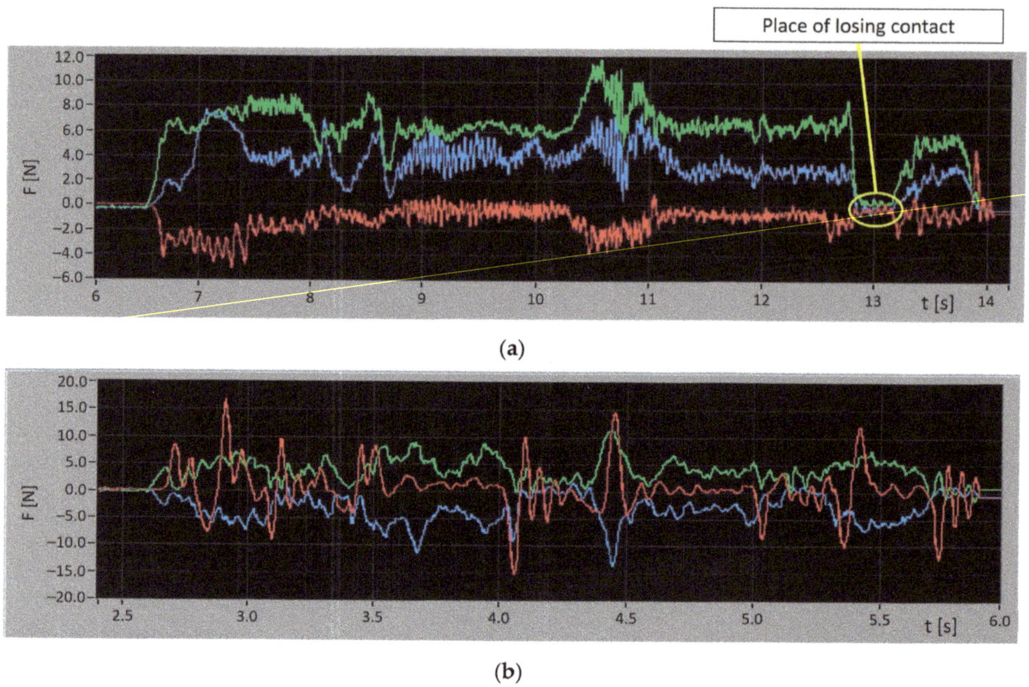

Figure 8. (**a**) Force registration in case of loss of contact; (**b**) force registration without losing contact.

For this purpose, the deviation from the nominal value was discretized with a set of points. The obtained points were approximated by a third-degree spline function using Matlab/Simulink software. The generated function was then used to determine the proposed quality indicators. The sum of squares of the lapses and the maximum absolute value of the derivative of the error function were used as quality indicators. Juxtaposition of the obtained indicator values allowed for the selection of suboptimal parameters. The resulting parameters in the form of the value of the tool–workpiece contact force and the speed of the TCP at a given nominal value of tool rotation were adopted as the best solution from the available set of solutions. A detailed description of the conducted research is included in [14].

The second problem defined above concerns the selection of the damping and the speed of the force increase. It is related to the quality of the machined surface. As a result of the research work, the indicated parameters were determined for the analyzed diffuser machining process. The experiment consisted in driving along one of the rectangular edges of the sample with a given contact force and recording the force value with variable parameters of the force control system. The course of the experiment is shown in the diagram in Figure 9.

In the first step, the influence of the damping coefficient D on the stabilization of the set force for different values of forces was investigated [25]. In the ABB documentation, the damping coefficient D is a definition of how large a contact force is required for the robot to move at a certain speed. The signals from the force sensor were recorded using the ABB Signal Test Viewer software, additionally exporting the recorded measurements to Matlab software in order to determine RMS errors. An example of a measurement chart is shown in Figure 10.

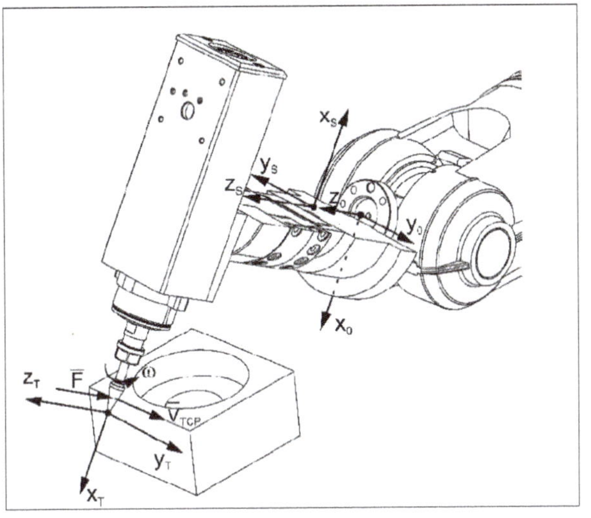

Figure 9. Course of the experiment.

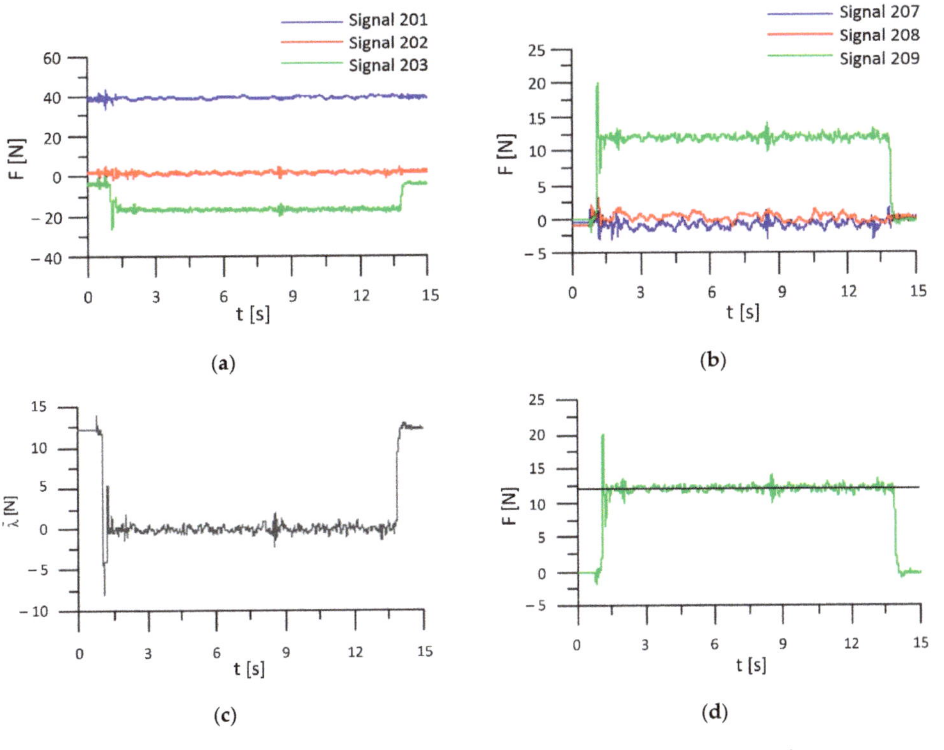

Figure 10. Signals corresponding to the values of forces in various reference systems recorded with the use of Signal Test Viewer software, at the set force 12 N and the damping coefficient D = 100%: (**a**) the signals in the force sensor reference system x_S, y_S, z_S; (**b**) signals in the tool coordinate system x_T, y_T, z_T; (**c**) force error; (**d**) set and actual force.

Figure 10 shows that the signal 209 (related to the direction of the Z axis) along the entire trajectory oscillated around a force command value of 12 N. This value is close to the reference value in this example test. It should be noted that in the initial phase of the robot's contact with the environment, there was a noticeable overshoot of signal 209 to a value close to 17.5 N. The force error diagram shows that throughout the entire measurement, when the manipulator tool was in contact with the workpiece, the force error diagram had quite large oscillations (error values of 2 N and the RMSE value of 1.873567). The second part of the experiment consisted in making and analyzing the measurements with regard to the value of the speed of force changes. A number of tests were performed. An example of a test is presented in Figure 11.

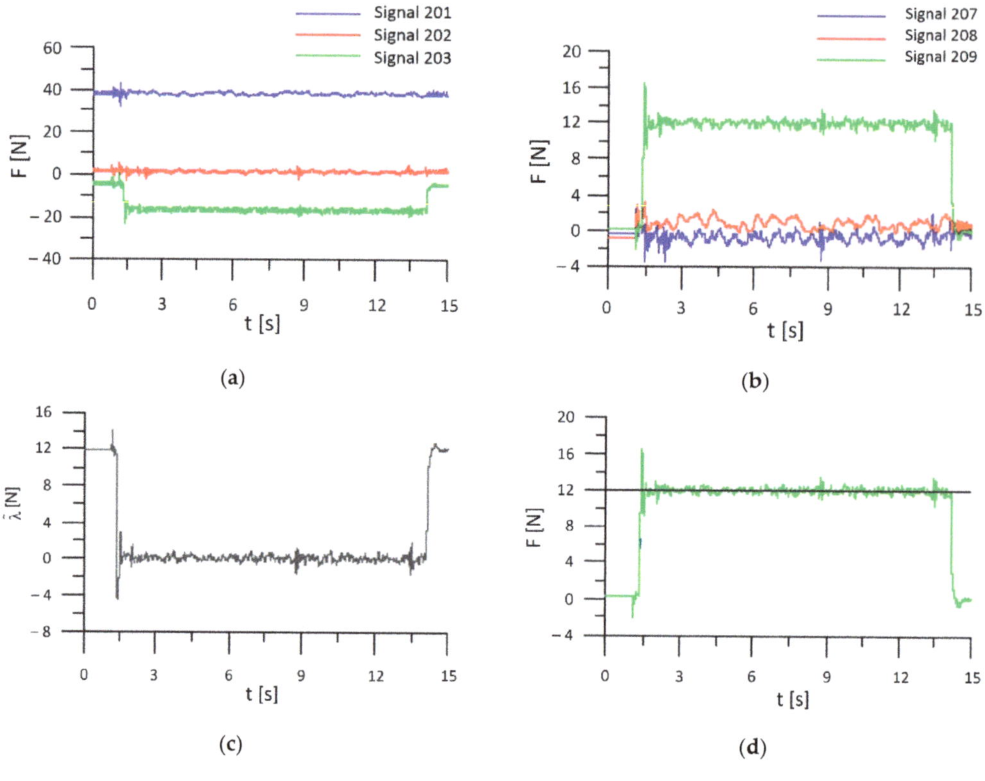

Figure 11. Signals corresponding to the values of forces in various reference systems recorded with the use of Signal Test Viewer software, at the set force 12 N and the rate of force increase FC = 75 N/s: (**a**) the signals in the force sensor reference system x_S, y_S, z_S; (**b**) signals in the tool coordinate system x_T, y_T, z_T; (**c**) force error; (**d**) set and actual force.

Figure 11 shows the lack of overshoot in the course of the real force (signal 209). However, this comes at the cost of a higher RMSE mean square error and standard fluctuation. The authors conducted research for selected values of the D coefficient of 150%, 100%, 75%, and 50%. Research was conducted for these coefficients. The results for the coefficient D = 100% are shown in Figures 10 and 11. The selected value of D = 100% was selected experimentally, taking into account the force stabilization and the speed of force change. The appropriate selection of the parameter is an important element in programming, if the manipulator is to perform cutting operations with high accuracy and in a stable manner (this corresponds to the quality of the machined surface) [25–27]. To sum up, the use of the Test Signal Viewer software allows for better use of the Force Control software feature and extends the possibilities of its implementation.

The method of determining the best parameters of damping and rate of force buildup was carried out on the basis of an experimental study, which consisted of making a pass along one edge of a rectangular sample. The value of the tool–sample contact force was assumed, and at different values of damping and rate of force buildup, the correctness of realization of the assumed force was measured.

Thus, the effect of the damping factor D and the force increase factor FC on stabilizing the set clamping force of F = 12 N was tested. Signals from the force sensor were recorded using the ABB Signal Test Viewer application, dedicated to retrieving signal values from ABB's robot controller. Figures 10 and 11 show selected results of the experiment implementation. Figures 10c and 11c show the course of the force value deviation.

Two indicators of the quality of force control implementation were used for the analysis: the root-mean-square error (RMSE) and the standard deviation according to the relationships:

$$\varepsilon_\lambda = \sqrt{\frac{1}{n}\sum_{k=1}^{n}\overline{\lambda}_k^2} \tag{1}$$

$$\sigma = \sqrt{\frac{1}{n}\sum_{k=1}^{n}\left(\overline{\lambda}_k - \overline{\overline{\lambda}}\right)^2} \tag{2}$$

where: ε_λ—root-mean-square error; σ—standard deviation; n—number of samples; k—sample no.; $\overline{\lambda}_k$—contact force error in the k-th measurement; $\overline{\overline{\lambda}}$—arithmetic mean of force value.

In the first part of the experiment, seven tests were performed by varying the damping parameter in the range of 25 ÷ 175%, with a constant force increase factor FC = 50 N/s.

The results of the analysis are presented in the form of a plot in Figures 12 and 13.

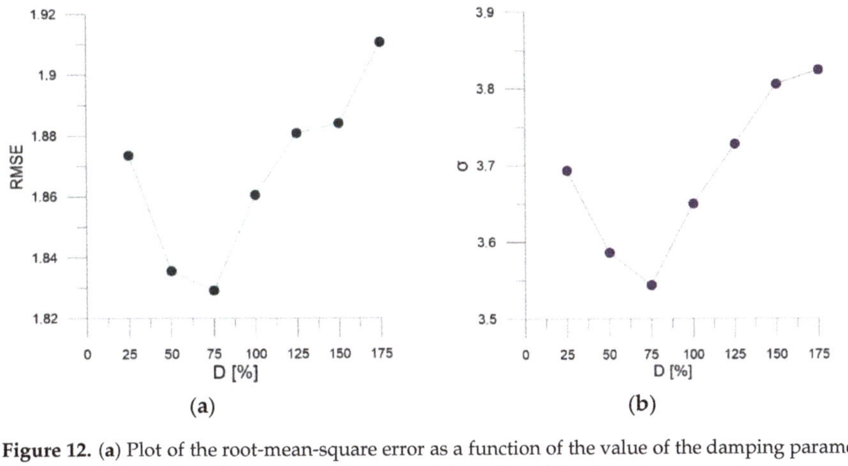

(a) (b)

Figure 12. (**a**) Plot of the root-mean-square error as a function of the value of the damping parameter; (**b**) plot of the standard deviation as a function of the value of the damping parameter.

Figure 12a presents a plot of the root-mean-square error as a function of the value of the damping parameter, while Figure 12b shows a plot of the standard deviation as a function of the value of the damping parameter.

In the second part of the experiment, tests were carried out for eight values of the force increase factor FC in the range of 10 ÷ 150 N/s, with a constant damping factor of D = 100%.

Figure 13a shows a plot of the root-mean-square error as a function of the value of the force increase parameter, and Figure 13b shows a plot of the standard deviation as a function of the value of the force increase parameter.

The smallest values of the root-mean-square error and standard deviation ratios were observed for a damping ratio of D = 75% and a force increase ratio of FC = 30 N/s.

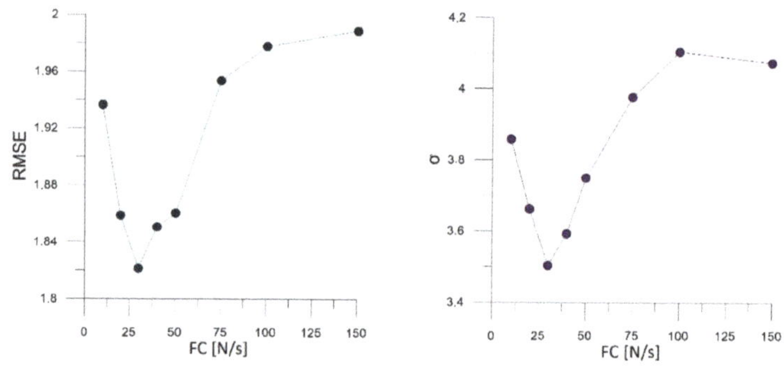

Figure 13. (**a**) A plot of the root-mean-square error as a function of the value of the force increase parameter; (**b**) a plot of the effect of changes in the force increase parameter on the standard deviation.

3.3. *Tune Master*

Robotization of technological processes related to machining has been carried out in the industry for many years. Robots are used primarily for the operation of CNC machines and the processing of so-called soft materials such as plastics, wood, and aluminum alloys. With the increase in the repeatability and accuracy of robots, the spectrum of their applications is growing. Robots can find application in the automation of manual processes such as deburring, and removal of mold lines on cast elements [28–32]. Due to the low rigidity compared to your CNC, robots will not replace them in the so-called heavy and highly accurate machining. Robots are more and more commonly used where high accuracy is not so important, e.g., in deburring cast iron castings.

As part of the work carried out, a robotic station was developed to machine castings of gearbox housings made as gray cast iron castings made in disposable sand molds. A view of the station model is shown in Figure 14. The station components are described in Figure 15. The idea of the station's operation is based on the use of the IRB1600 robot with a 2D laser scanner, which allows the measurement of fluctuations from the reference model. Then, the position offset of the cutting tool is calculated. The measured workpiece is transported with a five-axis IRBP B250 positioner to the machining zone, the so-called dirty zone. There, the IRB 4600 robot equipped with an electrospindle and tool storage performs the machining of workpieces.

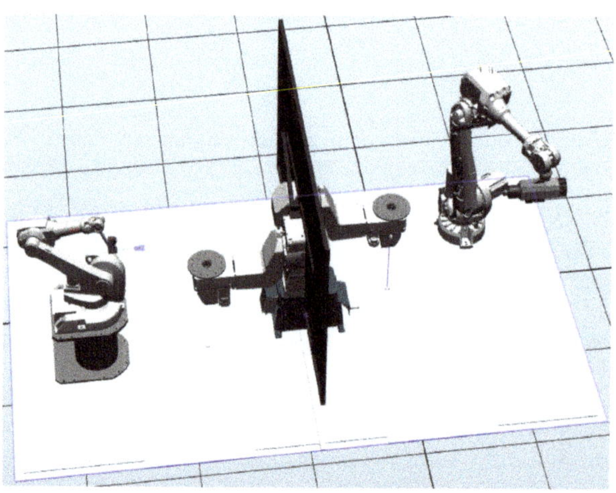

Figure 14. Station model.

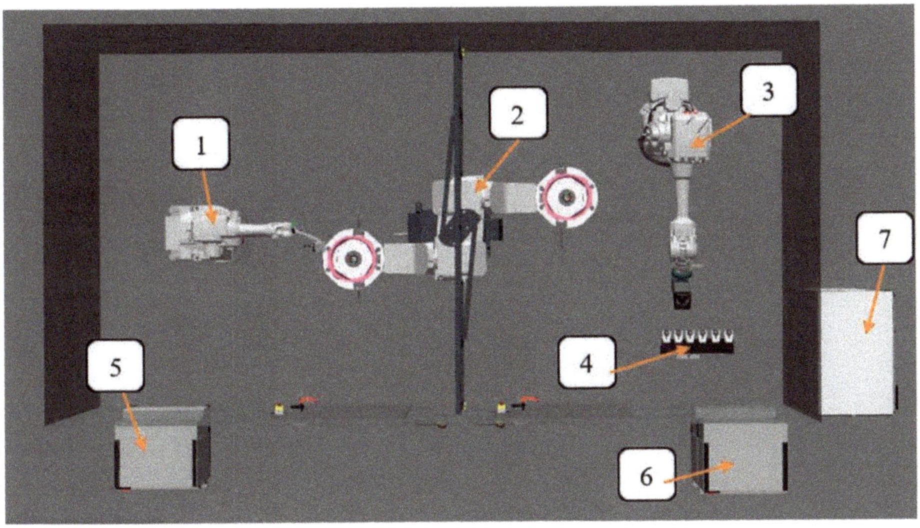

Figure 15. Station diagram: (1) IRB 1600 robot with Keyence measuring head; (2) IRBP B250 five-axis positioner; (3) IRB 4600 robot with electrospindle and force sensor; (4) tool storage; (5) IRB 1600 robot controller; (6) IRB 4600 robot controller; (7) electrical cabinet.

The machined workpiece is a part from the automotive industry in which the time of execution of machining processes is crucial. Therefore, it is important to perform machining as efficiently as possible. In the presented station, the FC SpeedChange software feature was used, which allows, by measuring the force in the cutting direction, for increasing or decreasing the speed of the tool movement. In order to monitor the forces in test work, so as to determine the tool speed change thresholds, parameters such as tool movement speeds and forces in the process must be known. Tune Master software was used to monitor these parameters. An exemplary view of working with the software during tests is presented in Figure 16.

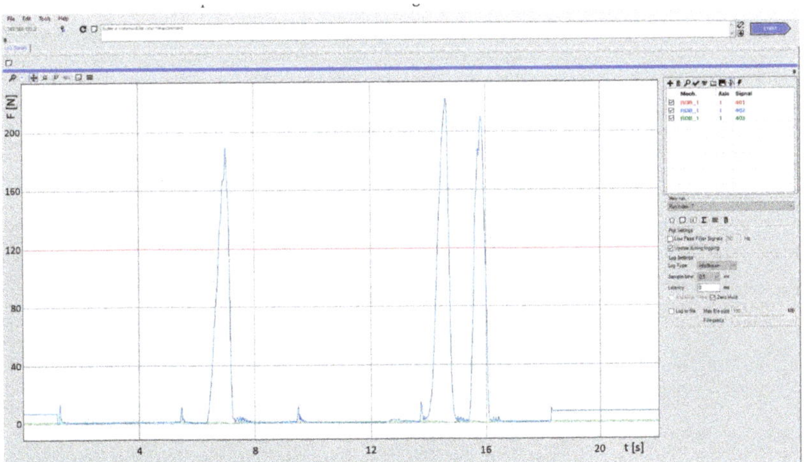

(a)

Figure 16. *Cont.*

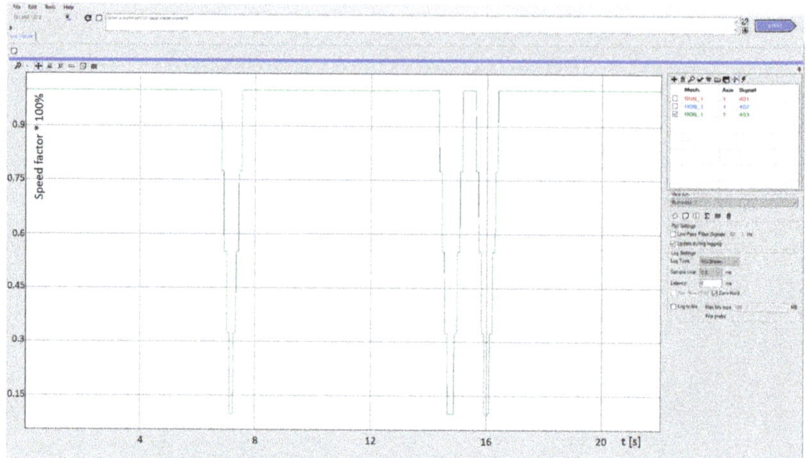

(**b**)

Figure 16. Tune Master Log Signal window view: (**a**) registered forces; (**b**) speed ration.

The tuned parameter for this solution was the value of the force at which the robot's TCP feed rate is reduced (Figure 16a, signal 401, red color). This is a very important limiting parameter. With cutting forces (Figure 16a, signal 402, blue color) below the accepted value, the TCP moved with the feed rate specified by the programmer, e.g., 100 mm/s. When the limit value was exceeded, the feed rate was reduced (value 1 corresponds to the speed set by the programmer, picture 16b, signal 403, green color). The limiting force parameter was tuned using Tune Master software. The selection of this parameter was carried out experimentally. The authors set the maximum feed rate specified by the cutting tool manufacturer. Contact forces were then recorded. If the contact force that occurred was safe for the tool then it was taken as a limit and set as a parameter in the software. If this parameter was selected, the cycle time was optimized. Efforts were made to achieve the highest possible TCP feed rate, i.e., the shortest possible machining time for the part. The most important criterion was not to allow damage to the tool. It was checked that the forces during cutting did not exceed the maximum forces allowed by the tool manufacturer. The software and optimization resulted in a machining cycle time of 750 s. On the basis of references from similar projects, it can be assumed that the cycle time without optimization would be 30% longer.

The use of the Tune Master software enabled the optimization and correct programming of the casting machining station, which was implemented in a Kutno foundry powered by Luma Automation. Photos of the work in the laboratory and the implementation in the production plant are presented in Figures 17 and 18.

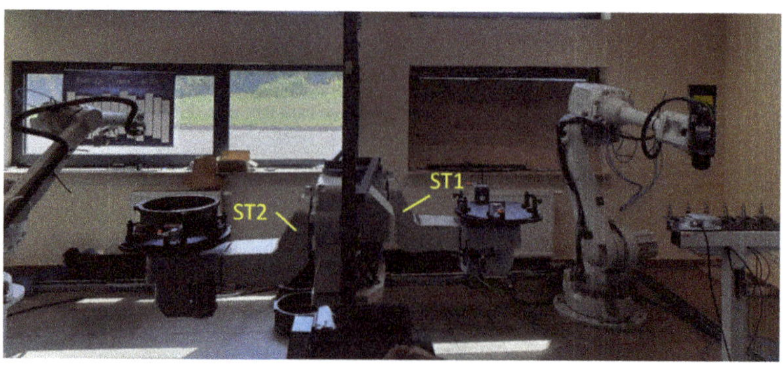

Figure 17. Robotic station during tests in the laboratory.

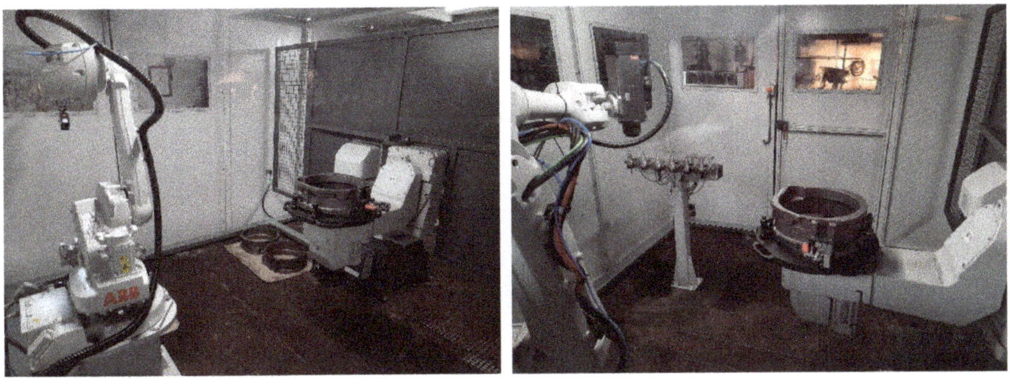

(**a**) (**b**)

Figure 18. Robotic station in the production plant: (**a**) view of the measuring part; (**b**) view of the machining part.

4. Conclusions

The paper presented examples of the use of three programs to monitor the operation of ABB robotic stations with the IRC5 controller. Their possibilities, functionalities and industrial examples of use were shown. The presented programs, thanks to the possibility of monitoring, recording, and data processing, can extend the possibilities of implementing robotic stations to more demanding technological processes. The use of the aforementioned IT tools in many cases allows for a significant reduction in the programming time of robots, and in some cases for the optimization of energy consumption, which brings financial gains over a wide time horizon.

In our practice, we have so far used the abovementioned software to carry out such tasks as the development of a robotic process of deburring the edges of parts with variable shape used in aircraft engines with the use of the automatic tool trajectory adaptation system. The work was carried out for the diffuser of the V2500 engine used in Airbus A320 and McDonnell Douglas MD-90 airplanes. Another example is the already mentioned station for measuring the thickness of the segments of the OGVs of the GP7000 engine produced by a wide consortium, including by Pratt and Whitney, used in the Airbus A380 aircraft. In addition, monitoring software was widely used in work on the automation of machining thin-walled aircraft gearbox housings made of light alloys. The research work carried out concerned the deburring of the machined sections of the aluminum ADT housing of the PW1500G engine used in the Airbus A220 aircraft. The aforementioned software was widely used in the selection of parameters for the robotic blade grinding process of the PT6 motor manufactured by Pratt and Whitney. The company also benefited from the work related to polishing the teeth of the FDGS gear wheels. The monitoring of brush wear in the process of deburring the blade locks of the CFM International LEAP engine used in the Airbus A320 and Boeing 737 aircrafts commissioned by Safran Aircraft Engines is another example of the use of monitoring programs. An example of implementation in the automotive industry is the already mentioned work on the development and commissioning of a technology for the production of highly accurate iron castings for the automotive sector, consisting of the machining of cast housings. Research work was commissioned by Luma Automation S.A.

In conclusion, the described possibilities of programs for monitoring robotic processes by ABB can become very valuable information, giving rise to new ideas and solutions, and showing new possibilities of implementing complex processes.

Author Contributions: Conceptualization, A.B., D.S. and K.K.; methodology, K.K., P.G. and D.S.; software, A.B., P.G. and K.K.; validation, M.M. and A.O.; formal analysis, M.M. and M.U.; investigation,

A.B.; resources, K.K.; data curation, M.U.; writing—original draft preparation, A.B.; writing—review and editing, M.M. and A.B.; visualization, P.G. and K.K.; supervision, A.B.; project administration, P.G.; funding acquisition, A.B. All authors have read and agreed to the published version of the manuscript.

Funding: This research received no external funding.

Data Availability Statement: Data are contained within the article.

Conflicts of Interest: The authors declare no conflict of interest.

References

1. Hovgard, M.; Lennartson, B.; Bengtsson, K. Applied energy optimization o multi-robot systems through motion parameter tuning. *CIRP J. Manuf. Sci. Technol.* **2021**, *35*, 422–430. [CrossRef]
2. Burghardt, A.; Kurc, K.; Szybicki, D.; Muszyńska, M.; Szczęch, T. Monitoring the parameters of the robot-operated quality control process. *Adv. Sci. Technol. Res. J.* **2017**, *11*, 232–236. [CrossRef]
3. Ramasamy, S.; Eriksson, K.; Peralippatt, S.; Perumal, B.; Danielsson, F. Optimized Online Path Planning Algorithms Considering Energy. In Proceedings of the 2021 26th IEEE International Conference on Emerging Technologies and Factory Automation (ETFA), Vasteras, Sweden, 7–10 September 2021. [CrossRef]
4. Blomdell, A.; Dressler, I.; Nilsson, K.; Robertsson, A. Flexible application development and high-performance motion control based on external sensing and reconfiguration of ABB industrial robot controllers. In Proceedings of the ICRA 2010 Workshop on Innovative Robot Control Architectures for Demanding (Research) Applications, Anchorage, AK, USA, 3–7 May 2010; pp. 62–66. Available online: https://pdfslide.net/documents/icra-2010-workshop-on-innovative-robot-control-a-liation-1kuka-roboter-gmbh.html?page=11 (accessed on 26 August 2022).
5. Wiśniewski, M.; Peta, K.; Pietrowiak, A.; Ciszak, O.; Grochalski, K. Analiza ścieżki narzędzia w operacji montażu z zastosowaniem programu Robot-Studio. *Technol. I Autom. Montażu* **2017**, *4*, 47–53.
6. Kaczmarek, W.; Panasiuk, J.; Tomaszuk, M. Komputerowe wspomaganie tworzenia zrobotyzowanych aplikacji cięcia laserem. *Mechanik* **2015**, *88*, 377–384. [CrossRef]
7. Muszyńska, M.; Szybicki, D.; Gierlak, P.; Kurc, K.; Burghardt, A.; Uliasz, M. Application of Virtual Reality in the Training of Operators and Servicing of Robotic Stations. In *Collaborative Networks and Digital Transformation*; PRO-VE 2019; IFIP Advances in Information and Communication, Technology; Camarinha-Matos, L., Afsarmanesh, H., Antonelli, D., Eds.; Springer: Cham, Switzerland, 2019; Volume 568. [CrossRef]
8. Szybicki, D.; Obal, P.; Kurc, K.; Gierlak, P. Programming of Industrial Robots Using a Laser Tracker. *Sensors* **2022**, *22*, 6464. [CrossRef] [PubMed]
9. Abreu, P.; Barbosa, M.R.; Lopes, A.M. Robotics virtual lab based on off-line robot programming software. In Proceedings of the 2013 2nd Experiment International Conference, IEEE, Coimbra, Portugal, 18–20 September 2013; pp. 109–113. [CrossRef]
10. Obal, P.; Burghardt, A.; Kurc, K.; Szybicki, D.; Gierlak, P. Monitoring the parameters of industrial robots. In Proceedings of the International Workshop on Modeling Social Media; Springer: Cham, Switzerland, 2018; Volume 548, pp. 230–238. [CrossRef]
11. Szybicki, D.; Burghardt, A.; Kurc, K.; Pietruś, P. Calibration and verification of an original module measuring turbojet engine blades geometric parameters. *Arch. Mech. Eng.* **2019**, *66*, 97–109. [CrossRef]
12. Krantz, M.; Andersson, R. Robotized Polishing and Deburring with Force Feedback Control. 2010. Available online: https://www.diva-portal.org/smash/get/diva2:331100/FULLTEXT01.pdf (accessed on 2 May 2010).
13. Padmanabhan, S.N.; Halil, Z.; Sun, Y.; Vu, T.T.; Yeo, S.H.; Wee, A. Experimental investigation of Robotic Surface Finishing Using Abrasive Disc. *Int. J. Mater. Mech. Manuf.* **2016**, *6*, 127–130. [CrossRef]
14. Burghardt, A.; Szybicki, D.; Kurc, K.; Muszyńska, M. Optimization of process parameters of edge robotic deburring with force control. *Int. J. Appl. Mech. Eng.* **2016**, *21*, 987–995. [CrossRef]
15. Burghardt, A.; Kurc, K.; Szybicki, D. Robotic automation of the turbo-propeller engine blade grinding process. *Appl. Mech. Mater.* **2016**, *817*, 206–213. [CrossRef]
16. Niu, B.; Chi, Y.; Zhang, H. Dynamic electrode force control of resistance spot welding robot. In Proceedings of the 2009 IEEE International Conference on Robotics and Biomimetics (ROBIO), Guilin, China, 19–23 December 2009. [CrossRef]
17. Soron, M.; Kalaykov, I. A robot prototype for friction stir welding. In Proceedings of the 2006 IEEE Conference on Robotics, Automation and Mechatronics, Bangkok, Thailand, 1–3 June 2006. [CrossRef]
18. Cook, G.E.; Crawford, R.; Clark, D.E.; Strauss, A.M. Robotic friction stir welding. *Ind. Robot.* **2004**, *31*, 55–63. [CrossRef]
19. Chen, M.; Li, Z.; Zhan, S.; Zhang, G. Dynamic Drive Chain Error Analysis of Industrial Robots with Cyber Sensing Technology. In Proceedings of the 2017 IEEE 7th Annual International Conference on CYBER Technology in Automation, Control, and Intelligent Systems (CYBER), Honolulu, HI, USA, 31 July 2017–4 August 2017. [CrossRef]
20. Zhang, B.; Gravel, D.; Zhang, G.; Wang, J.; Bell, A. Robotic force control assembly parameter optimization for adaptive production. In Proceedings of the 2011 IEEE International Conference on Robotics and Automation, Shanghai, China, 9–13 May 2011. [CrossRef]
21. Verl, A.; Valente, A.; Melkote, S.; Brecher, C.; Ozturk, E.; Tunc, L.T. Robots in machining. *CIRP Ann.* **2019**, *68*, 799–822. [CrossRef]

22. Gale, S.; Eielsen, A.A.; Gravdahl, J.T. Modelling and simulation of a flywheel based energy storage system for an industrial manipulator. In Proceedings of the 2015 IEEE International Conference on Industrial Technology (ICIT), Seville, Spain, 17–19 March 2015. [CrossRef]
23. Hislop, J.; Isaksson, M.; McCormick, J.; Hensman, C. Validation of 3-Space Wireless Inertial Measurement Units Using an Industrial Robot. *Sensors* **2021**, *21*, 6858. [CrossRef] [PubMed]
24. Gierlak, P.; Burghardt, A.; Szybicki, D.; Kurc, K. Eliminating the Inertial Forces Effects on the Measurement of Robot Interaction Force. In *Methods and Techniques of Signal Processing in Physical Measurements*; MSM; Lecture Notes in Electrical Engineering; Hanus, R., Mazur, D., Kreischer, C., Eds.; Springer: Cham, Switzerland, 2018; Volume 548. [CrossRef]
25. Burghardt, A.; Gierlak, P.; Kurc, K.; Szybicki, D. Automatic Detection of Industrial Robot Tool Damage Based on Force Measurement. *Teh. Vjesn.* **2020**, *27*, 1385–1393. [CrossRef]
26. Szybicki, D.; Burghardt, A.; Kurc, K.; Gierlak, P. Device for Contact Measurement of Turbine Blade Geometry in Robotic Grinding Process. *Sensors* **2020**, *20*, 7053. [CrossRef] [PubMed]
27. Burghardt, A.; Szybicki, D.; Gierlak, P.; Kurc, K.; Muszyńska, M. Robotic Grinding Process of Turboprop Engine Compressor Blades with Active Selection of Contact Force. *Teh. Vjesn.* **2022**, *29*, 15–22. [CrossRef]
28. Hendzel, Z.; Burghardt, A.; Gierlak, P.; Szuster, M. Conventional and fuzzy force control in robotised machining, Solid State Phenomena. *Trans. Tech. Publ.* **2014**, *210*, 178–185. [CrossRef]
29. Burghardt, A.; Szybicki, D.; Kurc, K.; Muszyńska, M.; Mucha, J. Experimental study of Inconel 718 surface treatment by edge robotic deburring with force control. *Strength Mater.* **2017**, *49*, 594–604. [CrossRef]
30. Burghardt, A.; Kurc, K.; Szybicki, D.; Muszyńska, M.; Nawrocki, J. Robot-operated quality control station based on the UTT method. *Open Eng.* **2017**, *7*, 37–42. [CrossRef]
31. Burghardt, A.; Kurc, K.; Szybicki, D.; Muszyńska, M.; Szczęch, T. Robot-operated inspection of aircraft engine turbine rotor guide vane segment geometry. *Tech. Gaz.* **2017**, *24*, 345–348. [CrossRef]
32. Burghardt, A.; Kurc, K.; Szybicki, D.; Muszyńska, M.; Nawrocki, J. Software for the robot-operated inspection station for engine guide vanes taking into consideration the geometric variability of parts. *Teh. Vjesn.* **2017**, *24*, 349–353. [CrossRef]

electronics MDPI

Article

Quaternion Attitude Control System of Highly Maneuverable Aircraft

Michał Gołąbek *, Michał Welcer, Cezary Szczepański, Mariusz Krawczyk, Albert Zajdel and Krystian Borodacz

Łukasiewicz Research Network—Institute of Aviation, Al. Krakowska 110/114, 02-256 Warszawa, Poland
* Correspondence: michal.golabek@ilot.lukasiewicz.gov.pl

Abstract: In the era of rapid advancements in manned and unmanned aviation and robotics, there is a need for high-performance, robust attitude control of highly maneuverable fixed-wing aircraft, both manned and unmanned (UAVs). This paper presents an extension to research on spacecraft attitude control. The article extends existing concepts and applies them to the control problem of aircraft operating in Earth's atmosphere. First, a general concept of quaternions is presented. Next, the attitude controller's architecture is discussed. The controller synthesis is described using quaternion algebra. The quaternion-based attitude controller is then compared with a classical Euler-based attitude controller. The methodology for comparison and performance evaluation of both controllers is described. Lastly, the results of the simulations and a comparison of the two controllers are presented and discussed. The presented control scheme outperforms classical methods based on Euler angles, particularly at the aircraft's high pitch and roll angles.

Keywords: attitude; control; aircraft; UAV; fixed-wing; quaternions; Euler angles; avionics

Citation: Gołąbek, M.; Welcer, M.; Szczepański, C.; Krawczyk, M.; Zajdel, A.; Borodacz, K. Quaternion Attitude Control System of Highly Maneuverable Aircraft. *Electronics* **2022**, *11*, 3775. https://doi.org/10.3390/electronics11223775

Academic Editors: Imre J. Rudas, Piotr Szymak, Stanisław Hożyń and Paweł Piskur

Received: 7 October 2022
Accepted: 16 November 2022
Published: 17 November 2022

1. Introduction

The objective of this paper is to present an attitude controller for fixed-wing aircraft operating over large ranges of pitch and roll angles. The presented controller provides desired control qualities over a wide range of flight conditions in terms of speed of response and performance robustness. The aircraft considered in this paper is the Extra 330SC, which is a single-seat aerobatic airplane, manufactured by Extra Flugzeugbau. The long-term goal justifying the need for such a controller is to design a complete autopilot system for unmanned, highly maneuverable fixed-wing strike and fighter aircraft. The attitude controller in this context can be thought of as one of the internal control loops of the autopilot. It means that input to the attitude controller comes from an external controller, such as a guidance system.

Often when speaking about an aircraft's attitude, Euler angles appear in mathematical description. The Euler angles approach is frequently chosen because of its intuitiveness. Although intuitive, Euler angles suffer from non-linearities and singularities, resulting in undesired effects such as gimbal lock. Moreover, an aircraft's control based on Euler angles results in unwanted effects of control coupling between a rudder and an elevator for high roll and pitch angles. This paper presents an effective way of dealing with shortcomings associated with Euler angles by replacing them with quaternions in attitude description and control.

A quaternion number system extends the complex number system and is applied to describe rotations in three-dimensional space. A more stable and often simpler attitude representation can be achieved with quaternions than with Euler angles. Quaternions' computational stability comes from the fact that only unit quaternions are required to describe rotations, reducing numeric errors when they are normalized. They lack the gimbal-lock effect inherent to Euler angles because of their relationship to the axis–angle representation. Their mathematical simpleness comes from the fact that no trigonometric functions are required, only arithmetic functions such as quaternion products (which are computationally efficient). Mathematical tools associated with quaternions also prove the advantage of quaternions over Euler angles. Linear interpolation between two rotations is straightforward in quaternion notation, while impossible in Euler angles notation. Moreover, integrating angular rates to quaternions is much easier than integrating angular rates to Euler angles. This article will present an efficient use of quaternions in an airplane attitude control problem.

Research in quaternion attitude control has been conducted in the area of spacecraft control, and its results can be found in [1–3]. This article aims to extend this control concept and apply it to the aircraft operating in Earth's atmosphere.

The quaternion number system extends the complex number system. A quaternion is a four-dimensional hyper-complex number that can describe transformations in three-dimensional space. Unit quaternions, being quaternions with a norm equal to one, are often used to describe rotations in three-dimensional space. Publications [4–6] explain the quaternion construct and associated quaternion algebra in more detail.

A quaternion consists of one real part and three imaginary parts. q_w is called the real part and q_x, q_y, q_z are called imaginary parts. Equations (1) and (2) present the two most popular quaternion notations.

$$q = q_w + q_x\boldsymbol{i} + q_y\boldsymbol{j} + q_z\boldsymbol{k} \tag{1}$$

$$q = \begin{bmatrix} q_w \; q_x \; q_y \; q_z \end{bmatrix}^T \tag{2}$$

The complex conjugate of a quaternion is defined as:

$$\overline{q} = \begin{bmatrix} q_w & -q_x & -q_y & -q_z \end{bmatrix}^T \tag{3}$$

The norm of a quaternion q is defined as:

$$|q| = \sqrt{q_w^2 + q_x^2 + q_y^2 + q_z^2} \tag{4}$$

The unit quaternion is a quaternion of the norm equal to one.

$$|q_{unit}| = 1 \tag{5}$$

The quaternion inverse is defined as:

$$q^{-1} = \frac{\overline{q}}{||q||^2} \tag{6}$$

For unit quaternion, its inverse is equal to its conjugate:

$$q_{unit}{}^{-1} = \overline{q}_{unit} \tag{7}$$

The quaternion product of quaternions p and q, also called the Hamilton product and denoted by $\otimes$, is defined as:

$$p \otimes q = \begin{bmatrix} p_wq_w - p_xq_x - p_yq_y - p_zq_z \\ p_wq_x + p_xq_w + p_yq_z - p_zq_y \\ p_wq_y - p_xq_z + p_yq_w + p_zq_x \\ p_wq_z + p_xq_y - p_yq_x + p_zq_w \end{bmatrix} \tag{8}$$

Quaternion multiplication is non-commutative in the same way as real-life rotations are non-commutative:

$$p \otimes q \neq q \otimes p \tag{9}$$

All quaternions used in this paper to represent rotation or attitude are unit quaternions. Please note that this does not apply to quaternion derivatives. Care must be taken when calculating a product of a quaternion and a quaternion derivative since this product will also not be unitary.

2. Materials and Methods

The entire attitude control process is performed in the quaternion form. Yaw, pitch and roll angles are not used to control the aircraft's attitude. The classical three controllers (each for a separate: yaw, pitch and roll angle) have been replaced with one controller operating on a single quaternion representing the aircraft's attitude. The proposed controller is comprised of two major parts working in a cascade. The first (external) part consists of a single quaternion "Q_P Attitude Controller" working similarly to a classical "P" (proportional) controller but being evaluated in quaternion space (Q_P stands for quaternion proportional). The second (internal) part consists of three PID controllers [7] responsible for angular rate control of the aircraft in its three axes. The inputs to the Q_P Attitude Controller are: a desired (setpoint) attitude in the form of a quaternion, denoted as q_{sp}, and a measured attitude in the form of a quaternion, denoted as q_{meas}. Internally, the controller calculates the attitude error, denoted as q_{err}, according to Equation (11). The output of the Q_P Attitude Controller are three angular rate setpoints of the aircraft around its three axes: x, y, z, denoted as: $\omega_{x_{sp}}, \omega_{y_{sp}}, \omega_{z_{sp}}$. The output of the Q_P Attitude Controller (after saturation) constitutes an input to the three angular rate controllers, which are PID controllers [7]. The PID controllers take two inputs. The first input is the setpoint angular rate, and the second is the measured angular rate from the feedback path. The PI part of the controller works on the angular rate error (setpoint minus measured value), and the D part works on the measured value from the feedback path. The outputs from the angular rate controllers are the desired angles of deflection of the aircraft's control surfaces denoted as: $\delta_{A_{sp}}, \delta_{H_{sp}}, \delta_{V_{sp}}$, which correspond to the deflection angles of the ailerons, elevator and rudder, respectively. Feedback signals in the system consist of three measured angular rates around the x, y and z axes denoted as: $\omega_{x_{meas}}, \omega_{y_{meas}}, \omega_{z_{meas}}$, and a quaternion representing an aircraft's measured attitude q_{meas}. All signals in the feedback path come from the aircraft's Attitude and Heading Reference System (AHRS) (Figure 1).

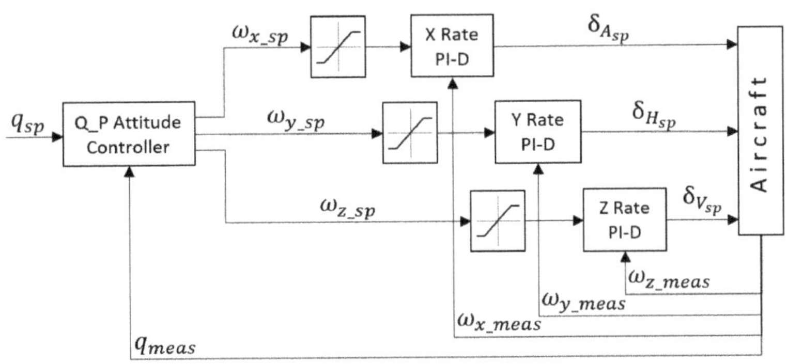

Figure 1. The schematic of the quaternion-based controller architecture.

It has to be noted that the presented quaternion attitude control system is not a complete autopilot system. The presented attitude controller constitutes an inner part of the autopilot. An outer part in a form of a guidance system (GS) is required for a complete autopilot. The guidance system is responsible for calculating the setpoint attitude of the aircraft based on assumed flight plan, in particular the assumed flight trajectory and flight parameters. The attitude control system then executes the setpoint attitude calculated by the GS. In this paper, the GS was replaced by man-generated attitude setpoints contained in test scenarios. In the future, we plan to make GS work fully automatically, which can be achieved, for example, by implementing the inverse dynamics methods described in [8,9].

The assumed coordinate system of the airplane is the standard aeronautical reference system [10], where the x-axis points to the nose of the aircraft, the y-axis points to the right wingtip of the aircraft and the z-axis points to the bottom of the airplane's fuselage (Figure 2).

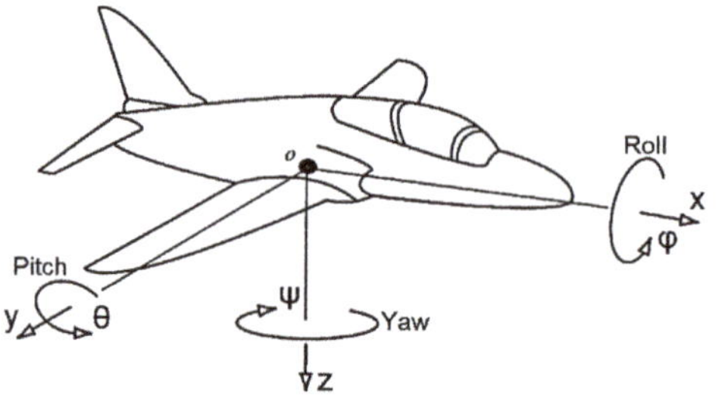

Figure 2. The assumed aircraft coordinate system [10].

This section describes the process of the Q_P Attitude Controller synthesis. We begin with a definition of the aircraft's attitude error. The attitude error is a difference between a setpoint attitude and a measured attitude. Typically, in a regular PID controller operating on real numbers, an error value is defined as a difference between the desired setpoint (SP) and the measured process variable (PV), i.e., e(t) = SP(t) − PV(t) [7]. In the proposed quaternion attitude controller, the attitude error quaternion q_{err} is calculated based on the following relationship (10) between the setpoint attitude quaternion q_{sp} and the measured attitude quaternion q_{meas}, assuming intrinsic rotations:

$$q_{sp} = q_{meas} \otimes q_{err} \tag{10}$$

Thus, after left-multiplication by $\overline{q}_{meas}$ we obtain:

$$q_{err} = \overline{q}_{meas} \otimes q_{sp} \tag{11}$$

Since two rotations can describe every attitude, we have to ensure that the attitude error q_{err} is represented by the shorter of the two possible rotations. To do this, we convert a quaternion q_{err} to axis–angle representation (12). In this representation, we check the angle of rotation θ_{err} around the axis u. If the angle is greater than 180 degrees (13), we invert the axis of rotation and calculate a new angle of the shorter rotation $\theta_{err_{short}}$ (14).

$$\begin{aligned} \theta_{err} &= 2{\cdot}\mathrm{acos}(q_{w_{err}}) \\ u_x &= \frac{q_x}{\sqrt{1-q_{w_{err}}^2}} \\ u_y &= \frac{q_y}{\sqrt{1-q_{w_{err}}^2}} \\ u_z &= \frac{q_z}{\sqrt{1-q_{w_{err}}^2}} \end{aligned} \tag{12}$$

If:

$$\theta_{err} > \pi \tag{13}$$

then:

$$\begin{aligned} \theta_{err_{short}} &= |\theta_{err} - 2\pi| \\ u_{x_{short}} &= -u_x \\ u_{y_{short}} &= -u_y \\ u_{z_{short}} &= -u_z \end{aligned} \tag{14}$$

else:

$$\begin{aligned} \theta_{err_{short}} &= \theta_{err} \\ u_{x_{short}} &= u_x \\ u_{y_{short}} &= u_y \\ u_{z_{short}} &= u_z \end{aligned} \tag{15}$$

Next, we need to convert the new attitude error from the axis–angle representation back to the quaternion form (16).

$$\begin{aligned} q_{w_{err_{short}}} &= \cos\left(\frac{\theta_{err_{short}}}{2}\right) \\ q_{x_{err_{short}}} &= u_{x_{short}} \cdot \sin\left(\frac{\theta_{err_{short}}}{2}\right) \\ q_{y_{err_{short}}} &= u_{y_{short}} \cdot \sin\left(\frac{\theta_{err_{short}}}{2}\right) \\ q_{z_{err_{short}}} &= u_{z_{short}} \cdot \sin\left(\frac{\theta_{err_{short}}}{2}\right) \end{aligned} \tag{16}$$

A faster check for the shorter rotation can be performed according to [2].
If:

$$q_{w_{err}} < 0 \tag{17}$$

then:

$$\begin{aligned} q_{w_{err_{short}}} &= -q_{w_{err}} \\ q_{x_{err_{short}}} &= -q_{x_{err}} \\ q_{y_{err_{short}}} &= -q_{y_{err}} \\ q_{z_{err_{short}}} &= -q_{z_{err}} \end{aligned} \tag{18}$$

else:

$$\begin{aligned} q_{w_{err_{short}}} &= q_{w_{err}} \\ q_{x_{err_{short}}} &= q_{x_{err}} \\ q_{y_{err_{short}}} &= q_{y_{err}} \\ q_{z_{err_{short}}} &= q_{z_{err}} \end{aligned} \tag{19}$$

Next, we assume that a quaternion derivative $\dot{q}_{sp}$ representing a desired (setpoint) rate of rotation required to achieve a desired (setpoint) attitude is proportional to the attitude error, i.e., $\dot{q}_{sp} \propto q_{err}$. After the introduction of proportional gain K_p the setpoint rate of rotation becomes (20):

$$\dot{q}_{sp} = K_p \cdot q_{err_{short}} \tag{20}$$

Next, we need to express the obtained setpoint rate of rotation as an angular rate vector ω_{sp}. The relationship between the quaternion derivative and angular rate vector in the body reference frame is given by [6] (21):

$$\omega = 2\overline{q} \otimes \dot{q} \tag{21}$$

Thus:

$$\omega_{sp} = 2\overline{q}_u \otimes \dot{q}_{sp} \tag{22}$$

where q_u is an unrotated unit quaternion (23):

$$q_u = [1 \;\; 0 \;\; 0 \;\; 0]^T \tag{23}$$

and its conjugate $\overline{q}_u$ (24):

$$\overline{q}_u = q_u \tag{24}$$

In extended form, the setpoint angular rate vector becomes (25):

$$\begin{bmatrix} \omega_{x_{sp}} \\ \omega_{y_{sp}} \\ \omega_{z_{sp}} \end{bmatrix} = 2 \cdot \begin{bmatrix} \overline{q}_{w_u} \cdot \dot{q}_{x_{sp}} + \overline{q}_{x_u} \cdot \dot{q}_{w_{sp}} + \overline{q}_{y_u} \cdot \dot{q}_{z_{sp}} - \overline{q}_{z_u} \cdot \dot{q}_{y_{sp}} \\ \overline{q}_{w_u} \cdot \dot{q}_{y_{sp}} - \overline{q}_{x_u} \cdot \dot{q}_{z_{sp}} + \overline{q}_{y_u} \cdot \dot{q}_{w_{sp}} + \overline{q}_{z_u} \cdot \dot{q}_{x_{sp}} \\ \overline{q}_{w_u} \cdot \dot{q}_{z_{sp}} + \overline{q}_{x_u} \cdot \dot{q}_{y_{sp}} - \overline{q}_{y_u} \cdot \dot{q}_{x_{sp}} + \overline{q}_{z_u} \cdot \dot{q}_{w_{sp}} \end{bmatrix} \tag{25}$$

Thus, after simplification (26):

$$\begin{bmatrix} \omega_{x_{sp}} \\ \omega_{y_{sp}} \\ \omega_{z_{sp}} \end{bmatrix} = 2 \cdot \begin{bmatrix} \dot{q}_{x_{sp}} \\ \dot{q}_{y_{sp}} \\ \dot{q}_{z_{sp}} \end{bmatrix} \tag{26}$$

and finally (27):

$$\begin{bmatrix} \omega_{x_{sp}} \\ \omega_{y_{sp}} \\ \omega_{z_{sp}} \end{bmatrix} = 2 \cdot K_p \cdot \begin{bmatrix} q_{x_{err_{short}}} \\ q_{y_{err_{short}}} \\ q_{z_{err_{short}}} \end{bmatrix} \tag{27}$$

The ω_{sp} vector is an output from the Q_P Attitude Controller and input to the three PID controllers responsible for the aircraft's angular rate control, as shown in Figure 1. The simplicity of the result makes the control computationally efficient and makes a direct physical interpretation of the result possible. A fact worth noting is that, unlike Euler-based attitude controllers, the quaternion-based controller will tend to generate a rotation resulting in the shortest path, provided that the aircraft under consideration can achieve the same angular rates in all three (body) axes.

The controller was fully implemented in the "C" language. An extensive quaternion math library was created to support the required quaternion algebra. Preliminary tests were conducted on a simple dynamic model and ran on a simulation in the "C" language. Further tests were conducted using Simulink R2020b combined with Laminar Research X-Plane 11 simulation environment, using a much more complex dynamic model. The controller's compiled "C" files were used as Simulink blocks during simulations. X-Plane flight dynamics and dynamic airplane models were used during the simulations.

To evaluate the performance of the attitude control, the quaternion-based controller (Figure 1) was compared to the conventional Euler-based controller (Figure 3). The Euler-based controller operates on three Euler angles to represent the aircraft's attitude. Therefore, the attitude control is performed on these three angles instead of the one quaternion as it is accomplished in the quaternion-based controller.

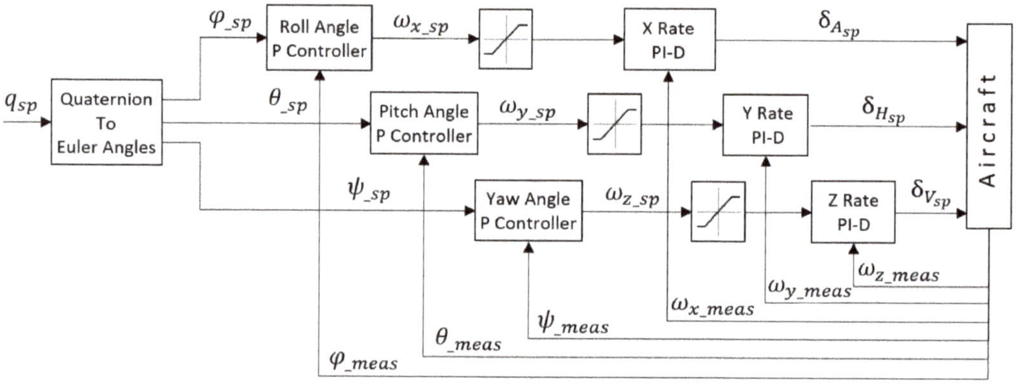

Figure 3. The schematic of the Euler-based controller architecture.

The described Euler-based controller is designed in the same cascaded architecture as the proposed quaternion-based controller. The outer control loop is the airplane's attitude control loop, and the inner control loop is the airplane's angular rate control loop. The outer loop consists of three proportional ("P") controllers. The inner loop consists of three PID controllers, a variation of a classical PID controller, as described in [7]. The outer control loop output (after saturation) constitutes an input to the inner control loop. φ_{sp}, θ_{sp}, ψ_{sp} denote setpoint roll, pitch and yaw angles, respectively. $\omega_{x_{sp}}$, $\omega_{y_{sp}}$, $\omega_{z_{sp}}$ denote setpoint angular rates in the airplane's x, y and z axes, respectively. $\delta_{A_{sp}}$, $\delta_{H_{sp}}$, $\delta_{V_{sp}}$ denote commanded deflections of ailerons, elevator and rudder, respectively. $\omega_{x_{meas}}$, $\omega_{y_{meas}}$, $\omega_{z_{meas}}$ on the feedback path denote measured angular rates in the airplane's x, y and z axes, respectively. φ_{meas}, θ_{meas}, ψ_{meas} on the feedback path denote the airplane's measured roll, pitch and yaw angles, respectively. Details of the classical, Euler-based, cascaded controller can be found in [11]. Computations required to convert from quaternion to Euler angles are described in [5].

Both controllers were implemented in Simulink and coupled with X-Plane 11 simulation environment and then tested against each other during several test scenarios. The test scenarios were designed in such a way as to reflect real-life turn maneuvers of an airplane performed at various roll angles. For example, the turns were simulated at the following roll angles: 30, 60 (Figure 4), 80 and 90 degrees (knife-edge turn).

Figure 4. Example of a simulated maneuver—a turn at a 60-degree roll angle.

The test scenarios were designed in the following way: each aircraft's distinct (key) setpoint attitude during the maneuver was given a setpoint "pose" number. The Euler angles (psi, theta, phi) were determined for each distinct pose. Between the following poses, time differences (delta T) were determined. Test scenarios created this way (Figure 5) were inputs to the Simulink/X-Plane simulation.

	A	B	C	D	E	F
1	Pose #	Time [sec]	Delta T [sec]	Yaw [deg]	Pitch [deg]	Roll [deg]
2	0	0,0		0,0	2,0	0,0
3	1	2,0	2,0	0,0	2,0	0,0
4	2	12,0	10,0	90,0	2,0	60,0
5	3	22,0	10,0	179,0	2,0	60,0
6	4	32,0	10,0	-90,0	2,0	60,0
7	5	42,0	10,0	0,0	2,0	0,0
8	6	102,0	60,0	0,0	2,0	0,0

Figure 5. Example of the test scenario—a turn at a 60-degree roll angle.

Each test scenario was processed in Simulink in the following fashion. First, each setpoint pose attitude was converted from Euler angles representation to quaternion representation. Then, intermediate poses were calculated using Spherical Linear Interpolation (SLERP) between all distinct key poses from the scenarios. In this way, a smooth transition between key setpoint poses was obtained. A smooth attitude setpoint was therefore achieved. This smooth, interpolated attitude setpoint was finally fed to the inputs of the attitude controllers (quaternion-based and Euler-based).

An Extra 330 SC airplane (Figure 6) was chosen to be simulated. A dynamic model of the airplane for the X-Plane 11 was purchased from the official X-Plane Store. Part number of the product: Vertigo-330SC. The fact of its use by one of the Red Bull Air Race professional pilots, during his virtual trainings, validates the accuracy of the purchased dynamic model.

Figure 6. Extra 330 SC during the simulated turn.

Both attitude controllers utilized equal proportional gains. In addition, both controllers used the same internal loop responsible for angular rate control, so the same PID gains were automatically assured between the two controllers. Therefore, it is important when comparing the performance of the two controllers.

To improve the robustness of the control, the outputs of the angular rate controllers were correlated with the square of the aircraft airspeed, which comes directly from the lift force equation [10].

3. Results

This section presents the results of the performed simulations. The simulations consisted of turns performed at 30-, 60-, 80- and 90-degree roll angles by the quaternion-based and the Euler-based controllers. For each simulated turn, two types of plots are presented—the setpoint tracking performance plot and the tracking error plot. The setpoint tracking performance plots illustrate the setpoint attitude and the measured attitude achieved by the two controllers, presented in the form of Euler angles (ψ, θ, φ). The tracking error plots illustrate tracking errors in yaw, pitch and roll for both controllers.

First, the turns at a 30-degree roll angle were simulated, with the results as shown in (Figures 7 and 8).

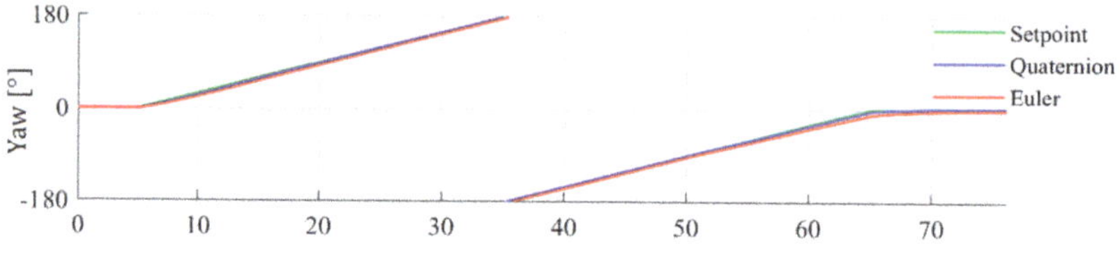

Figure 7. *Cont.*

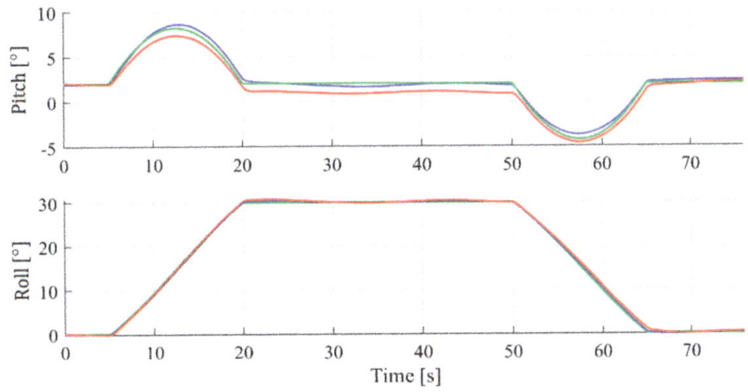

Figure 7. Setpoint tracking performance—turn at a 30-degree roll angle.

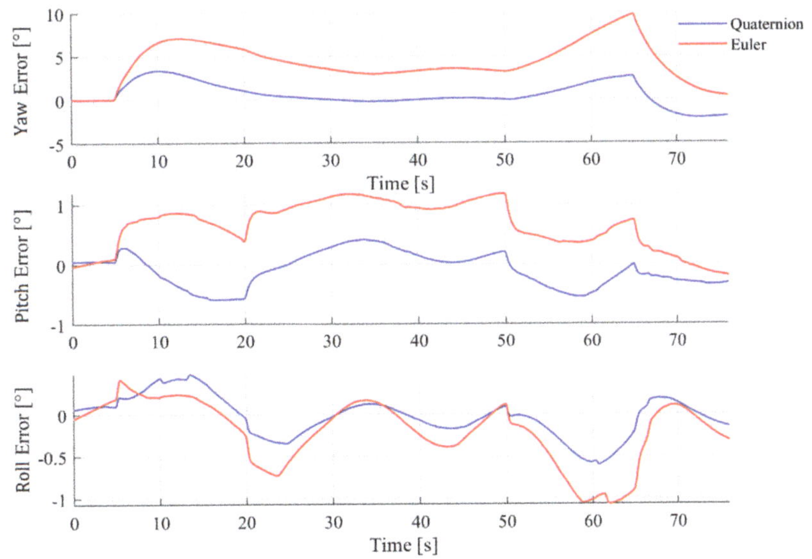

Figure 8. Tracking error—turn at a 30-degree roll angle.

At the 30-degree roll angle, both controllers perform the turn very well. The quaternion-based controller has a slight edge in tracking yaw and pitch angles, though.

Next, the turns at a 60-degree roll angle were simulated, with the results as shown in (Figures 9 and 10).

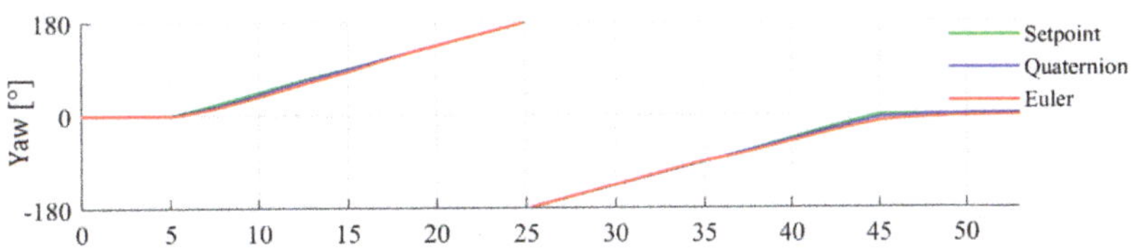

Figure 9. *Cont.*

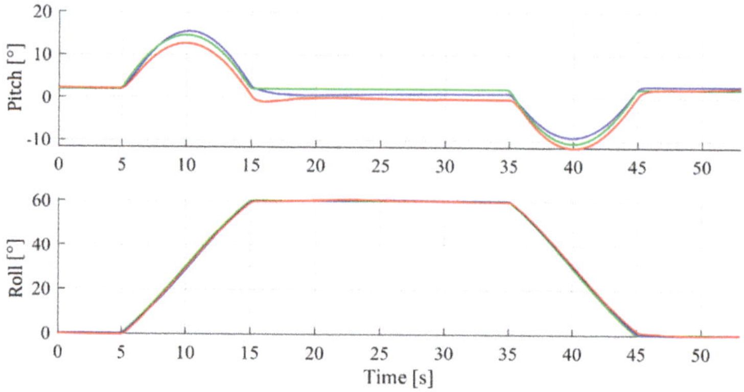

Figure 9. Setpoint tracking performance—turn at a 60-degree roll angle.

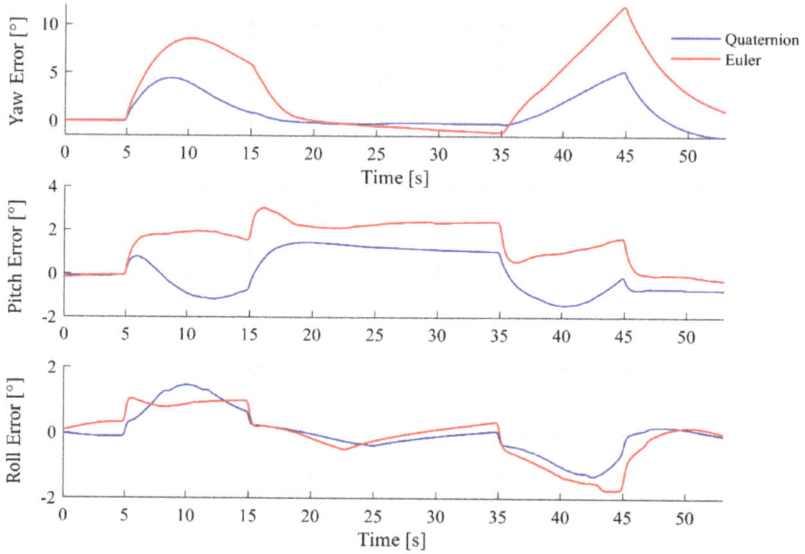

Figure 10. Tracking error—turn at a 60-degree roll angle.

At the 60-degree roll angle, both controllers also perform well. Once again, the quaternion-based controller has a slight edge in tracking yaw and pitch angles.

Next, the turns at an 80-degree roll angle were simulated, with the results as shown in (Figures 11 and 12).

At the 80-degree roll angle, the difference in the performance of the two controllers becomes apparent. The quaternion-based controller performs significantly better with a much tighter tracking of yaw and pitch angles. The Euler-based controller performs worse than the quaternion one due to the couplings: rudder–pitch, and elevator–yaw in the aircraft's control. The Euler-based controller does not "understand" that at high roll angles, the rudder controls the pitch angle, and the elevator controls the yaw angle. Due to its architecture and mathematical properties of quaternions, the quaternion-based controller "understands" this phenomenon and is able to perform proper and efficient control of the aircraft.

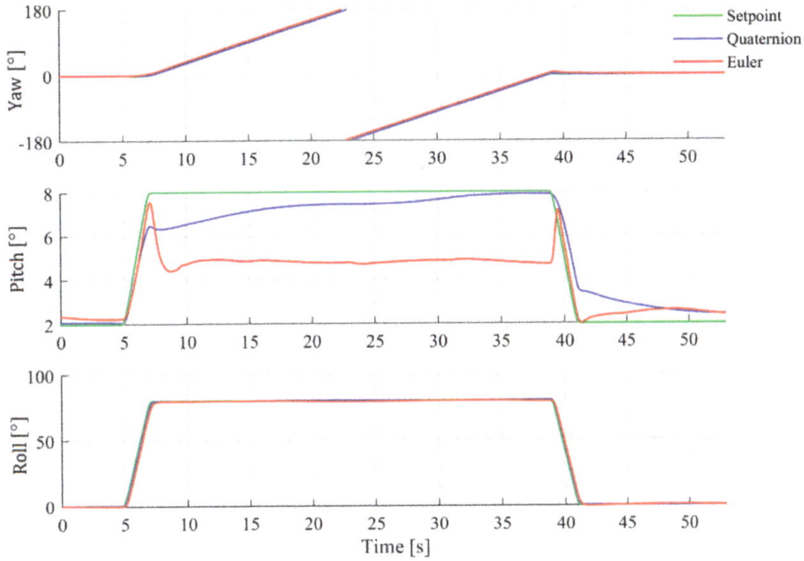

Figure 11. Setpoint tracking performance—turn at an 80-degree roll angle.

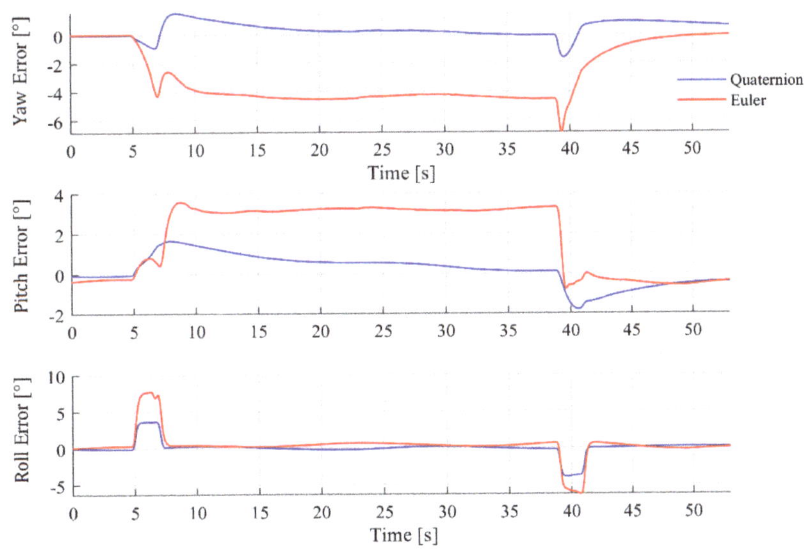

Figure 12. Tracking error—turn at an 80-degree roll angle.

Lastly, the turns at a 90-degree roll angle were simulated, with the results as shown in (Figures 13 and 14).

At the 90-degree roll angle, the difference in the performance of the two controllers becomes even more significant. The quaternion-based controller continues to perform better and better as the roll angle increases, compared to the Euler-based controller. It is due to stronger rudder–pitch and elevator–yaw couplings as the roll angle increases.

Because the dynamic model was purchased and there was no insight into it, the experimental stability analysis of the controller was performed. The quaternion-based controller response was evaluated at various attitudes and various setpoints resulting in attitude errors spanning from 0 to 180 degrees in all three Euler angles. The experimental stability analysis did not show any unstable areas in the system.

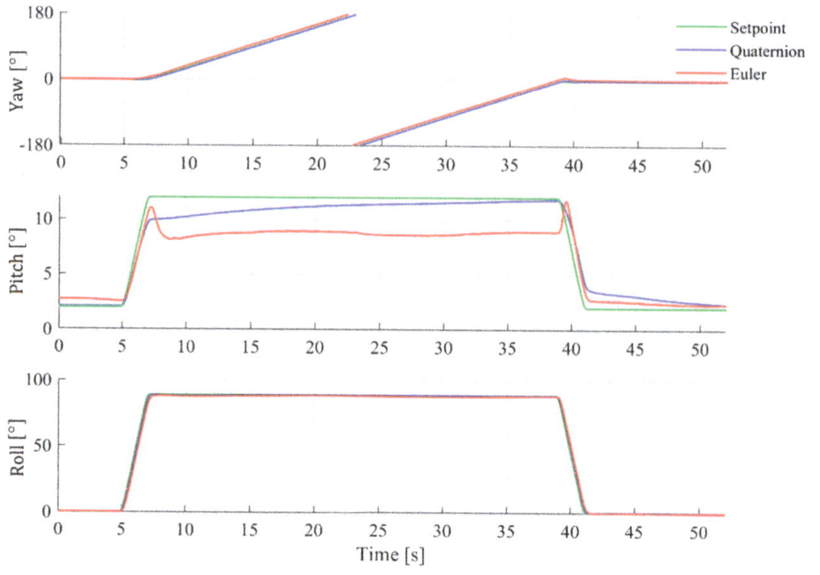

Figure 13. Setpoint tracking performance—turn at a 90-degree roll angle.

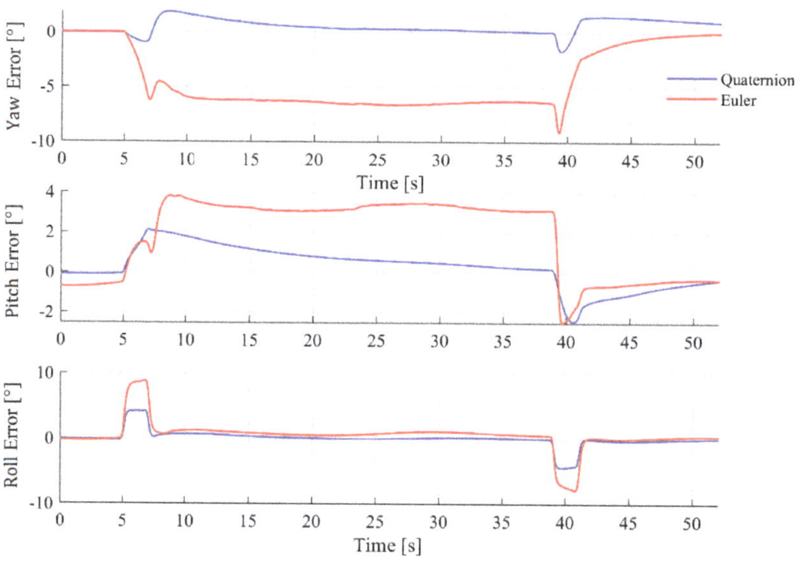

Figure 14. Tracking error—turn at a 90-degree roll angle.

The proposed controller outperforms the Euler angle-based controller. The difference in performance between the two controllers becomes larger as the pitch and roll angles of the aircraft's maneuver increase. Where Euler-based attitude control is sufficient at low pitch and roll angles, it becomes lacking at high angles, mostly due to the elevator and rudder couplings. The major advantage of the quaternion approach reveals itself at high roll and pitch angles. The major difference in the setpoint tracking is observed in the yaw and pitch channels. There is no relevant difference in the setpoint tracking for the roll channel between the two controllers. This is because the roll is the last rotation in Euler angles and is equivalent to the aircraft's rotation around its x-axis.

4. Conclusions

The controller can be successfully employed to control manned and unmanned aerial vehicles operating at high pitch and roll angles. In addition, a slightly modified version of the controller can be used to control cruising missiles and space rockets. Some preliminary tests were conducted in this direction, employing a Matlab simulation of an ILR-33 Amber sounding rocket, with great results.

Quaternion attitude control, counterintuitively, can be less complex than Euler angle control since a single quaternion represents a single rotation. In contrast, a single set of Euler angles represent three consecutive rotations, adding to the complexity of the control.

Author Contributions: Conceptualization, M.G.; methodology, M.G., M.K. and C.S.; software, M.G.; validation, M.G. and M.W.; formal analysis, M.K., C.S. and K.B.; investigation, M.G.; resources, M.G.; data curation, M.W.; writing—original draft preparation, M.G.; writing—review and editing, M.K., C.S. and A.Z.; visualization, M.G. and M.W.; supervision, M.K., C.S. and A.Z.; project administration, M.K., C.S. and A.Z.; funding acquisition, M.K., C.S. and A.Z. All authors have read and agreed to the published version of the manuscript.

Funding: This research received no external funding.

Data Availability Statement: Not applicable.

Conflicts of Interest: The authors declare no conflict of interest.

References

1. Wie, B.; Barba, P.M. *Quaternion Feedback for Spacecraft Large Angle Maneuvers*; Ford Aerospace & Communications Corporation: Palo Alto, CA, USA, 1984.
2. Markley, F.L.; Crassidis, J.L. *Fundamentals of Spacecraft Attitude Determination and Control*; Springer: Berlin/Heidelberg, Germany, 2019.
3. Chaturvedi, N.A.; Sanyal, A.K.; McClamroch, N.H. Rigid-Body Attitude Control. *IEEE Control. Syst.* **2011**, *31*, 30–51. [CrossRef]
4. Kuipers, J.B. *Quaternions and Rotation Sequences: A Primer with Applications to Orbits, Aerospace, and Virtual Reality*; Princeton University Press: Princeton, NJ, USA, 1999.
5. Diebel, J. Representing attitude: Euler angles, unit quaternions, and rotation vectors. *Matrix* **2006**, *58*, 1–35.
6. Graf, B. Quaternions and dynamics. *arXiv* **2008**, arXiv:0811.2889.
7. Ogata, K. *Modern Control Engineering*; Prentice Hall: Englewood Cliffs, NJ, USA, 2010.
8. Blajer, W.; Kosma, Z.; Krawczyk, M.; Graffstein, J. Prediction of the dynamic characteristics and control of aircraft in prescribed trajectory flight. *J. Theor. Appl. Mech.* **2001**, *39*, 79–103.
9. Blajer, W.; Graffstein, J.; Krawczyk, M. UAV program motion and control in prescribed mission. *J. Theor. Appl. Mech.* **1998**, *36*, 963–978.
10. Cook, M.V. *Flight Dynamics Principles*; Butterworth-Heinemann: Oxford, UK, 2013.
11. Krawczyk, M.; Szczepanski, C.; Zajdel, A. Simulation and Testing of Flight Stabilization System Using Trimmers. In *Automation 2021: Recent Achievements in Automation, Robotics and Measurement Techniques*; Springer: Berlin/Heidelberg, Germany, 2021; pp. 185–196, ISBN 978-3-030-74893-7.

electronics

MDPI

Article

Analysis of Impulse Responses Measured in Motion in a Towing Tank

Agnieszka Czapiewska [1], Andrzej Luksza [2], Ryszard Studanski [2,*] and Andrzej Zak [3]

1 Faculty of Electronics, Telecommunications and Informatics, Gdansk University of Technology, Narutowicza St. 11/12, 80-233 Gdansk, Poland
2 Faculty of Electrical Engineering, Gdynia Maritime University, Morska St. 81-87, 81-255 Gdynia, Poland
3 Faculty of Mechanical and Electrical Engineering, Polish Naval Academy, Smidowicza St. 69, 81-127 Gdynia, Poland
* Correspondence: r.studanski@we.umg.edu.pl; Tel.: +48-58-558-6479

Abstract: The growing interest in developing autonomous underwater vehicles (AUVs) and creating underwater sensor networks (USNs) has led to a need for communication tools in underwater environments. For obvious reasons, wireless means of communication are the most desirable. However, conducting research in real conditions is troublesome and costly. Moreover, as hydroacoustic propagation conditions change very significantly, even during the day, the assessment of proposed underwater wireless communication methods is very difficult. Therefore, in the literature, there are considered simulators based on real measurements of underwater acoustic (UWA) channels. However, these simulators make an assumption that, during the transmission of elementary signals, the impulse response does not change. In this article, the authors present the results of the measurements realized in a towing tank where the transmitter could move with a precisely set velocity and show that the analyzed channel was non-stationary, even during the time of the transmission of a single chirp signal. The article presents an evaluation method of channel stationarity at the time of the chirp transmission, which should be treated as novelty. There is also an analysis of the impulse responses measured in motion in a towing tank.

Keywords: UWA channel; impulse response; underwater communication; signal propagation; channel stationarity; chirp signal

Citation: Czapiewska, A.; Luksza, A.; Studanski, R.; Zak, A. Analysis of Impulse Responses Measured in Motion in a Towing Tank. *Electronics* **2022**, *11*, 3819. https://doi.org/10.3390/electronics11223819

Academic Editor: Olivier Sename

Received: 18 October 2022
Accepted: 16 November 2022
Published: 20 November 2022

1. Introduction

The enormous progress in the field of wireless telecommunications has created expectations regarding the provision of high-quality wireless data transmission in the water environment. This comes from a growing interest in the utilization of autonomous underwater vehicles (AUVs) and underwater sensor networks (USNs) for science, military, and industrial applications [1,2]. The underwater acoustic (UWA) channel is a transmission medium and has a number of limitations in comparison to the radio or optical channel in the air. The main limitations are the relatively small range due to the strong attenuation of the signals, especially for higher frequencies, and the instability of the propagation conditions. These conditions are particularly important in the case of ensuring the transmission, e.g., in shallow waters, in wrecks, etc., where there are numerous reflections of the signal from the surface, bottom, and underwater obstacles [3,4]. In many cases, wireless supervision of working robots, e.g., in wrecks, requires high data throughput with the robot in motion. Movement in conditions of a strong multipath causes rapid changes in propagation conditions, i.e., a short time of channel stationarity. Obtaining high bit rates requires transmission in a wide band. Narrowing the bandwidth may be favored by the use of multivalent modulation, but then the requirements of stability of the propagation conditions increase.

Conducting research in real conditions is particularly troublesome and costly for logistical reasons [4]. Additionally, due to the large variability of hydrometeorological conditions, the results of the tests carried out at different times of the day and year may differ significantly [5]. Conducting research on various underwater wireless communication systems makes it much easier to have sets of sequences of recorded impulse responses of the UWA channel [6]. However, it should be noted that the most common assumption in underwater transmission simulation studies is that the channel is invariable for the duration of the elementary signal. However, in the case of the mutual movement of the transmitter and the receiver, this approach is an oversimplification.

The primary method of estimating a communication channel is to determine the channel impulse response parameters. Hence, the main aim of the study was to evaluate the parameters of the impulse response of the UWA channel measured in motion in the towing tank.

In the literature on the subject, there are publications concentrated on the measurement of impulse responses in UWA channels. Most frequently, the simulation results are presented, such as in [7], where the results of simulation studies on the pilot-based channel estimation with a minimal mean square error (MMSE) for the SM OFDM communication system are presented, and [8] where simulation studies for low frequencies (25 Hz) are presented. In [9], it was shown that the fact of reducing the depth of the water with the wave propagation direction has an impact on the form of the received signal. In [9], there was discussed the influence of the transducer movement in underwater acoustic communication on the basis of simulation analysis. In this article, it is noted that, due to the relative speed of the transmitter and receiver, there is a Doppler frequency shift, resulting in a change in the amplitude of the impulse response of the channels. The simulations indicate that the Doppler effect resulting from the relative velocity contributes to a large variation in the amplitude of the UWA channel impulse response. As the relative velocities or Doppler spread increases, the amplitude variation and phase variation of the arrival structure also increase. There are articles in which, in addition to simulation tests, experiments under real conditions are also presented, such as in [10], where various channel estimators are presented based on simulation tests, as well as experiments carried out in shallow water at various distances from several dozen to several hundred meters but under statics conditions. Channel modeling is important for the estimation of transmission quality. In [11] and [12], the authors present methods of underwater channel modeling. The channel impulse response can also be utilized for the estimation of underwater platform motion parameters [13].

In [14,15], the results of the measurements of the UWA channel impulse responses in real conditions, with the use of correlation methods using chirp and BPSK signals, are presented. Several articles have pointed out that the shape of the impulse response is also greatly influenced by the mutual motion of the transmitter and receiver, as well as by the motion of the water. For example, in [16], there is a discussion of the difficult propagation conditions in the multipath UWA channel due to changes caused by sea currents, as well as slow changes in temperature and/or salinity. It was noted that, even at the speed of 0.5 m/s between the receiver and transmitter, the changes in the propagation conditions would be particularly dynamic. Nevertheless, the article presents examples of impulse responses from measurements in static conditions, encouraging, at the same time, to conduct the research in motion. The article [17] shows the impact of daily changes on the quality of transmission. Given the disruptive nature of the acoustic channel, where an acoustic link between two nodes may present only a small packet loss for several hours, later, for a few hours, the link presents a high packet loss and then returns to stability again. In such a situation, it is not trivial to understand that when a drop in the performance of an acoustic network, it is caused by a DoS attack or by bad channel conditions. The increase in packet loss can be caused by several factors; for example, the increase in noise can be caused by a ship traveling close to the network deployment, by the presence of strong rain and wind, or by the presence of shadow zones caused by a temperature drop and the consequent change of the sound speed profile. The article [18] presents the results of the research on the UWA channel impulse response carried out in shallow waters in the area of Tanjung Balau, Johor, Malaysia. The transmitter and receiver were submerged at different depths and separated by different distances. Chirp pulses with a linear frequency modulation were used in the research. The cross-correlation between the transmitted and received signals represents the estimation of the impulse response of the channel (the multipath profile). The results show that the amplitude of the successive paths will not decrease sharply, and vice versa, as the distance between the sender and receiver increases. Moreover, the time difference between the different paths will be small as the distance increases. In other words, the successive paths will coincide in time. It should be noted that, although it has been pointed out that the mutual movement of the transmitter and the receiver will have a large impact on the form of the impulse response, the tests in real conditions are still carried out statically, and such results are presented. It is also emphasized in the literature that having an impulse response form allows the creation of channel simulation models. For example, in [19], the parameters of the UWA channel in shallow water were determined, and with their help, a channel model variable in time and frequency was developed. However, the measurements were also carried out in static conditions; therefore, the channel model does not take into account the changes in the impulse responses caused by the movement of the receiver in relation to the transmitter. In [20], an attempt to create a model of the swimming pool response suitable for simulation experiments, with the detection and localization of the emergency signal, was presented. Here, the analysis was carried out for stationary conditions, as well. It should also be emphasized that having information about the current form of the channel impulse response is essential for the creation of modern communication systems. In [21], it was indicated that the coherence bandwidth is one of the key transmission parameters used for designing the physical layer of a data transmission system to minimize the influence of time dispersion on the received signal. It can be calculated on the basis of the channel impulse response measured with the use of the correlation method and frequency-modulated signals or pseudorandom binary sequences.

According to the above, the authors state that they have not encountered any scientific reports presenting the results of research on impulse responses in a UWA channel obtained in motion and even more in difficult propagation conditions characterized by a strong multipath.

Underwater channels are characterized by a long memory time, often greater than the coherence time. Therefore, on the basis of successively emitted signals at intervals greater than the memory time, the coherence time cannot be determined. It is expected that this time will decrease with increasing the mutual speed between the transmitter and receiver.

The main contribution of this paper is to make measurements and analyses and to share the estimates of the impulse responses obtained in difficult propagation conditions in motion. It is based on the logarithmic chirp signals transmitted into the water from a transducer and received by four hydrophones. The measuring stand allows measurements in the frequency range from 5 kHz to 170 kHz. The towing tank in which the measurements were carried out provided difficult propagation conditions due to the strong multipath and the possibility of moving the transmitter in relation to the receivers at a precisely set speed. For the purposes of analyzing the obtained data, a method was developed to assess the non-stationarity of the propagation channel for the duration of the measurement signal. To the best of the authors' knowledge, no research group has published any measurements of this type. This paper will facilitate the work of those researchers who are interested in wireless underwater communication in harsh environments. In conclusion, the main distinctive contributions of our research, which make it innovative, are:

1. Research in a wide range of center frequencies and wide bands;
2. The development of a method for evaluating the stationarity of the UWA channel for a time duration of the measurement signal;
3. Provide complex impulse responses measured in motion in difficult propagation conditions.

The article is organized as follows: In Section Two, the measurement conditions have been described, especially the dimensions of the towing tank, the deployment of the transducers, and the configuration of the measuring stand. Section Three includes a proposition of a method for evaluating the stationarity of the UWA channel, as well as simulation results confirming the correctness of the adopted solution and an analysis of the signals measured in the towing tank. Section 4 presents the results of research focused on impulse responses measured in movement in the towing tank. The mathematical dependencies are given, and an analysis of the influence of the parameters of the generated signal on the number of replicas and root mean square delay spread was carried out. The final section contains a summary of the carried out research, as well as the general conclusions resulting from the measurements.

2. Description of Measurement Conditions

The research was conducted in the towing tank of the Faculty of Mechanical Engineering and Ship Technology of the Gdańsk University of Technology, presented in Figure 1. The height of the water column was 3 m, the width of the towing tank was 4 m, and its length was 40 m. In the central part of the towing tank, in the line of the transmitter's movement, four hydrophones were placed 1.5 m from the right wall (looking in the direction of the movement), at a depth of 1 m, in pairs, at a distance of 0.35 m between each other and 1.98 m between the pairs. The transmitting hydrophone was attached to the moving platform and immersed 1 m deep at a distance of 1.60 m from the right wall. The prepared structure allowed the transmitter to move at a speed of 1.5 m/s, with an accuracy of 0.01 m/s.

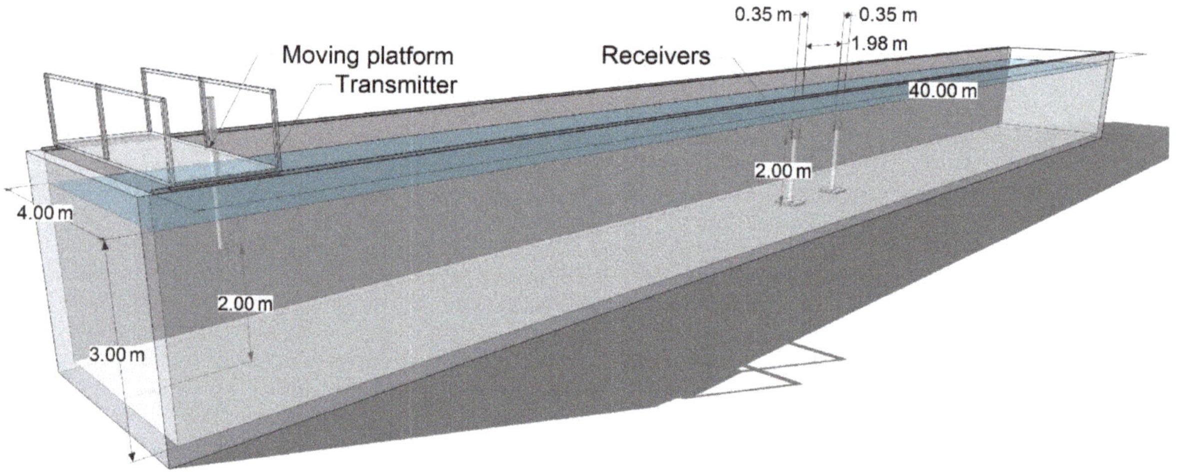

Figure 1. Deployment of the measuring devices.

In the transmitting part of the measurement stand, the signal was digitally formed in a Matlab environment and sent via an NI USB-6366 DAC to an ETEC PA1001 amplifier and further to a Reson TC4013 hydrophone. The receiving part consisted of four independent lines with a Reson TC4013 hydrophones-1 piece, a Reson TC4034-1 piece, and Reson TC4014-2 pieces. The lines with the Reson TC4013 and Reson TC4034 hydrophones were additionally equipped with a Reson VP1000 amplifier. The NI USB-6366 card connected to the computer on which the signals were recorded was used as an analog-to-digital converter.

For the purposes of analyzing the recorded signals, the speed of the sound propagation in the water was measured using an STD/CTD SAIV AS SD204 probe. During the measurements, this speed was constant and equal to 1477 m/s.

In the hydroacoustic channel, the transmitted signal propagates along L deterministic paths. The impulse response, $h(\tau)$, characterizes the multipath propagation, which means that it describes the l-th path into which the generated signal splits in response to random inhomogeneities of the medium, i.e., the presence of obstacles. Every replica, because of the specific distance from the transmitter to the receiver, will cause a delay of the signal in the reception point by time τ_l. Additionally, during the relative movement of the transmitter and receiver, we have to take into account the Doppler effect. Let parameter η_l describe the Doppler shift in the l-th path.

The equation describing the channel model can be written as follows [22,23]:

$$y(t) = \sum_{l=1}^{L} x(\eta_l t - \tau_l)h(\tau_l) + n(t), \tag{1}$$

where $x(t)$ is the transmitted signal, $y(t)$ is the received signal, and $n(t)$ is the additive noise.

The Doppler scattering of the signal is accounted for in (1) by summing the terms with various values of the Doppler parameter, η_l. Parameters τ_l and η_l and the impulse $h(\tau)$ vary within a measurement session. In ideal stationary conditions, this parameter should not change in time. In static real conditions, they are assumed to be fixed within the time interval T_s, corresponding to the processing of each separate signal.

In our experiments, we used logarithmic chirp signals, where the frequency of the signal varies exponentially as a function of time [24]:

$$f_i(t) = f_0 k^t, \tag{2}$$

where

$$k = \left(\frac{f_1}{f_0}\right)^{\frac{1}{T}}, \tag{3}$$

where f_0 is the starting frequency, f_1 is the ending frequency, and T is the sweep time.

It must be noted that $|f_0 - f_1| = B$ is a band occupied by the chirp, and $f_C = f_0 + \frac{f_1 - f_0}{2}$ is the central frequency of the chirp.

The corresponding time-domain function for the phase of a logarithmic chirp is the integral of the frequency, so we can finally write:

$$\phi(t) = \phi_0 + 2\pi f_0 \left(\frac{k^t - 1}{\ln(k)}\right), \tag{4}$$

where ϕ_0 is the initial phase (at $t = 0$).

The corresponding time-domain function for a logarithmic chirp is the sign of the phase:

$$x(t) = \sin(\phi(t)), \tag{5}$$

We assume that if $f_0 < f_1$, i.e., the frequencies will change from low to high, we will talk about a chirp-up; otherwise, if $f_0 > f_1$, i.e., the frequencies will change from high to low, we will talk about a chirp-down.

To estimate the impulse response of a UWA channel, we can use a matched filter. A matched filter is obtained by correlating a known delayed transmitted signal (template) with a received signal. This is equivalent to convolving the received signal with a conjugated time-reversed version of the template $x^*(-t)$. It allows us to detect the presence of the template in the received signal. According to the above estimate of an impulse response of a UWA channel, $h(\tau)$ can be expressed as follows [25]:

$$h(\tau) = \int_{-\infty}^{+\infty} x^*(-t + \tau)y(t)dt, \tag{6}$$

The matched filter is a linear filter that maximizes the output signal-to-noise ratio.

3. Evaluation of the Stationarity of the UWA Channel

Logarithmic chirp signals, which were emitted at the same time and on the same central frequency, were used to assess the stationarity of the channel. The measure of stationarity is the value of the cross-correlation of the impulse response determined for the chirp-up and chirp-down signals normalized to the root of the product of the maximum auto-correlation values of the impulse responses obtained from the chirps up and down. This coefficient can be described by the equation:

$$\hat{R}_{UD,coeff} = \frac{\max(\hat{R}_{UD}(m))}{\sqrt{\hat{R}_{UU}(0)\,\hat{R}_{DD}(0)}}, \tag{7}$$

where $R_{UU}(0)$ is the maximum of the autocorrelation for the estimation of the impulse response determined by a chirp-up; $R_{DD}(0)$ is the maximum of the autocorrelation for the estimation of the impulse response determined by a chirp-down; $R_{UD}(m)$ is the cross-correlation for the estimation of the impulse response determined by a chirp-up and a chirp-down.

A channel is considered stationary when the similarity coefficient, $\hat{R}_{UD,coeff}$, is greater than 0.5.

3.1. Simulation Research

The validation of the proposed solution of evaluating the stationarity of the hydroacoustic channel was carried out with the use of a UWA channel available in the Matlab computing environment library. The documentation of the UWA channel simulator shows that for the duration of the transmitted signal, the channel is a stationary one, i.e., the parameters determined for the individual paths do not change over time. Using this model, we can confirm that the proposed method allows for the determination of stationarity. The simulations were performed for the case of the transmitter moving in relation to the receiver. The simulation main parameters were as follows:

- The distance between the receiver and transmitter was 900 m;
- The channel depth was 6 m;
- The underwater sound propagation speed was 1470 m/s;
- The receiver immersion depth was 1 m;
- The transmitter immersion depth was 2 m;
- The source velocity was 15 m/s.

The estimates of the impulse responses obtained with the utilization of the chirp-up signal with different carrier frequencies tend to be radically different, as can be clearly seen in Figure 2.

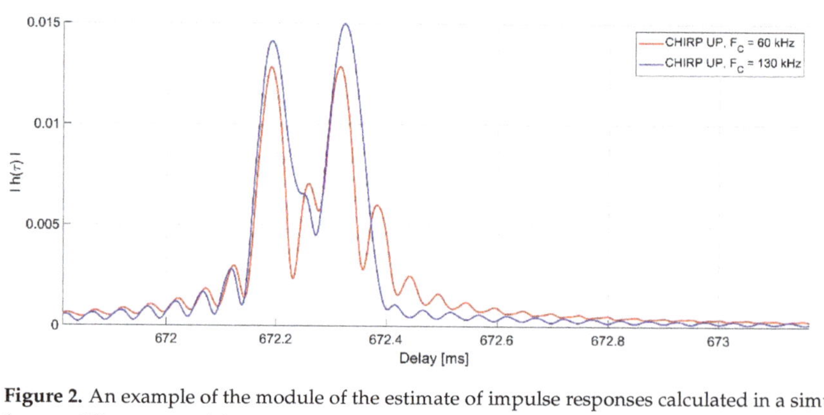

Figure 2. An example of the module of the estimate of impulse responses calculated in a simulation for two different central frequencies in shallow water for a transmitter moving at a speed of 15 m/s.

In this case (shown in Figure 2), the determined value of the similarity coefficient is 0.952. Under the same conditions, chirp-up and chirp-down signals were sent at the same time on the same center frequency (see Figure 3). In this case, the similarity coefficient is 0.993.

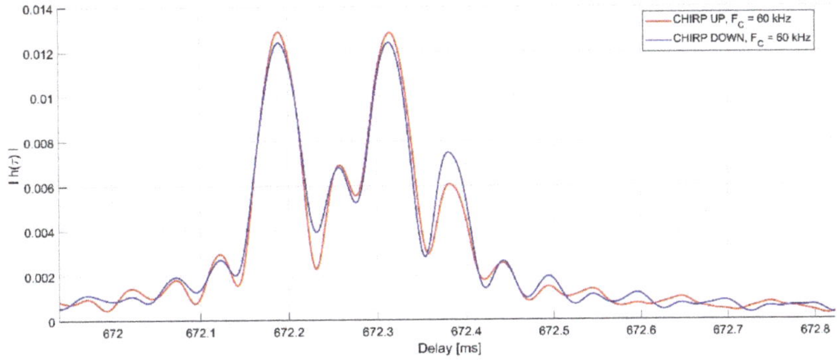

Figure 3. An example of the module of the estimate of impulse responses calculated in a simulation for chirp-up and chirp-down signals sent at the same time on the same center frequency in shallow water for a transmitter moving at a speed of 15 m/s.

It should be noted that although the channel did not change, a lower similarity coefficient was obtained in the case where two different center frequencies were used. On this basis, it was assumed that, in further research, we will use chirp-up and -down signals transmitted at the same time and on the same central frequency.

3.2. Evaluation of the Stationarity of the Measurement Channel

The stationarity of the measurement channel was evaluated in motion for the following transmitter speeds: 0.25 m/s, 0.5 m/s, 1 m/s, and 1.5 m/s. The measurement signal was generated at three different center frequencies: the F_C i.e., 50 kHz, 95 kHz and 140 kHz in band B equals 2 kHz, 10 kHz, 20 kHz, and 40 kHz, and the duration time, T, equals 30 ms or 60 ms.

Figure 4 shows the average (from four receivers) values of the measured similarity coefficients as a function of the distance from the receiver for the transmitter speed of $V = 1$ m/s and band $B = 20$ kHz and the duration of the measurement signal $T = 60$ ms.

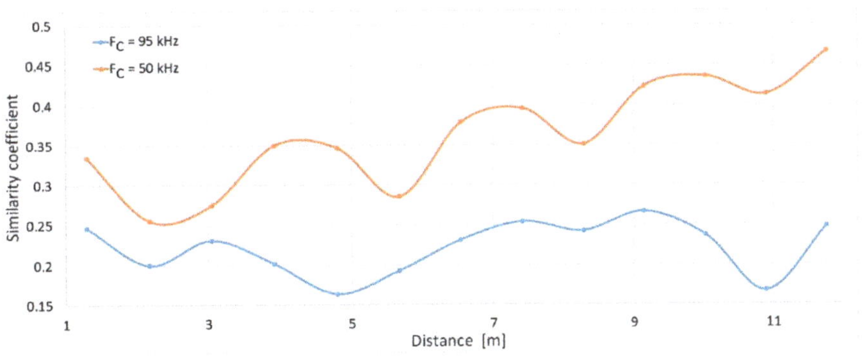

Figure 4. Measured similarity coefficients as a function of the distance between transmitter and receiver for the band $B = 20$ kHz and chirp duration of $T = 60$ ms at the transmitter velocity of $V = 1$ m/s.

The obtained results indicate that at the velocity $V = 1$ m/s for the chirp band $B = 20$ kHz and the duration $T = 60$ ms at a distance from 1 m to 12 m from the receiver, the channel was non-stationary, both for the center frequency $F_C = 50$ kHz and $F_C = 95$ kHz.

Figure 5 shows the average, from the four receivers, values of the measured similarity coefficients as a function of the distance from the receiver for the transmitter speed as before, i.e., $V = 1$ m/s, but the band was smaller and equals $B = 10$ kHz, and the duration of the chirp signal, T, was 30 ms.

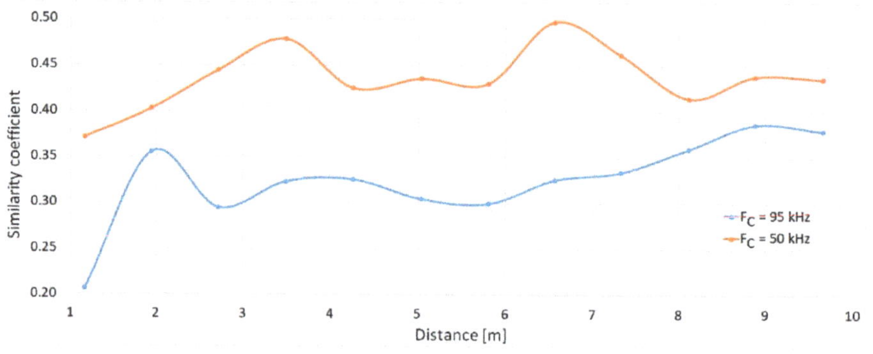

Figure 5. Measured similarity coefficients as a function of the distance from the receiver for the band $B = 10$ kHz and chirp duration $T = 30$ ms at the transmitter velocity $V = 1$ m/s.

The measurement results presented in Figure 5 show that despite the narrower band $B = 10$ kHz and a shorter chirp duration for two times, the channel was also non-stationary; however, the achieved similarity coefficients in Figure 5 are greater compared to those in Figure 4.

Figure 6 shows the average (from four receivers) values of the measured similarity coefficients as a function of the distance from the receiver for the transmitter velocity of $V = 0.5$ m/s in the band of $B = 20$ kHz and the duration of the chirp of $T = 30$ ms.

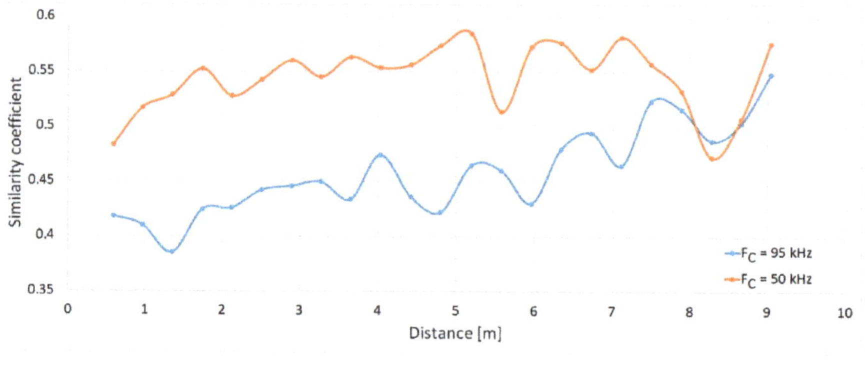

Figure 6. Measured similarity coefficients as a function of the distance from the receiver for the band $B = 20$ kHz and chirp duration $T = 30$ ms at the transmitter velocity $V = 0.5$ m/s.

From the obtained results, it can be concluded that by reducing the velocity of a moving transmitter, the values of the similarity coefficient are greater than those measured at higher velocities. From the above, it should be considered that the stationarity of the channel cannot be unambiguously associated with the distance between the transmitter and the receiver. Therefore, for each velocity, the similarity coefficients from all distances from the transmitter were averaged (Figure 7).

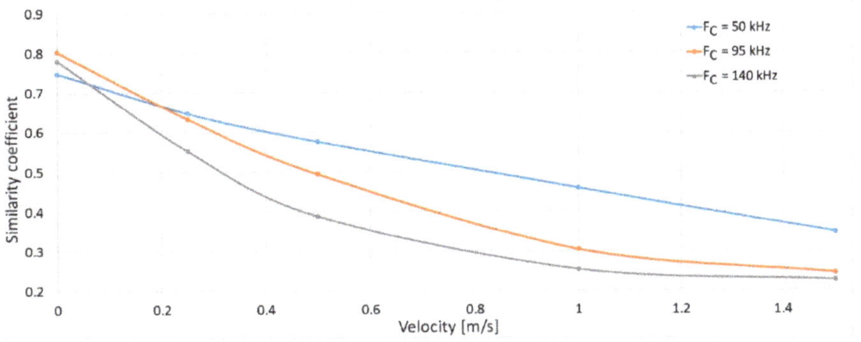

Figure 7. Similarity coefficients as a function of the transmitter velocity for the band $B = 20$ kHz and chirp duration $T = 30$ ms for the various central frequencies, F_C.

When comparing the values of the similarity coefficient for different velocities of the transmitter, it can be seen that for $B = 20$ kHz, only at the speed of $V = 0.25$ m/s and in static conditions (the transmitter did not move) for all of the analyzed center frequencies, the stationarity of the UWA channel was found.

Figure 8 shows the assessment of the influence of the measurement signal bandwidth on the value of the similarity coefficient.

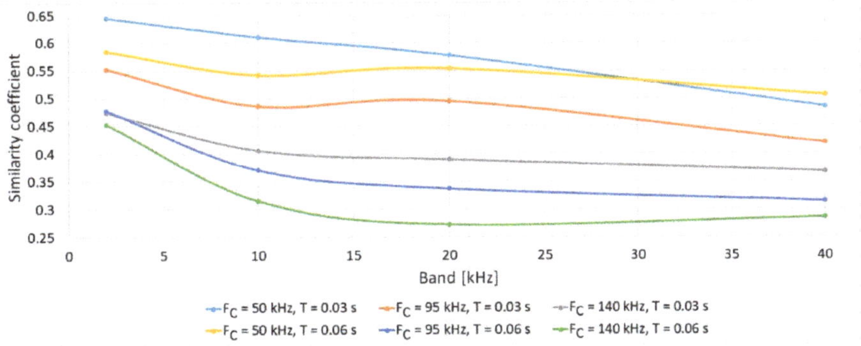

Figure 8. Effect of bandwidth B for different center frequencies, F_C, and chirp duration, T, on the value of the similarity factor for the transmitter speed, $V = 0.5$ m/s.

Based on the research results presented in Figure 8, it can be seen that the narrower the bandwidth occupied by the signal and the shorter its duration, the easier it is to ensure channel stationarity. Also, as the center frequency of the signal increases, the channel becomes non-stationary.

The analysis of the results presented in Figures 4–8 shows, as should be expected, that the higher the speed of the transmitter's movement in relation to the receiver, the longer the signal duration, and the greater the bandwidth, the more difficult it is to ensure the channel stationarity.

4. Research Results

The analysis of the impulse response parameters measured in motion was performed using the measurement stand and under the conditions described in Section 2. Single logarithmic chirp signals were transmitted to obtain complex impulse responses. It should be noted that two signals were not sent simultaneously, i.e., up and down, at the same time and at the same center frequency, as it was conducted in the evaluation of the stationarity of the UWA channel (Section 3). Thus, the impulse response was not overloaded. That is, on a given center frequency, the sequence of up/down/up-chirps were successively transmitted with an interval of 100 ms between them. Then, such a set of measuring signals was repeated with a time interval of 500 ms. Figure 9 shows an example of the registered impulse responses in the first measurement channel during the transmitter's approach to the receiver.

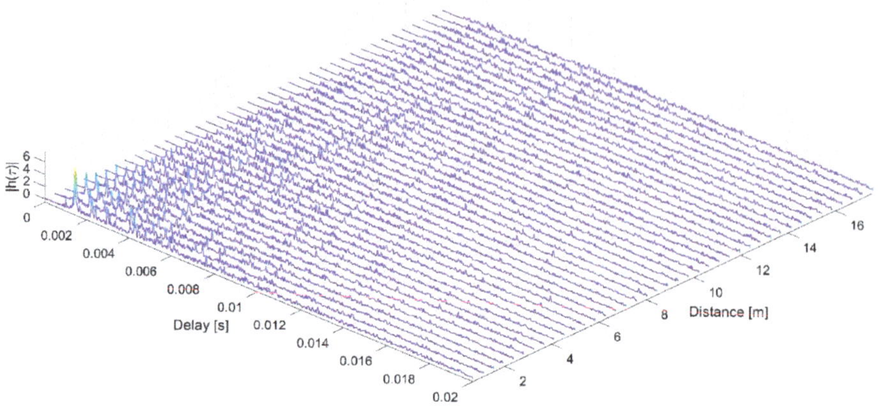

Figure 9. An example of changes in the modules of the impulse response estimates depending on the distance from the signal source for: $F_c = 60$ kHz, $v = 0.5$ m/s, $T = 60$ ms, and $B = 20$ kHz.

The obtained result clearly shows that as the transmitter gets closer to the receiver, the strength of the received signal increases, and the delay between the signal arriving directly from the direct route and the replicas increases. This is because the direct path of signal propagation from the transmitter to the receiver is shortened, e.g., when the transmitter passes the receiving hydrophone, then, in the adopted measurement configuration, the first replica, reflected from the water surface, must cover a distance of 2 m.

By analyzing the impulse responses measured in motion, the mean square delay spread, τ_{rms}, and the number of replicas were determined.

The root mean square delay spread, τ_{rms}, was determined according to the equation [26]:

$$\tau_{rms} = \sqrt{\frac{\sum_{i=1}^{N} \tau_i^2 p(\tau_i)}{P_m} - \overline{\tau}^2}, \tag{8}$$

where $p(\tau_i)$ is the power of the i-th replica, N is the number of replicas, and $\overline{\tau}$ is the average delay of the received replicas, which can be calculated according to the following equation [26]:

$$\overline{\tau} = \frac{\sum_{i=1}^{N} \tau_i p(\tau_i)}{P_m}, \tag{9}$$

where P_m is the total power of all signal replicas in a single impulse response, described by equation [26]:

$$P_m = \sum_{i=1}^{N} p(\tau_i). \tag{10}$$

Only those replicas whose level was not lower than −15 dB in relation to the signal with the highest power for the impulse response were taken for analysis, which is in line with the recommendation of ITU-R P.1407-7 [26]. All recorded signals were subjected to resampling before determining the impulse response. The resampling was performed on the basis of the measured speed of the sound in the water and the current speed of the transmitter. The aim of this procedure was to minimize the influence of the Doppler effect on the form of the final impulse response estimates.

4.1. *Analisis of Number of Replicas*

The assessment of the number of replicas meeting the conditions in accordance with Recommendation ITU-R P.1407-7 was carried out as a function of the transmitter velocity, V, the bandwidth occupied by the chirp signal, B, the center frequency, F_C, the chirp duration, T, and the distance between the transmitter and receiver.

Figure 10 shows the relation between the number of replicas and the distance for different bandwidths, B, and the duration of the chirp, T, at the transmitter's speed of $V = 1.5$ m/s.

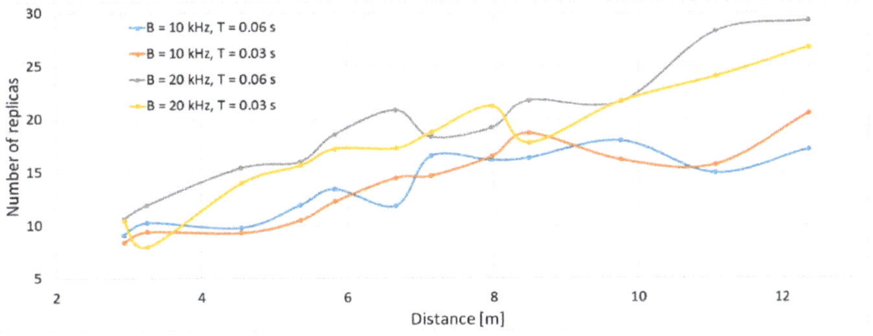

Figure 10. Comparison of the number of replicas as a function of the transmitter–receiver distance for different chirp bandwidths, B, and the duration, T.

The obtained result shows that the number of replicas does not depend on the chirp duration but depends on the bandwidth of the measurement signal. This relationship results from the fact that the wider the bandwidth, B, the greater the resolution of the obtained impulse response estimate, and as a result, the greater the number of distinguished replicas. Therefore, the results for different chirp durations can be averaged. Another example of this relationship, this time for a speed of $V = 1$ m/s, is shown in Figure 11.

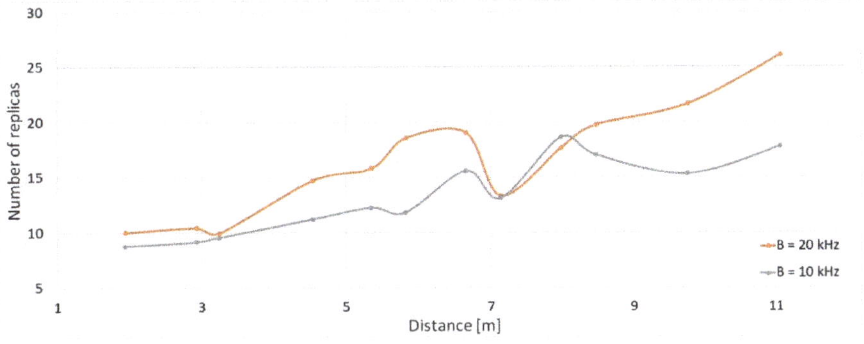

Figure 11. Comparison of the number of replicas as a function of the distance between the transmitter and the receiver for different chirp bandwidths, B.

An analysis of the influence of the movement speed of the transmitter relative to the receiver on the number of replicas was also carried out. The result is shown in Figure 12.

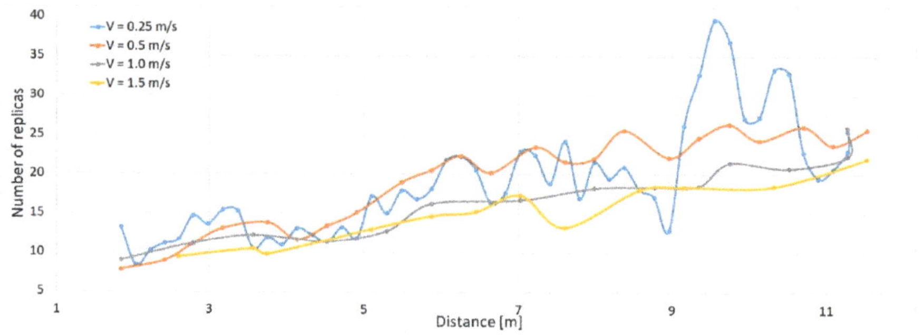

Figure 12. Number of replicas for different transmitter velocities as a function of its distance from the receiver for $B = 20$ kHz.

The obtained result shows that, in practice, the number of essential replicas in the channel does not depend on the speed.

4.2. Analisis of Root Mean Square Delay Spread

In the next stage of the analysis, the values of the root mean square delay spread, τ_{rms}, were compared for different transmission speeds of the transmitter and different bandwidths of the measurement signal, center frequency, and duration of the measurement signal as a function of the distance between the transmitter and the receiver. The results are shown in Figures 13–15.

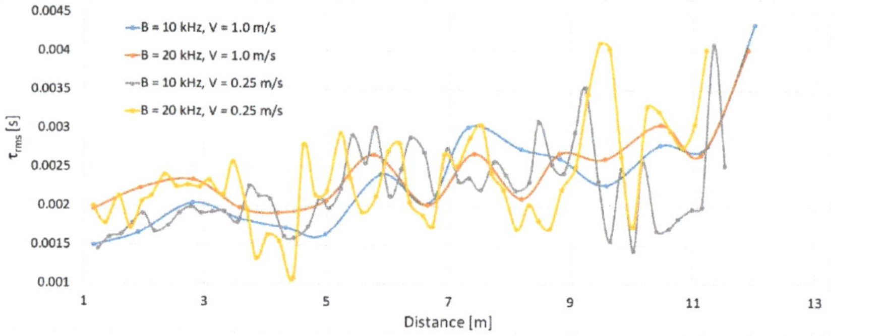

Figure 13. Influence of the distance from the receiver on the delay spread for chirps with a center frequency of $F_C = 60$ kHz, a duration of $T = 30$ ms, and different bandwidths, B, for transmitter velocities of $V = 1$ m/s and $V = 0.25$ m/s.

The results shown in Figure 13 show that despite the increase in the number of replicas with the bandwidth, the value of this parameter depends only on the distance.

The presented results show that, regardless of the movement speed of the transmitter relative to the receiver, the delay spread, τ_{rms}, does not depend on the bandwidth of the measurement signal, B, the center frequency, F_C, and the duration of the measurement signal, T. It should also be noted that the delay spread, τ_{rms}, depends only on the distance between the transmitter and the receiver.

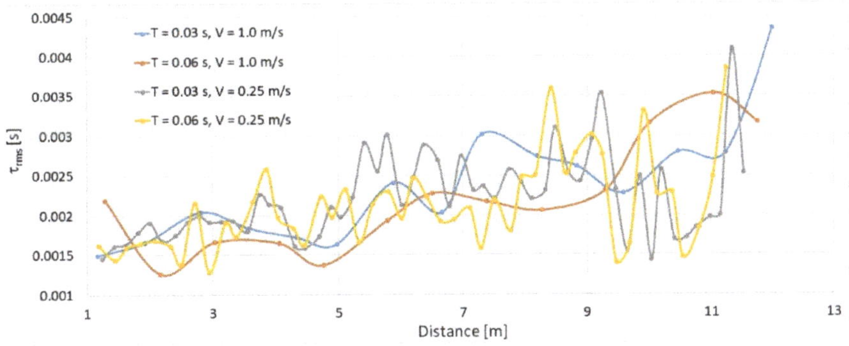

Figure 14. Influence of the distance from the receiver on the delay spread for chirps with a center frequency of $F_C = 60$ kHz, a bandwidth of $B = 10$ kHz, and different durations, T, for the transmitter velocities of $V = 1$ m/s and $V = 0.25$ m/s.

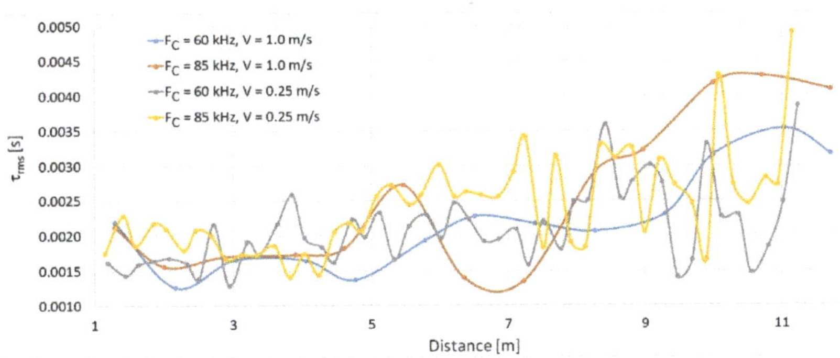

Figure 15. Influence of the distance from the receiver on the delay spread for chirps with a bandwidth of $B = 10$ kHz, a duration of $T = 60$ ms, and different center frequencies, F_C, for the transmitter velocities of $V = 1$ m/s and $V = 0.25$ m/s.

5. Conclusions

The knowledge of the impulse response of the UWA channel is important due to the influence of propagation phenomena on the form of the received signal. With this information, simulation tests can be carried out. This will allow for, among other things, quick and low-cost validation of the operation of underwater systems, which utilize elastic wave propagation. Most of the simulators of acoustic wave propagation in water available today provide data for static scenarios. In the opinion of the authors, this does not correspond to scenarios in motion. Therefore, one of the objectives of the research was to find out about the propagation conditions when the transmitter is moved relative to the receiver. For the authors, this research is an introduction to work on wireless underwater communication systems operating in difficult propagation conditions, e.g., in wrecks or between moving objects.

The aim of the presented research was to determine and analyze the estimates of impulse responses in a towing tank characterized by a strong multipath. In addition, the authors had the possibility to carry out measurements between objects during the movement of the transmitter with a precisely set velocity. As a result of the mutual movement of the receiver and the transmitter, the impulse response changed with the distance between the devices. As a result, a non-stationary channel was obtained, which was confirmed by the proposed method of channel stationarity evaluation. The measurements were conducted for selected speeds of the transmitter {0.0, 0.25, 0.5, 1.0, and 1.5} [m/s], selected central frequencies {50, 60, 85, 90, and 140} [kHz], selected bandwidths {2, 10, 20, 40} [kHz], and chirp duration times {0.03 and 0.06} [s].

Based on the tests carried out in the towing tank, it was found that, despite the fact that they were carried out in a non-stationary channel, one of the most important parameters describing the channel was that the root mean square delay spread did not change as a function of the transmitter movement velocity and the parameters of the generated chirp signal, but only depended on the distance between the transmitter and the receiver.

The complex estimates of the impulse responses obtained as a result of the measurements have been made available as Supplementary Materials to this article.

Supplementary Materials: The impulse responses measured in motion in a towing tank used in this study are openly available in FigShare at https://doi.org/10.6084/m9.figshare.21586596.v1.

Author Contributions: Conceptualization, R.S. and A.C.; methodology, R.S., A.C. and A.Z.; software, A.Z. and A.C.; validation, R.S., A.C., A.L. and A.Z.; formal analysis, A.Z. and R.S.; investigation, R.S., A.C. and A.L.; data curation, R.S.; writing—original draft preparation, R.S.; writing—review and editing, A.C, A.L. and A.Z.; visualization, R.S., A.C. and A.Z; funding acquisition, A.L. All authors have read and agreed to the published version of the manuscript.

Funding: This research was supported financially by the Ministry of Science and Higher Education program, the "Regional Excellence Initiative" of 2019–2022, project No. 006/PID/2018/19. The sum of the financing was 11,870,000 PLN.

Acknowledgments: We would like to thank the Faculty of Mechanical Engineering and Ship Technology of the Gdansk University of Technology for the opportunity to conduct research on the towing tank.

Conflicts of Interest: The authors declare no conflict of interest.

References

1. Karim, S.; Shaikh, F.K.; Chowdhry, B.S.; Mehmood, Z.; Tariq, U.; Naqvi, R.A.; Ahmed, A. GCORP: Geographic and cooperative opportunistic routing protocol for underwater sensor networks. *IEEE Access* **2021**, *9*, 27650–27667. [CrossRef]
2. Wu, J.; Song, C.; Ma, J.; Wu, J.; Han, G. Reinforcement Learning and Particle Swarm Optimization Supporting Real-Time Rescue Assignments for Multiple Autonomous Underwater Vehicles. *IEEE Trans. Intell. Transp. Syst.* **2021**, *23*, 6807–6820. [CrossRef]
3. Mizeraczyk, J.; Studanski, R.; Zak, A.; Czapiewska, A. A Method for Underwater Wireless Data Transmission in a Hydroacoustic Channel under NLOS Conditions. *Sensors* **2021**, *21*, 7825. [CrossRef] [PubMed]
4. Campagnaro, F.; Signori, A.; Otnes, R.; Goetz, M.; Sotnik, D.; Komulainen, A.; Nissen, I.; Favaro, F.; Guerra, F.; Zorzi, M. A Simulation Framework for Smart Adaptive Long and Short-range Acoustic Networks. In Proceedings of the OCEANS 2021, San Diego, CA, USA, 20–23 September 2021. [CrossRef]
5. Lv, Z.; Bai, Y.; Jin, J.; Wang, H.; Ren, C. Analysis of wave fluctuation on underwater acoustic communication based USV. *Appl. Acoust.* **2021**, *175*, 107820. [CrossRef]
6. Van Walree, P.A.; Socheleau, F.X.; Otnes, R.; Jenserud, T. The Watermark Benchmark for Underwater Acoustic Modulation Schemes. *IEEE J. Ocean. Eng.* **2017**, *42*, 1007–1018. [CrossRef]
7. Soman, A.M.; Nakkeeran, R.; Shinu, M.J. Pilot Based MMSE Channel Estimation for Spatial Modulated OFDM Systems. *Int. J. Electron. Telecommun.* **2021**, *67*, 685–691.
8. Zhang, Y.; Wang, Q.; Zhu, H.; Chai, Z.; Cui, Z.; Yang, F. Influence of upslope seabed on low frequency acoustic propagation characteristics in shallow sea. In Proceedings of the 2022 3rd International Conference on Geology, Mapping and Remote Sensing (ICGMRS), Zhoushan, China, 22–24 April 2022; pp. 810–813.
9. Kusuma, I.P.A.I.; M. Margareta, Z.B. Analysis of Channels Impulse Response Due to Transducer Movements in Underwater Acoustic Communication. In Proceedings of the 2019 IOP Conference Series: Materials Science Engineering, Surabaya, Indonesia, 23–25 July 2019; Volume 462, p. 012021.
10. Berger, C.; Zhou, S.; Preisig, J.; Willett, P. Sparse Channel Estimation for Multicarrier Underwater Acoustic Communication: From Subspace Methods to Compressed Sensing. *IEEE Trans. Signal Process.* **2010**, *58*, 1708. [CrossRef]
11. Onasami, O.; Feng, M.; Xu, H.; Haile, M.; Qian, L. Underwater Acoustic Communication Channel Modeling Using Reservoir Computing. *IEEE Access* **2022**, *10*, 56550–56563. [CrossRef]
12. Lucas, E.; Wang, Z. Performance Prediction of Underwater Acoustic Communications Based on Channel Impulse Responses. *Appl. Sci.* **2022**, *12*, 1086. [CrossRef]
13. Almanza-Medina, J.E.; Henson, B.; Shen, L.; Zakharov, Y. Motion estimation of underwater platforms using impulse responses from the seafloor. *TechRxiv* **2022**. [CrossRef]

14. Studanski, R.; Zak, A. Results of impulse response measurements in real conditions. *J. Mar. Eng. Technol.* **2017**, *16*, 337–343. [CrossRef]
15. Schmidt, J.; Kochańska, I.; Schmidt, A. Measurement of Impulse Response of Shallow Water Communication Channel by Correlation Method. *Hydroacoustics* **2017**, *20*, 149–158.
16. Stojanovic, M.; Preisig, J. Underwater acoustic communication channels: Propagation models and statistical characterization. *IEEE Commun. Mag.* **2009**, *47*, 84–89. [CrossRef]
17. Signori, A.; Campagnaro, F.; Nissen, I.; Zorzi, M. Channel-Based Trust Model for Security in Underwater Acoustic Networks. *IEEE Internet Things J.* **2022**, *9*, 20479–20491. [CrossRef]
18. Al_Aboosi, Y.Y.; Sha'ameri, A.Z. Experimental Multipath Delay Profile of Underwater Acoustic Communication Channel in Shallow Water. *Indones. J. Electr. Eng. Comput. Sci.* **2016**, *2*, 351–358. [CrossRef]
19. Naderi, M.; Pätzold, M.; Zajić, A.G. The design of measurement-based underwater acoustic channel simulators using the INLSA algorithm. In Proceedings of the OCEANS 2015, Genova, Italy, 18–21 May 2015; pp. 1–6.
20. Misiurewicz, J.; Bruliński, K.; Klembowski, W.; Kulpa, K. Multipath propagation of acoustic signal in a swimming pool. In Proceedings of the 2021 Signal Processing Symposium (SPSympo), Łódź, Poland, 20–23 September 2021; pp. 197–201.
21. Kochanska, I.; Schmidt, J.H.; Schmidt, A.M. Study of probe signal bandwidth influence on estimation of coherence bandwidth for underwater acoustic communication channel. *Appl. Acoust.* **2021**, *183*, 108331. [CrossRef]
22. Zakharov, Y.V.; Kodanev, V.P. Doppler Scattering Adapted Reception in a Hydroacoustic Cornmunication Channel. *Acoust. Phys.* **1995**, *41*, 219–223.
23. Zhao, Y.; Yu, H.; Wei, G.; Ji, F.; Chen, F. Parameter estimation of wideband underwater acoustic multipath channels based on fractional Fourier transform. *IEEE Trans. Signal Process.* **2016**, *64*, 5396–5408. [CrossRef]
24. Smith, S.W. *The Scientist and Engineer's Guide to Digital Signal Processing;* California Technical Pub: Sacramento, CA, USA, 2002.
25. Turin, G.L. An introduction to matched filters. *IRE Trans. Inf. Theory* **1960**, *6*, 311–329. [CrossRef]
26. Recommendation ITU-R P.1407-7 Multipath Propagation and Parameterization of its Characteristics (08/2019). Available online: https://www.itu.int/rec/R-REC-P.1407/en (accessed on 1 January 2022).

electronics

MDPI

Article

Selection of the Depth Controller for the Biomimetic Underwater Vehicle

Michał Przybylski

Faculty of Mechanical and Electrical Engineering, Polish Naval Academy, 81-127 Gdynia, Poland; m.przybylski@amw.gdynia.pl; Tel.: +48-261-262-524

Abstract: The aim of this paper is to select a depth controller for innovative biomimetic underwater vehicle drives. In the process of optimizing depth controller settings, two classical controllers were used, i.e., the proportional–integral–derivative (PID) and the sliding mode controllers (SM). The parameters of the regulators' settings were obtained as a result of optimization by three methods of the selected quality indicators in terms of the properties of the control signal. The starting point for the analysis was simulations conducted in the MATLAB environment for the three optimization methods on three types of indicators for three different desired depth values. The article describes the methods and quality indicators in detail. The paper presents the results of the fitness function obtained during the optimization. Moreover, the time courses of the vehicle position relative to the desired depth, the side fin deflection angles, the calculated parameters of the control signals, and the observations and conclusions formulated in the research were presented.

Keywords: biomimetic underwater vehicle; depth controller; genetic algorithms; particle swarm optimization; Pareto optimization

1. Introduction

In the 21st century, there has been dynamic development of mobile robots. One of the fields is underwater robots, where we can distinguish between ROV (Remote Operated Vehicle) and AUV (Autonomous Underwater Vehicle). Vehicles based on ROV technology, which dates back to the 1960s, have been ideally developed and are successfully used in all kinds of underwater operations [1–3]. A significant limitation of the mobility of these vehicles is the use of cable tether, which leads scientists and engineers to develop AUVs [4]. Autonomy, advanced control and positioning systems allow the realization of many civilian and military tasks. The development of bionics resulted in a new trend in the construction of underwater mobile robots, whose main idea is to imitate underwater animals. These vehicles are called biomimetic underwater vehicles (BUVs) [5,6], which mimic construction and motion kinematics. Most often, the prototype vehicle is inspired by the shape and movement of aquatic creatures, although there are designed based on manta rays, penguins and many others. Nevertheless, it is essential to carefully analyze the animal movement and to develop an appropriate simplified mathematical model, which will be used as close as possible to the BUV [7]. Artificial fins, similar in shape and appearance to the real ones, propel the BUV. Such propulsion is called wave or undulating propulsion, and it can be placed in different parts of the BUV's hull, depending on the type of design, its maneuverability and the speeds it can achieve. An electric motor usually drives it to generate sufficient vehicle thrust through oscillating motion. A sinusoidal function usually describes oscillatory motion, but different parts can be used depending on the type and purpose of the undulating propulsion. To describe oscillatory motion, usually use parameters such as the neutral position, which is the zero position for oscillation, the amplitude of oscillation, which defines the maximum deformation of the fins, and the frequency of oscillation [8].

Citation: Przybylski, M. Selection of the Depth Controller for the Biomimetic Underwater Vehicle. *Electronics* **2023**, *12*, 1469. https://doi.org/10.3390/electronics12061469

Academic Editor: Hamid Reza Karimi

Received: 7 February 2023
Revised: 9 March 2023
Accepted: 14 March 2023
Published: 20 March 2023

Control of underwater vehicles has been investigated quite thoroughly. However, still many difficulties occur, among other things, due to environmental disturbance, highly nonlinear behavior of vehicles, the complexity of the vehicle hydrodynamics or the application of new innovative propulsion systems. The articles [9–12] present various systems for controlling the depth of underwater vehicles. They concern both controllers tuned classically and using artificial intelligence algorithms. However, the works presented above concern, to a large extent, ROV or AUV vehicles using a classic screw drive propeller. Biomimetic underwater vehicles can be equipped with an artificial swim bladder [13] similar to a fish bladder, which controls buoyancy and depth. Another solution used to control the depth of biomimetic underwater vehicles [14,15] is changing the locomotion primitives of the fins, usually by changing frequency, amplitude or side fins' phase shift. In more advanced applications, changing the depth can be done by adjusting the angles of attack of the vehicle's control surfaces, their stiffness [16], or surface area of side fins [17] while the vehicle is moving at a certain speed through its wave propulsion.

The main contribution of this paper is to present depth controllers' different tuning methods for innovative BUV wave propulsion drives. The proposed methods use artificial neural networks in the tuning process and the experimental verification system of controller gains to evaluate its performance. In the presented literature, one can notice the lack of use of neural network-based methods for tuning depth controllers based on bioinspired wave propulsion drive. This research could have significant implications for the development of biomimetic autonomous underwater vehicles for various applications such as exploration of the ocean ecosystems, environmental monitoring, or even search and rescue missions. The rest of the paper is organized as follows. The mathematical model of the vehicle is based on the Fossen model [18] with modifications to include the new wave drive used. Commonly used underwater vehicles are driven mainly by a set of screw propellers, whereby the thrust generated by the propellers can be calculated using specific mathematical formulas [19]. In the case of a new wave propulsion system that mimics the action of fish fins, the forces and moments of force acting on the vehicle were calculated using the author's method presented in the literature [20]. Then the applied depth controllers and their control methods are presented. The following section offers the research problem and the results obtained in the simulation process. The final section formulates conclusions and plans for future research.

2. Materials and Methods

This chapter presents the simulation model and methods used in the tuning process of the depth controller settings. It contains two main subsections. The first concern describes biomimetic underwater vehicles with an innovative propulsion system. A mathematical model based on the above-mentioned vehicle is described in the further part of this subsection. While the second main subsection deal with the description of controllers, optimization methods and fitness functions used in the simulation process.

2.1. Control Object

This subsection describes the construction of a biomimetic underwater vehicle (mini-Cyber Seal) with particular emphasis on the new propulsion system. In the next part, the mathematical model of the vehicle is presented, where special attention is paid to the proposed solutions for modelling the vector of vehicle forces and moments.

2.1.1. Mini CyberSeal

The study used a physical model of the BUV mini CyberSeal vehicle shown in Figure 1, is a smaller prototype of the larger vehicle. The purpose of the downsized BUV was to test the performance of the larger vehicle's counterpart's propulsion system and to test a new type of control. Unlike its predecessors, the mini CyberSeal's propulsion system had two tail fins instead of one. They generate the major thrust and are additionally responsible for changing course. In addition, a side propulsion system is mounted in front of the fuselage, which generates additional thrust and changes the depth. The rear and side fins are made of polycarbonate and rubber.

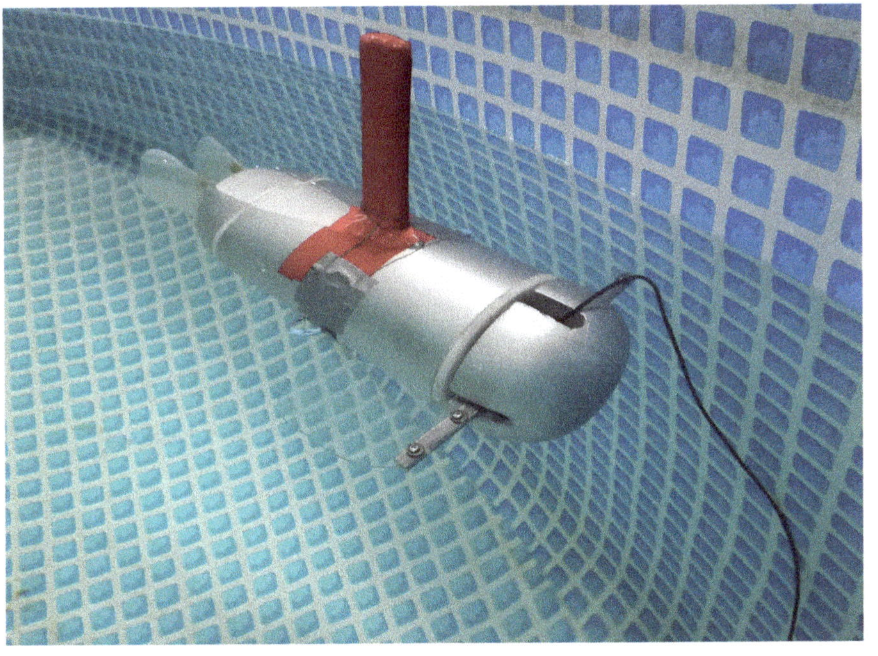

Figure 1. Model of the BUV mini CyberSel [own source].

The rear fins can move at a frequency of up to 3 Hz within a range of about 80 degs outward, and up to 12 degs to the vehicle's inside axis of symmetry. The side fins can also move at up to 3 Hz over a range of ±45 degress. All electronic components, sensors, servos and a 7.2 V 10 Ah li-ion battery were enclosed in a sealed tube made of POM-C material. The servos with a 1:1 gearbox, responsible for the movement of the fins, are controlled via POLOLU-1353 miniMaestro servo controllers via RS232. The base station communicates via a WiFi network supported by a TP-Link TL-WR702N access point and a WIZNET Wiz-145SR server port providing four serial ports. The vehicle has an internal artificial buoyancy bladder, which is responsible for the vehicle's static depth adjustment or buoyancy control. The mini CyberSeal vehicle has been equipped with an OS-5000 digital compass from Ocean Server and an A-10 depth sensor from Wika with a measurement range of 0–1 bar. The sensors mentioned above make it possible to read current depth values and vehicle motion parameters, i.e., angle of an inclination concerning individual axes of the coordinate systems.

2.1.2. Mathematical Model

The model captures the underwater vehicle's rigid body dynamics, hydrostatics and hydrodynamic effects. Critical issues for the modelling and simulation of BUVs are model complexity, ease of implementation and accuracy of prediction. Therefore, for the mathematical description, some simplifications are adopted for the vehicle: it has three planes of symmetry, moves at low speed in a viscous fluid, and has six degrees of freedom. When analyzing the motion of an underwater vehicle, two reference systems are defined:

(1) a stationary xyz coordinate system associated with the Earth,
(2) a moving $x_o y_o z_o$ coordinate system associated with the underwater vehicle.

The moving coordinate system is commonly called the "vehicle reference system", and its origin corresponds to the geometric center of the vehicle. The different axes of this coordinate system correspond to the following:

(1) x_o—the longitudinal axis directed from the stern to the bow,
(2) y_o—transverse axis directed to the starboard side,
(3) z_o—vertical axis directed towards the bottom.

Changes in the position of the moving $x_o y_o z_o$ coordinate system are described relative to the adopted xyz coordinate system associated with the Earth. Due to the low velocity of the vehicle, the acceleration of points on the Earth's surface due to its spin is neglected, and the xyz system is considered stationary. It is suggested that angular and linear velocities be described in the reference system associated with the vehicle while the vehicle's orientation is described in a stationary coordinate system. The quantities describing the vehicle's movement are defined according to the SNAME notation in Table 1 .

$$\mathbf{M}\dot{\nu} + \mathbf{D}(\nu)\nu + g(\eta) = \tau \tag{1}$$

where:

ν—vector of linear and angular velocities, i.e., $\nu = [u, v, w, p, q, r]$;

η—vector of vehicle position and Euler angle coordinates in the stationary system;

M—inertia matrix (equal to the sum of the rigid body mass matrix $\mathbf{M_{RB}}$ and the associated masses matrix $\mathbf{M_A}$);

$\mathbf{D}(\nu)$—hydrodynamic damping matrix;

$g(\eta)$—matrix of restoring forces (gravity forces P and buoyancy forces B);

τ—vector of forces and moments acting on the vehicle.

Table 1. Notation used in describing the movement of underwater vehicles.

Degrees of Freedom	Name of Movement	Forces and Moments	Angular and Linear Velocities	Position and Euler Angles
1	Movement in the direction of the x_o axis	X	u	x
2	Movement in the direction of the y_o axis	Y	v	y
3	Movement toward the z_o axis	Z	w	z
4	Rotation about the x_o axis	K	p	ϕ
5	Rotation about the y_o axis	M	q	θ
6	Rotation about the z_o axis	N	r	ψ

Given the assumptions mentioned above, a nonlinear model of motion in six degrees of freedom is adopted for simulating the movement of the mini CyberSeal. The action of the vehicle is described by six differential equations, which, presented in matrix form, have the following format: The right side of Equation (1) represents the vector of forces and moments of force acting on the vehicle generated by the vehicle's propulsion system (2).

$$\tau = [X, Y, Z, K, M, N] \tag{2}$$

where:

X, Y, Z —forces acting on the vehicle in the longitudinal, transverse and vertical symmetry axis, respectively;

K, M, N—moments of forces acting in relation to the longitudinal, transverse, and vertical symmetry axis, respectively.

The vector of forces and moments of forces generated by the wave propulsion can be calculated by considering the propulsion system set-up in each design. Figure 2 shows the mini CyberSeal propulsion model consisting of two counter-phased tail fins and two independently controlled side fins. The thrust produced by each fin should be conveyed to the center of gravity O (Figure 2) using simple vector transformation formulas:

$$X = X_{tl} + X_{tp} + X_l + X_p \quad (3)$$

$$Y = Y_{tl} + T_{tp} \quad (4)$$

$$Z = Z_l + Z_p \quad (5)$$

$$K = 0 \quad (6)$$

$$M = M_l + M_p \quad (7)$$

$$N = N_{tl} - N_{tp} + N_l - N_p \quad (8)$$

where:

tl, *tp*, *l* and *p*—subscripts referring to the action of the left rear fin, right rear fin, left-side fin and right-side fin, respectively.

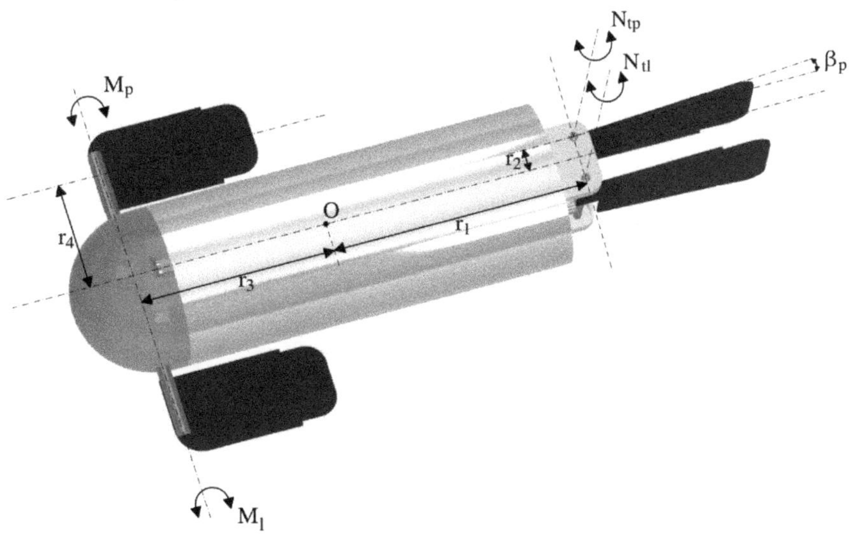

Figure 2. Mini CyberSeal propulsion model [own source].

The individual components of the vector, e.g., X_{tl}, Y_{tl}, N_{tl} can be calculated using the position of these fins with respect to the centre of gravity according to the equations:

$$X_{tl} = \cos(\beta_l) * T_{tl} \quad (9)$$

$$X_{tp} = \cos(\beta_p) * T_{tp} \quad (10)$$

$$Y_{tl} = \sin(\beta_l) * T_{tl} \quad (11)$$

$$Y_{tp} = \sin(\beta_p) * T_{tp} \quad (12)$$

$$N_{tl} = r_2 * X_t l + r_1 * Y_{tl} \quad (13)$$

$$N_{tp} = r_2 * X_t p + r_1 * Y_t p \quad (14)$$

$$X_l = \cos(\alpha_l) * T_l \quad (15)$$

$$Z_l = \sin(\alpha_l) * T_l \quad (16)$$

$$M_l = r_3 * Z_l \tag{17}$$

$$N_l = r_4 * X_l \tag{18}$$

More information on the mathematical dependencies of the novel drive used is contained in the [21]. As shown in Figure 1 or Figure 3, the Mini CyberSeal has two side fins and two tail fins, which generate time-varying thrust.

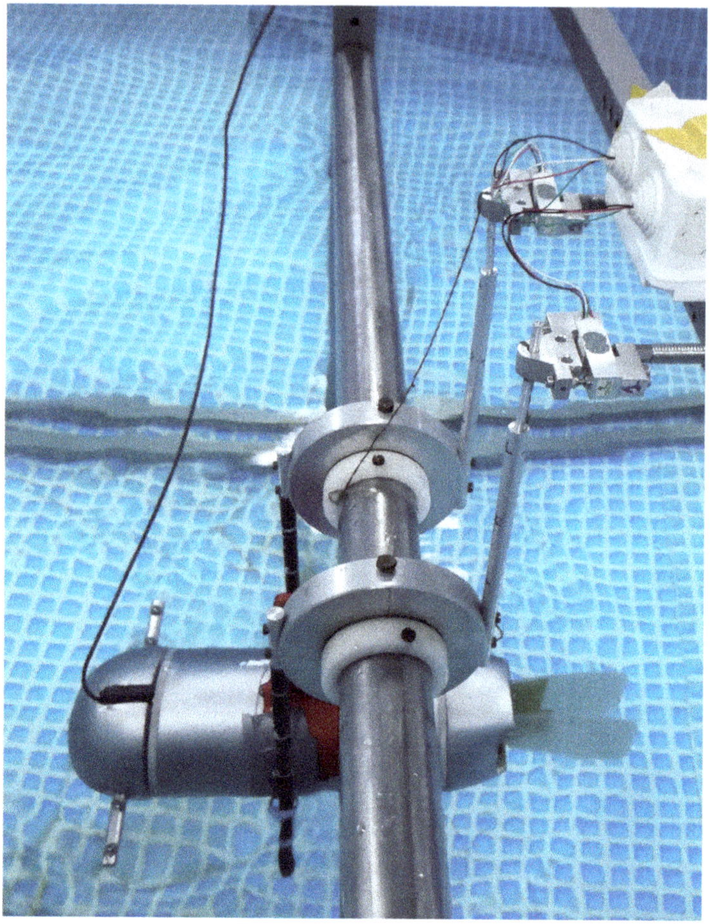

Figure 3. Mini CyberSeal thrust measurement stand [own source].

The value of the thrust T_{tl}, T_{tp}, T_l, T_l depends on the control parameters, including the frequency and amplitude of the deflection of each fin. The thrust values generated by each fin for different frequencies and amplitudes were determined experimentally. Finally, the vehicle's speed is variable and dependent on the frequency of fin oscillations. The method of measuring the thrusts generated by the mini CyberSeal used in the mathematical equations is presented in the literature [20]. The thrust T generated by the fin is the sum of two components:

$$T = T_{av} + T_{osc} \tag{19}$$

where:

T_{av}—constant thrust component at a specific fin oscillation frequency;

T_{osc}—variable component modelled by a sinusoidal wave with a specific amplitude (at a specific fin oscillation frequency).

At the same time, the test stand is shown in Figure 3. In addition, the right side of Equation (1) considers the effect of environmental disturbances such as wind, waves and sea currents, which significantly impact the BUV. The left side of Equation (1) describes the forces and moments of force caused by physical phenomena, such as rigid body inertia and the inertia of masses accompanying viscous fluid, the hydrodynamic drag exerted by water, and the balance of gravity and buoyancy forces. Using the mathematical relationships included in the literature [18], the matrix parameters describing the left side of Equation (1) can be calculated.

2.2. Methods

This subsection deals with the main part of the work concerning the selection of depth controllers' settings. The first subsection describes the controllers used, i.e., PID and SM. The following subsections describe the optimization methods and the fitness functions based on which the aforementioned methods realized optimization of used controllers settings.

2.2.1. Depth Controllers

Two classic controllers were used as depth controllers for the mini CyberSeal vehicle. The first one is about a PID controller that calculates the error value $e(k)$ as the difference between the set depth value and the value received from the depth sensor and applies a correction based on proportional, integral and derivative terms (denoted P, I, and D, respectively). Hence the name [22,23]. The action of the PID controller is described by the following formula presented in the discrete form:

$$u(k) = k_p e(k) + k_i \sum_{k=1}^{k_{max}} e(k) + k_d \Delta e(k) \tag{20}$$

where:

$u(k)$ is a control signal in k step of simulation;

$e(k)$ is an error signal in k step of simulation;

$\Delta e(k)$ is a change of error signals in k step of simulation, i.e., $e(k) - e(k-1)$;

k_p, k_i and k_d are constant quantities called gain factors.

The second controller used is the sliding mode controller (SM) [24,25], where sliding mode control is achieved by controlling nonlinear systems, which changes the dynamics of a nonlinear system by applying a discontinuous control signal, which forces the system to "slide" along upstream of the expected behavior of the system. It is calculated using the following formulas, also presented in the discrete form:

$$s(k) = \frac{\lambda e(k) + \Delta e(k)}{\varphi} \tag{21}$$

$$\text{if} \quad |s(k)| > 1, \quad \text{then} \quad s(k) = sign(s(k)) \tag{22}$$

$$u(k) = k_s s(k) \tag{23}$$

where:

$u(k)$ is a control signal in k step of simulation;

$s(k)$ is a normalized control signal in k step of simulation;

$e(k)$ is an error signal in k step of simulation;

$\Delta e(k)$ is a change of error signal in k step of simulation, i.e., $e(k) - e(k-1)$;

λ, φ, k_s is a constant settings of SM controller.

2.2.2. Optimization Methods

To tune controller settings, used the Global Optimization toolbox of the MATLAB environment [26,27]. The following three optimization methods were used: Genetic Algorithm (GA), Particle Swarm Optimization (PSO) and Pareto Simulation (PSA).

The genetic algorithm, first formalized as an optimization method by Holland, is a global optimization technique for multi-dimensional, nonlinear, and noisy problems and a stochastic search technique based on the mechanism of natural selection and natural genetics [28,29]. A genetic algorithm (GA) solves both constrained and unconstrained optimisation problems based on a natural selection process that mimics biological evolution. The algorithm repeatedly modifies a population of individual solutions. At each step, the genetic algorithm randomly selects individuals from the current population and uses them as parents to produce the children for the next generation. Changes are introduced into the offspring through mutation, crossover and other genetic operators. The procedure ends when satisfactory genotypes (a set of traits of an individual) are obtained, which are matched by phenotypes with a high fitness function (an individual from the population). Over successive generations, the population "evolves" toward an optimal solution. An initial population of 40 individuals was generated using a MATLAB random generator during optimization. Individuals in the current generation are estimated using one of the three fitness functions described in the following subsection. After calculating the fitness function, the reproduction algorithm creates children for the next generation. The following operators are used in reproduction: fitness rank scaling, Stochastic uniform selection function, Crossover fraction equal to 0.8, and Gaussian mutation function. The GA optimization was stopped when the maximum number of 100 generations was reached and/or when no change in the best fitness function value for new generations was detected during the next 50 steps [30].

The PSO algorithm is based on a simplified social model closely tied to swarming theory. It solves a problem by having a population of candidate solutions, here dubbed particles, and moving them around the search space according to simple mathematical formula over the particle's position and velocity. Each particle's movement is influenced by its local best-known position. Still, it is also guided toward the most notable positions in the search space, updated as better places are found by other particles [31]. A physical analogy might be a swarm of bees searching for food sources. In this analogy, each bee (referred to as a particle here) uses its memory and knowledge gained by the swarm to find the best available food sources. This is expected to move the swarm toward the best solutions. Based on the literature [32,33], the following PSO parameters were assumed: (1) MaxStallIterations (relative change in the value of the best objective function): 20, (2) MinNeighborsFraction (setting both the initial neighborhood size for each particle and the minimum neighborhood size): 1, (3) SwarmSize: 200. As in the case of GA, one of the critical problems is to properly define the fitness function to get the correct optimization rates. The fitness functions used were analogous to GA and are presented in the following subsection.

Pareto optimization (PSA) is a field of multi-objective decision-making that deals with mathematical optimization problems involving more than one objective function for simultaneous optimization. Multi-objective optimization has found application in many scientific areas, including engineering, economics and logistics, where optimal decisions must be made during trade-offs between two or more conflicting objectives. When we have several objective functions that we want to optimize simultaneously, these solvers find optimal trade-offs between competing objective functions. This method can also be applied to a single-objective problem. PSA uses pattern search on a set of points to iteratively search for non-dominated points. It should fulfil all constraints and linear constraints in each iteration. Theoretically, the algorithm converges to points near the true Pareto front [34]. In the algorithm used, in the beginning, the PSA creates an initial set of 200 randomly selected points and then checks whether these points are feasible concerning the bounds and linear constraints. If impossible, the algorithm projects the initial points into a linear subspace of linearly feasible points by solving a linear programming problem and removing duplicate points. The PSA then divides the points into two sets named "archive" and "iterative". The archive set contains non-dominated points associated with a mesh size less than 10^{-6} and satisfying all constraints within 10^{-6}. PSA checks the location of each point in the 'iterative' set. Success is achieved if the polled points yield at least one dominated point. PSA then extends the probing in successful directions multiple times, doubling the 6e grid to find a dominant point. If any non-dominated point is obtained, the grid size is halved. The algorithm stops when: (1) the mesh size exceeds the value (+Unity), (2) the fitness function decreases to the value (-Unity), (3) it reaches the maximum number of iterations equal to 400.

2.2.3. Fitness Function

It is necessary to determine proper fitness functions to obtain appropriate optimization results for the depth controller. For this paper, three functions commonly used in mathematics were formulated [35,36]. The first is Integral Absolute Error (ISA), the sum of the absolute values of the error signals $e(k)$ in all simulation steps. Its task is to select the controller parameters so that the error rate between the desired depth and the present depth value is as low as possible throughout the simulation period. The ISA in discrete form is shown in Equation (24).

$$f_{fit1} = \sum_{k=1}^{k_{max}} |e(k)| \tag{24}$$

The second fitness function is an Integral of Squared Error (ISE) in the discrete form Equation (25). The ISE integrates the square of the error over time. ISE will penalize for significant errors more than smaller ones (since the square of a large error will be much bigger). Control systems specified to minimize ISE will tend to eliminate significant errors quickly but will tolerate minor errors persisting for an extended period. Often this leads to fast responses but with considerable, low amplitude oscillation.

$$f_{fit2} = \sum_{k=1}^{k_{max}} e(k)^2 \tag{25}$$

The third proposed fitness function is based on combining two direct control quality indexes and is presented in the following Equation (26).

$$f_{fit3} = \sum_{i=1}^{i_{max}} t_r(i) + k_M \sum_{i=1}^{i_{max}} M_p(i) \tag{26}$$

It takes into consideration rising times t_r in [s] and first overshoots M_p in [rad] for all i_{max} changes of the desired course. Because of the small value of M_p in [rad] compared to t_r in [s] additional gain factor of the sum of first overshoots was introduced $k_M = 25$.

3. Research Problem and Results

The research problem of this work is to find a solution for the most effective depth control of the mini CyberSeal vehicle. With a mathematical model of the vehicle, two types of controllers (PID, SM), three different methods for optimizing controller settings, and three fitness functions, an attempt was made to select a suitable controller and its parameters. All regulators, methods and functions should be examined in the tuning process and verified for each combination (Figure 4). The tuning process was carried out for three different immersion changes: (1) 0.2 [m]—shallow immersion, (2) 0.5 [m]—deep immersion, (3) and then two immersion changes—first 0.5 [m] then second 0.2 [m] depth change after 20 s of simulation. In contrast, the verification process was based on the tuning process for each combination of 25 randomly selected depths from 0 to 0.7 [m] (in 0.1 [m] increments). The average value of the fitness function was conducted from 25 verification tests.

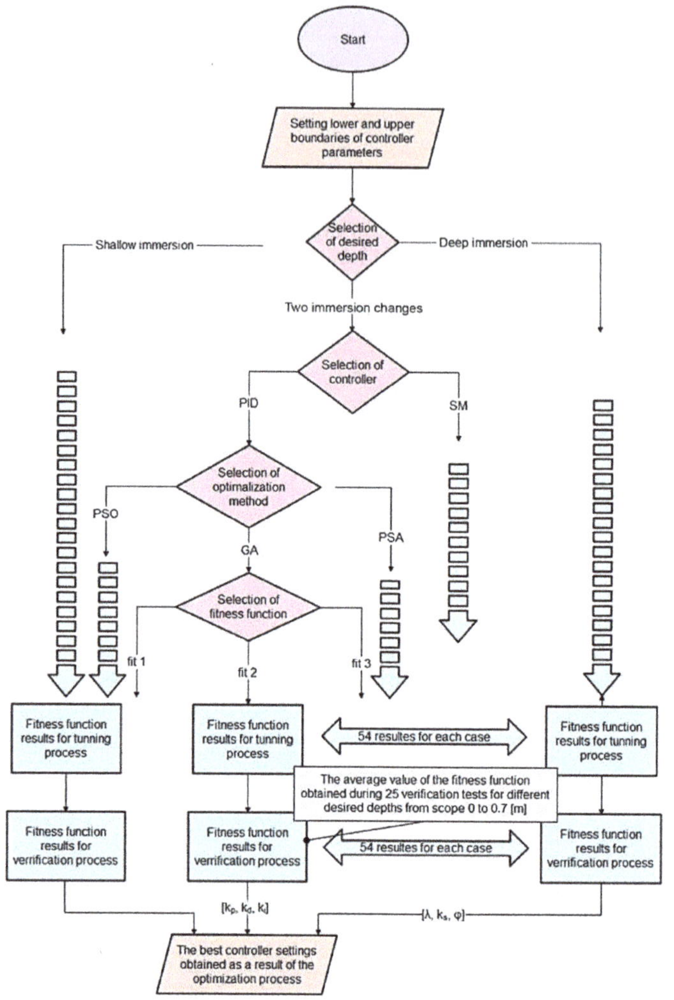

Figure 4. Flowchart optimization process to receive the best controllers settings.

Results and Discussion

The test results are shown in Tables 2–4 for the corresponding fitness functions 1–3. The table includes three different combinations of depth change for both the testing (T) and verification (V) processes. Also included in the table are combinations of controller type and its optimization method, e.g., SM-GA means SM controller was optimized by GA method, PID-PSA means PID controller was optimized by PSA method, etc. The initial study's objective was to find the optimal barriers for the controllers to avoid going into local minima. The research was carried out by an expert using a designed model of the CyberSeal, designed depth controllers and various tuning methods. As a result of the preliminary study, the upper and bottom barriers assume the following values: (a) for PID controller $[k_p, k_d, k_i]$—bottom barrier equals [200, 10,000, −1] and upper barrier equals [600, 40,000, 1], (b) for SM controller $[\lambda, k_s, \varphi]$—bottom barrier equals [−5, −100, −2] and upper barrier equals [5, 100, 2]. The main objective of the primary research was to compare two classic controllers, three methods optimized by fitness functions in response to changing the desired immersion value. The only quality control criterion optimized during tuning depth controllers was the minimized value of three different fitness functions. It means that all phrases describing the "best results", "most effective", etc., which are used later in the article, refer to the smallest fitness function values.

Analyzing the results contained in Table 2 obtained for fitness function no. 1, the following conclusions can be drawn: (1) in the verification process, the best results for both controllers and all methods were obtained for a large change in immersion value, (2) in the tuning process, the best results for both controllers for all methods were received for a slight change in the immersion value, (3) PID-GA obtained the best result in the verification process. In contrast, SM-PSO and SM-PSA got the best result in the tuning process considering all of the depth changes, (4) optimization methods present similar efficiency for all of the desired depth changes, (5) the smallest values of the f_{fit1} were obtained during tuning and verification for the SM controller than for the PID controller.

Table 2. Values of fitness function no. 1 for tuning (T) and verifying (V) BUV's depth controllers for three changes of desired immersion: shallow, deep and two following depth changes.

Controller Type	Shallow (T)	Immersion (V)	Deep (T)	Immersion (V)	For Two (T)	Changes (V)
PID-GA	30.5	48.6	73.3	55.5	128.6	54.2
SM-GA	29.2	88	71.3	54.3	126.2	96.2
PID-PSO	29.8	67.4	75.4	57.2	124.6	94.2
SM-PSO	30.1	92.1	65.6	55.2	116.5	95.1
PID-PSA	33.7	77,2	72.8	58.5	126.8	108.2
SM-PSA	32.1	116.1	67.6	96.8	120.1	99.7

Considering the results shown in Table 3 for fitness function no. 2, the following conclusions can be formulated: (1) in the verification process, the best results for both controllers and all methods were obtained for two changes in immersion value, (2) similar to earlier, the best results for both controllers for all methods in the tuning process were obtained for a slight change in the immersion value, (3) PID-GA obtained the best result in the verification process, while SM-PSO and SM-PSA obtained the best result in the tuning process taking into account all of the depth changes, (4) optimization methods present similar efficiency in the tuning process for all of the desired depth change.

Table 3. Values of fitness function no. 2 for tuning (T) and verifying (V) BUV's depth controllers for three changes of desired immersion: shallow, deep and two depths changes.

Controller Type	Shallow (T)	Immersion (V)	Deep (T)	Immersion (V)	For Two (T)	Changes (V)
PID-GA	11	36.4	84.2	53.2	141.8	30.8
SM-GA	11.2	75.1	85.6	81.2	167.8	80.1
PID-PSO	10.8	70.5	78.9	88.8	138.9	241.2
SM-PSO	11.4	165	72.2	78.2	129.8	77.2
PID-PSA	11.21	140.3	83.8	242	139.2	65.2
SM-PSA	11.2	199.8	76.8	358	134.4	92.6

Table 4 shows the results for fitness function no. 3. Analyzing its results, one can deduce: (1) as earlier, the best tuning process of both controllers was obtained for shallow immersion, while the best results for the verification process were received for two depth changes, (2) the smaller f_{fit3} was obtained during tuning and verification process by PID controller than SM, (3) PID-PSO received the smallest value of f_{fit3} in tuning for shallow immersion and verifying process for two depth changes, (4) optimization methods present similar efficiency for all of the desired depth change.

Table 4. Values of fitness function no. 3 for tuning (T) and verifying (V) BUV's depth controllers for three changes of desired immersion: shallow, deep and two following depths changes.

Controller Type	Shallow (T)	Immersion (V)	Deep (T)	Immersion (V)	For Two (T)	Changes (V)
PID-GA	8.21	14.1	9.26	15.9	20.1	9.11
SM-GA	8.9	22.41	14.76	24.2	26.2	15.6
PID-PSO	8.08	23.2	9.34	14.5	18.9	9.74
SM-PSO	9.01	24.6	11.45	24.31	27.6	30.41
PID-PSA	8.41	28.3	9.98	14.8	19	9.41
SM-PSA	9.5	25.5	12.32	15.81	31.2	15.81

Analyzing the above results in Tables 2–4, it can be concluded that the best results were obtained by PID-GA using objective function no. 3 for two depth changes. The result of f_{fit3} indicates that the selected controller settings in the optimization process performed best in the verification process for random desired depth (for 25 random changes in depth from 0 to 0.7 m). From Figures 5–7 show the simulation results for the controller settings obtained for the PID-GA method, using the f_{fit3} function and two depth changes. Figures 5–7 show, respectively, the results for each desired depth, i.e., 0.2 m, 0.5 m and two depth changes, first of 0.5 and then of 0.2. The simulation was carried out for 35 s. Considering the third case, the second desired depth signal occurred in 20 s of simulation. The timing was chosen so that the following depth changes occurred when there were no fluctuations in depth after the first change. The depth change was realized by the parallel swing of the side fins by the angle set by the control signal. In all the simulations, the oscillation frequency of the vehicle's side fins was constant and equal to 2 [Hz]. In the simulation, the side fin deflection angles were limited to 60 [deg] to avoid too excessive a vehicle trim angle. Each figure shows plots of the various parameters as a function of time: (1) graph of the dependence of the current vehicle depth on the desired, (2) values of the angular deflections of the side fins responsible for changing the depth of the vehicle based on the control signal along with the value of the vehicle trim (pitch), (3) error of depth over time. Analyzing the graphs presented, it can be noticed that selected controller gains are appropriate, and their selection is important for achieving good performance and stability of the system. The above graphs show that the object obtains satisfactory stability not only for the depths for which it was optimized (Figure 7) but also for other desired depths (Figures 5 and 6). The controller, in each case, reaches the required immersion quickly, and there is minimal overshoot. The controller's gains are not too large, so the system avoids exhibiting chattering, which is rapid switching between different control modes, which can lead to wear and tear on the actuators. The vehicle achieves stability within 12 s for all desired depths. To confirm the correctness of the adopted solution, i.e., tuning the controller for a specific value and then checking its settings for 25 random values, the simulation results for other controller settings are presented in Figure 8. For this figure, the controller settings were obtained for the PID-GA method, using the f_{fit3} function and small depth changes. Comparing the simulation results for the same depth change (in Figures 5 and 8), but obtained for two different controller settings, it can be observed that the controller tuned for the specific depth reached the desired depth faster, i.e., 8 s (Figure 8). This is confirmed by the results in Table 4, wherein the tuning process for a small change in depth, the PID-GA obtained the value of the fit function 8.21, while for two changes in depth is 20.1. However, better results were obtained in the verification process for two depth changes for the entire depth spectrum (from 0 to 0.7 m). The value of f_{fit3} for two depth changes was better by 50 per cent than for a small depth.

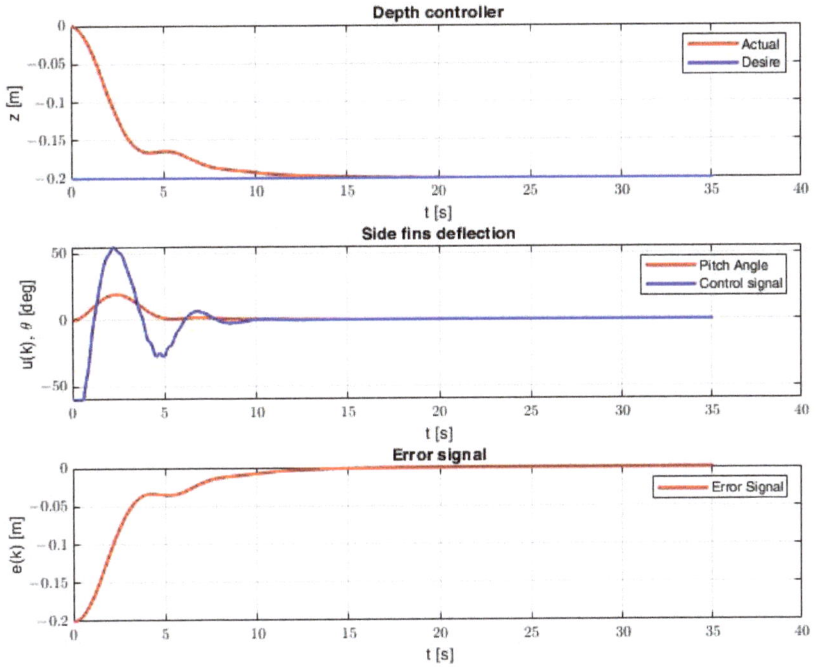

Figure 5. Changes of immersion and side fins deflection in time in response to desired depth: 0.2 [m] obtained for PID controller settings using GA method and fitness function no. 3 for two depth changes.

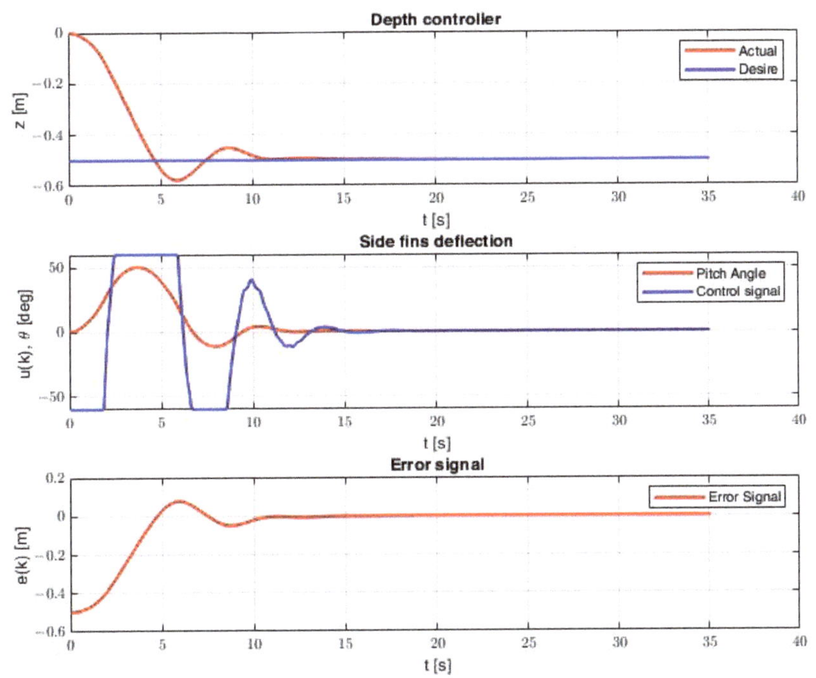

Figure 6. Changes of immersion and side fins deflection in time in response to desired depth: 0.5 [m] obtained for PID controller settings using GA method and fitness function no. 3 for two depth changes.

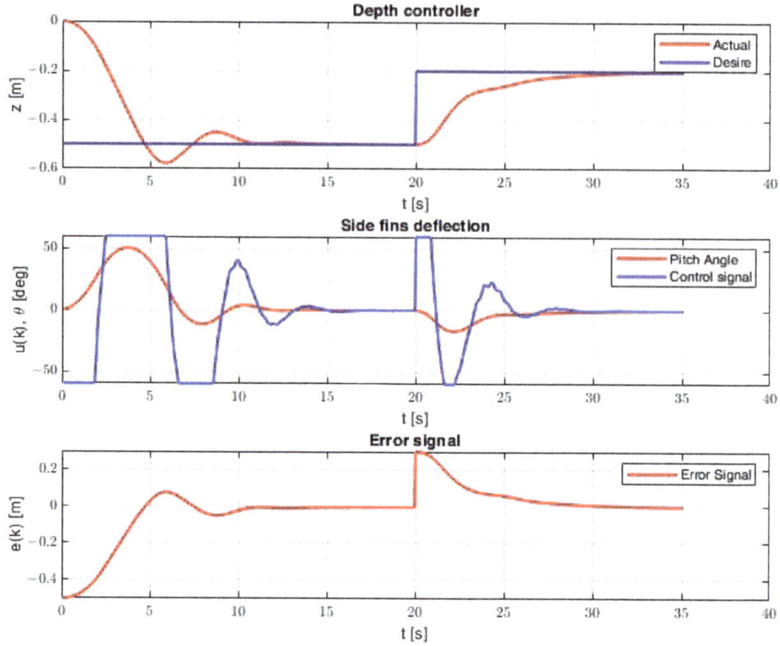

Figure 7. Changes of immersion and side fins deflection in time in response to subsequent desired depths: 0.5, 0.2 [m] obtained for PID controller settings using GA method and fitness function no. 3 for two depth changes.

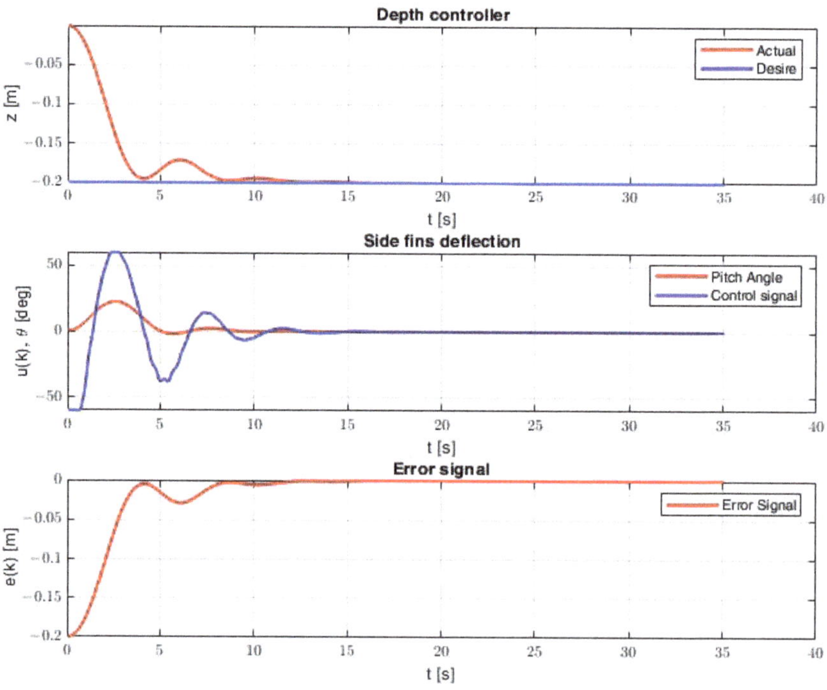

Figure 8. Changes of immersion and side fins deflection in time in response to desired depth: 0.2 [m] obtained for PID controller settings using GA method and fitness function no. 3 for small immersion.

4. Conclusions

This paper presents the tuning process of two classical depth controllers (PID, SM) of the biomimetic underwater vehicle with undulating propulsion. The tuning of the parameters of the controllers was carried out by applying three optimization methods and three fitness functions (GA, PSO, PSA), which provided a quality criterion. The optimization process was conducted for three different desired depth values, as shown in Figure 4.

Summarizing the research results obtained, several conclusions can be reached. The best results were not received with a slight change in immersion value. The results shown in Tables 2–4 confirm that despite the best values of the fitness function in the tuning process, the values for the verification process are several times higher. It can be recognized that the values of controller setting gains work efficiently only for small values. At the same time, they are not necessarily optimal for larger values used in the verification process. Also, the values of the fit function for different controller and optimization methods for the tuning and verification process are similar. Therefore, it is expected that the tuning of the controllers should be carried out for a larger desired depth or two or more changes of depth.

Furthermore, it can be deduced that, in most cases, better control quality was obtained for the PID controller than for the sliding controller. The differences can be seen in Table 4, where in all cases, both in the simulation and verification process, the fitness function obtained better values for the PID controller. The three optimization methods used to adjust the controller settings achieve comparable efficiency. Although the PSO and PSA methods achieved better tuning, the GA method achieved the best results in the verification process for each of the three different fitness functions.

Comparing the quality indicators, we can see that the ISE (f_{fit2}), which strongly penalizes any large deviation, obtained the worst results for a large change or two depth changes. In contrast, the best results were obtained for fitness function no. 3 (f_{fit3}) where used direct quality indicators, i.e., the rise time and the value of the first overshoots. It is because the fitness function obtained the most repetitive results for the tuning and verification process, and the values of fitness function no. 3 were the best compared to the other functions (f_{fit1}) and (f_{fit2}) using the classic indicators.

The presented results could be more comprehensive. An interesting issue would be further systematic testing of new multi-criteria indicators. In this way, it would be possible to use, for example, an LMS indicator or the like to limit sudden changes in the control signal and a more aggressive indicator such as ISE or ISE to increase the speed and accuracy of the process control. As part of future research, it is also planned to implement the applied regulators on a real object (mini CyberSeal) and verify them in natural conditions.

Funding: This research received no external funding

Institutional Review Board Statement: Not applicable.

Informed Consent Statement: Not applicable.

Data Availability Statement: Not applicable.

Conflicts of Interest: The author declares no conflict of interest.

Abbreviations

The following abbreviations are used in this manuscript:

AUV	Autonomous Underwater Vehicle
BUV	Biomimetic underwater vehicles
GA	Genetic algorithm
ISA	Integral Absolute Error
ISE	Integral of Squared Error
LMS	Least Median of Squares

PID	Proportional–integral–derivative controller
POM-C	Polyacetal (copolymer)
PSO	Particle Swarm Optimization
PSA	Pareto Simulation
ROV	Remote Operated Vehicle
SM	Sliding Mode controller

References

1. Ahmed, Y.; Yaakob, O.; Sun, B. Design of a new low cost ROV vehicle. *J. Teknol.* **2014**, *69* , 7. [CrossRef]
2. Hożyń, S.; Żak, B. Stereo Vision System for Vision-Based Control of Inspection-Class ROVs. *Remote Sens.* **2021**, *13*, 5075. [CrossRef]
3. Śmierzchalski, R.; Kapczyński, M. Autonomous Control of the Underwater Remotely Operated Vehicle in Collision Situation with Stationary Obstacle. *Pol. Marit. Res.* **2022**, *29*, 45–55. [CrossRef]
4. Kot, R. Review of Collision Avoidance and Path Planning Algorithms Used in Autonomous Underwater Vehicles. *Electronics* **2022**, *11*, 2301. [CrossRef]
5. Piskur, P. Strouhal Number Measurement for Novel Biomimetic Folding Fins Using an Image Processing Method. *J. Mar. Sci. Eng.* **2022**, *10*, 4. [CrossRef]
6. Praczyk, T.; Szymak, P.; Naus, K.; Pietrukaniec, L.; Hożyń, S. Report on research with biomimetic autonomous underwater vehicle—Low-level control. *Marit. Tech. J.* **2018**, *212*, 105–123. [CrossRef]
7. Shang, L.; Wang, S.; Tan, M.; Cheng, L. Swimming locomotion modelling for biomimetic underwater vehicle with two undulating long-fins. *Robotica* **2012**, *30*, 913–923. [CrossRef]
8. Piskur, P.; Szymak, P.; Przybylski, M.; Naus, K.; Jaskólski, K.; Żokowski, M. Innovative Energy-Saving Propulsion System for Low-Speed Biomimetic Underwater Vehicles. *Energies* **2021**, *34*, 8418. [CrossRef]
9. Hernández-Alvarado, R.; García-Valdovinos, L.G.; Salgado-Jiménez, T.; Gómez-Espinosa, A.; Fonseca-Navarro, F. Neural Network-Based Self-Tuning PID Control for Underwater Vehicles. *Sensors* **2016**, *16*, 1429. [CrossRef]
10. Song, X.; Liu, F.; Zou, Z.; Zhu, Y.M.; Yin, J.; Xu, F. Nonlinear Underwater Robot Controller Design with Adaptive Disturbance Prediction and Smoother. *Int. J. Comput. Intell. Syst.* **2011**, *4*, 634–643. [CrossRef]
11. Teran-Picon, J.C.; Camacho, A.J.O. Genetics algorithms as a tuning tool for PID and sliding mode controllers. *Rev. Téc. Ing. Univ. Zulia* **2005**, *28*, 200–209.
12. Wu, B.; Li, S.; Wang, X. Discrete-Time Adaptive Sliding Mode Control of Autonomous Underwater Vehicle in the Dive Plane. In *Intelligent Robotics and Applications*; Springer: Heidelberg, Germany, 2009; pp. 157–164.
13. Kang, B.; Lee, Y.; Piao, T.; Ding, Z.; Wang, W.D. Robotic soft swim bladder using liquid–vapor phase transition. *Mater. Horizons* **2021**, *8*, 939–947. [CrossRef]
14. Chemori, A.; Kuusmik, K.; Salumäe, T.; Kruusmaa, M. Depth Control of the Biomimetic U-CAT Turtle-Like AUV with Experiments in Real Operating Conditions. In Proceedings of the 2016 IEEE International Conference on Robotics and Automation (ICRA), Stockholm, Sweden, 16–21 May 2016; pp. 4750–4755. [CrossRef]
15. Morawski, M.; Talarczyk, T.; Malec, M. Depth control for biomimetic and hybrid unmanned underwater vehicles. *Tech. Trans.* **2021**, *118*. [CrossRef] [PubMed]
16. Hożyń, S. An Automated System for Analysing Swim-Fins Efficiency. *NašE More Znan. čAsopis More Pomor.* **2020**, *67* (Suppl. S3), 10–17. [CrossRef]
17. Piskur, P.; Szymak, P.; Kitowski, Z.; Flis, L. Influence of fin's material capabilities on the propulsion system of biomimetic underwater vehicle. *Pol. Marit. Res.* **2020**, *4*, 179–185. [CrossRef]
18. Fossen, T. *Handbook of Marine Craft Hydrodynamics and Motion Control*; Wiley: Hoboken, NJ, USA, 2011.
19. Katzschmann, R.K.; DelPreto, J.; MacCurdy, R.; Rus, D. Exploration of underwater life with an acoustically controlled soft robotic fish. *Sci. Robot.* **2018**, *3*, eaar3449. [CrossRef]
20. Szymak, P.; Przybylski, M. Thrust Measurement of Biomimetic Underwater Vehicle with Undulating Propulsion. *Marit. Tech. J.* **2018**, *213*, 69–82. [CrossRef]
21. Przybylski, M. Mathematical Model of Biomimetic Underwater Vehicle. In Proceedings of the 33rd International ECMS Conference on Modelling and Simulation, Caserta, Italy, 11–14 June 2019; Volume 33, pp. 343–350. [CrossRef]
22. Liang, J.; Huang, W.; Zhou, F.; Liang, J.; Lin, G.; Xiao, E.; Li, H.; Zhang, X. Double-Loop PID-Type Neural Network Sliding Mode Control of an Uncertain Autonomous Underwater Vehicle Model Based on a Nonlinear High-Order Observer with Unknown Disturbance. *Mathematics* **2022**, *10*, 3332. [CrossRef]
23. Li, H.; He, B.; Yin, Q.; Mu, X.; Zhang, J.; Wan, J.; Wang, D.; Shen, Y. Fuzzy Optimized MFAC Based on ADRC in AUV Heading Control. *Electronics* **2019**, *8*, 608. [CrossRef]
24. Szymak, P. Selection of Method for Underwater Robot Control. *Machatronic Syst. Mater. Mechatron. Syst. Robot.* **2010**, *164*, 149–154. [CrossRef]
25. Vu, Q.V.; Dinh, T.A.; Nguyen, T.V.; Tran, H.V.; Le, H.X.; Pham, H.V.; Kim, T.D.; Nguyen, L. An Adaptive Hierarchical Sliding Mode Controller for Autonomous Underwater Vehicles. *Electronics* **2021**, *10*, 2316. [CrossRef]
26. Global Optimization Toolbox.Solve Multiple Maxima, Multiple Minima, and Nonsmooth Optimization Problems. Available online: https://www.mathworks.com/products/global-optimization.html (accessed on 12 July 2022).

27. Przybylski, M.; Szymak, P.; Kitowski, Z.; Piskur, P. Comparison of Different Course Controllers of Biomimetic Underwater Vehicle with Two Tail Fins. In *Advanced, Contemporary Control*; Springer: Berlin/Heidelberg, Germany, 2020; Volume 1196, pp. 1507–1518.
28. Lim, C.-H.; Yoon, Y.-S.; Kim, J.-H. Genetic algorithm in mix proportioning of high-performance concrete. *Cem. Concr. Res.* **2004**, *34*, 409–420. [CrossRef]
29. Goldberg, D.E. *Genetic Algorithms in Search, Optimization and Machine Learning*; Addison Wesley: Boston, MA, USA, 1989.
30. Jurczyk, K.; Piskur, P.; Szymak, P. Parameters Identification of the Flexible Fin Kinematics Model Using Vision and Genetic Algorithms. *Pol. Marit. Res.* **2020**, *27*, 39–47. [CrossRef]
31. Venter, G.; Sobieszczanski-Sobieski, J. Particle Swarm Optimization. *AIAA J.* **2003**, *41*, 1583–1589. [CrossRef]
32. Pedersen, M.E. Good Parameters for Particle Swarm Optimization. *Hvass Lab.* **2010**, 1551–3203. Available online: https://www.semanticscholar.org/paper/Good-Parameters-for-Particle-Swarm-Optimization-Pedersen/a4ad7500b64d70a2ec84bf57cfc2fedfdf770433 (accessed on 13 March 2023).
33. Praczyk, T. Hill-Climb-Assembler Encoding: Evolution of Small/Mid-Scale Artificial Neural Networks for Classification and Control Problems. *Electronics* **2022**, *11*, 2104. [CrossRef]
34. Tušar, T.; Filipic, B. Visualization of Pareto Front Approximations in Evolutionary Multi-objective Optimization: A Critical Review and the Prosection Method. *IEEE Trans. Evol. Comput.* **2015**, *19*, 2. [CrossRef]
35. Bogusz, K.; Rajkowski, B.; Domański, P. Porównanie różnych sposobów optymalizacji nastaw regulacji procesów przemysłowych z uwzględnieniem wpływu wskaźników oceny ich jakości. *Pomiary Autom. Robot.* **2019**, *23*, 27–39. [CrossRef]
36. Pawliński, M.; Plamowski, S.; Domański, P. Wpływ wybranych wskaźników jakości regulacji na parametry sygnału sterującego w układzie z regulatorem PID. *Pomiary Autom. Robot.* **2019**, *23*, 31–42. [CrossRef]

electronics

Article

Dynamics of Separation of Unmanned Aerial Vehicles from the Magnetic Launcher Cart during Takeoff

Edyta Ładyżyńska-Kozdraś [1,*], Anna Sibilska-Mroziewicz [1], Krzysztof Sibilski [2], Danyil Potoka [1] and Andrzej Żyluk [2]

1 Institute of Micromechanics and Photonics, Warsaw University of Technology, Sw. A. Boboli 8, 02-525 Warsaw, Poland
2 Air Force Institute of Technology, Księcia Bolesława 6, 01-494 Warsaw, Poland
* Correspondence: edyta.ladyzynska@pw.edu.pl

Abstract: Most aircraft launchers exhibit a rapid acceleration of the launching aircraft, often exceeding ten times the acceleration due to gravity. However, only magnetic launchers offer flexible control over the propulsion force of the launcher cart, enabling precise control over the aircraft's acceleration and speed during its movement on the launcher. Consequently, extensive research is being conducted on magnetic launchers to ensure the repeatability of launch parameters, protect against aircraft overloads, and ensure operator safety. This article describes the process of modeling and analyzing the dynamical properties of a launch cart of an innovative prototype launcher, which employs a passive magnetic suspension with high-temperature superconductors, developed under the GABRIEL project. The developed mathematical model of the magnetic catapult cart was employed to conduct numerical studies of the longitudinal and lateral movement of the cart, as well as the configuration of the UAV–cart system during UAV takeoff under variable atmospheric conditions. An essential aspect of the research involved experimentally determining the magnetic levitation force generated by the superconductors as a function of the gap. The results obtained demonstrate that the analyzed catapult design enables safe UAV takeoff. External factors and potential vibrations resulting from uneven mass distribution in the UAV–cart system are effectively balanced by the magnetic forces arising from the Meissner effect and the flux pinning phenomenon. The primary advantage of the magnetic levitation catapult, in comparison to commercial catapults, lies in its ability to provide a reduced and consistent acceleration throughout the entire takeoff process.

Keywords: magnetic launcher; levitation force; UAV; dynamics modeling; numerical simulation

Citation: Ładyżyńska-Kozdraś, E.; Sibilska-Mroziewicz, A.; Sibilski, K.; Potoka, D.; Żyluk, A. Dynamics of Separation of Unmanned Aerial Vehicles from the Magnetic Launcher Cart during Takeoff. *Electronics* **2023**, *12*, 2883. https://doi.org/10.3390/electronics12132883

Academic Editors: Piotr Szymak, Stanisław Hożyń and Paweł Piskur

Received: 13 May 2023
Revised: 26 June 2023
Accepted: 27 June 2023
Published: 29 June 2023

1. Introduction

The designers of modern unmanned aerial vehicles (UAVs) must solve a series of technical problems related to the critical aspects of their operation in order to meet the needs of the rapidly developing market. One of the key aspects is the UAV takeoff procedure. The takeoff of a large portion of fixed-wing UAVs requires separate devices called launchers or aircraft catapults. Currently, rocket systems, rubber, pneumatic, and hydraulic launchers are most commonly used for this purpose. They are powered by various means, including steam, compressed air, energy stored in elastic elements (such as rubber bands), or even rocket boosters. The operation of these devices involves high acceleration for the launching UAV, often exceeding 10 times the Earth's gravitational acceleration, and a lack of control over the aircraft's trajectory during the launch. Magnetic launchers, which allow for significantly higher speeds compared to traditional catapult solutions while also enabling contactless operation, are an attractive alternative to the currently used launchers.

The first interest in electromagnetic catapults came from the navy, and during World War II, Westinghouse constructed the initial aircraft catapult system utilizing electromagnetic interactions [1]. However, due to the high costs of production and maintenance, as well as their large size and weight, these systems were not commercially viable. Electromagnetic technology was displaced by steam launchers in the past, and it was only at the turn of the century that the concept was revisited. Scientists were prompted by the miniaturization of electrical devices, which allowed for the construction of increasingly cheaper and smaller equipment, thus reducing the size and operating costs of electromagnetic launchers. There are several types of magnetic launchers, classified based on their design and motion generation methods. Rail launchers and coil launchers are particularly noteworthy. Among coil launchers, synchronous and asynchronous (inductive) launchers can be distinguished based on the method of generating magnetic force. Such launchers have been employed in recently commissioned US Navy aircraft carriers and in high-speed MAGLEV trains like the Inductrack. Rail launchers can utilize linear induction motors as the propulsion for the launch cart. Rail launchers are typically composed of two conductive rails made of electrically conductive material. A movable connector, also made of a conductor, is placed between the rails. The contact between the two rails and the connecting element must ensure good current flow. As a result of electric charge flow, following Ampere's and Biot–Savart's laws, a magnetic field is generated around the rails, which interacts with the charges moving in the connector. The movable element experiences a magnetic force that pushes it along the rails. The method described above is used in the high-speed MAGLEV trains of the Inductrack type. However, the Inductrack technique has a drawback as levitation forces occur after reaching a certain minimum speed. Additionally, due to their cost, they are difficult to accept in unmanned aircraft launch solutions. The solution presented in this article is significantly simpler and more cost-effective than the Inductrack approach.

The subject of the presented research is an innovative prototype of a catapult utilizing passive magnetic suspension with high-temperature superconductors [2,3]. The catapult prototype was constructed as part of the GABRIEL project (Integrated Ground and On-board system for Support of the Aircraft Safe Take-off and Landing) [4,5]. The design of the prototype catapult features magnetic tracks attached to a stationary base, a levitating takeoff platform, and a linear drive (Figure 1a). During the UAV takeoff, it is mounted on the takeoff platform, which is an integral part of the cart made of a duralumin frame and supported on four supports (Figure 1b), in which high-temperature YBCO superconductors are placed (Figure 1c). After cooling the YBCO with liquid nitrogen, the takeoff cart lifts off and levitates above the source of the magnetic field.

(a) (b) (c)

Figure 1. Elements of a magnetic catapult utilizing the Meissner effect: (**a**) magnetic tracks with a launch platform installed on them; (**b**) a levitating support for the launch cart; and (**c**) high-temperature superconductors placed in the support.

One undeniable advantage of this solution is its simple design, which facilitates the transportation and installation with just two personnel required for operation. Moreover, this solution incorporates all the advantages of electromagnetic launchers, allowing for the adjustment of the launch cart speed and acceleration to minimize the applied overload on the aircraft during takeoff. An additional benefit of this solution is the ability to utilize the system for the landing of the UAV, a feature that conventional or coil launchers cannot guarantee.

At the end of the GABRIEL project, tests were carried out on the designed unmanned aerial vehicle in Aachen for takeoff and landing. The tests were successful [6], but technical issues were encountered, such as the slowly damped vibrations of the cart and the loss of its stable position after disconnecting the linear drive. These issues served as inspiration for further research on the innovative magnetic catapult project.

This article presents numerical studies of the motion of the cart and the catapult system during the UAV launch under varying atmospheric conditions based on a developed model of the dynamics of the launch cart. An important aspect of the research was the experimental determination of the magnetic levitation force as a function of the gap generated by the superconductors.

2. The Model of the Launch Cart for a Magnetic Launcher

In this study, a mathematical model of the launch cart for the magnetic launcher was developed. This model takes into account key factors, such as the characteristics of the environment and the individual components of the launcher, which include the magnetic tracks and the cart that launches the UAV. To further describe and understand the system, various coordinate systems were introduced, which are shown in Figure 2.

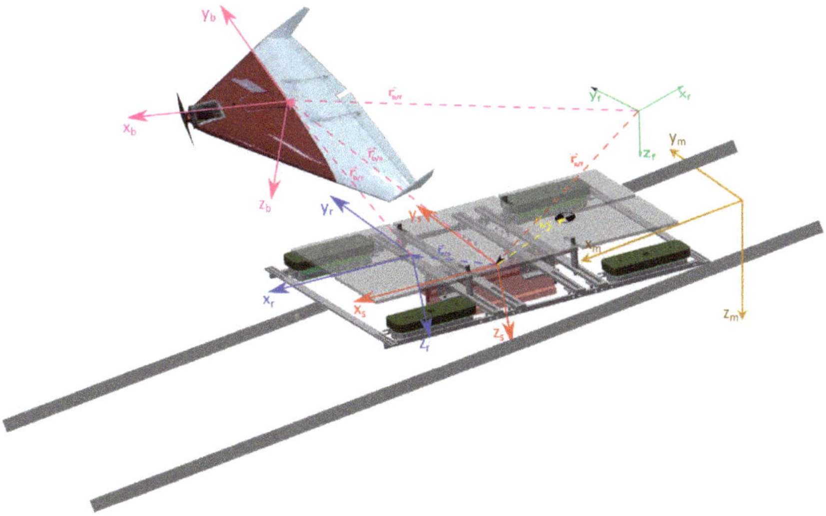

Figure 2. Model of the magnetic launcher system along with the adopted reference systems.

An inertial system ($O_f x_f y_f z_f$) was firmly tied to the surface of the Earth, providing a fixed reference point. The magnetic system ($O_m x_m y_m z_m$) is used to describe the distribution of the magnetic field and the levitation force. This system is aligned with the symmetry axis of the launcher tracks. The cart system ($O_s x_s y_s z_s$) is aligned with the longitudinal symmetry axis of the cart and illustrates the direction of the cart's movement. More detailed descriptions of these reference systems, along with a thorough analysis of the physical model and external influences, can be found in the referenced paper [7].

The dynamics of the launch cart for a magnetic launcher, which moves in three-dimensional space, is described by a system of mutually coupled nonlinear ordinary differential equations. These equations were derived for the launch cart modeled as a rigid body with six degrees of freedom, using the principle of momentum and angular momentum conservation. The mathematical model of the launch cart's motion dynamics was thoroughly developed in [8], and its main stages were discussed in the article [9]. The formalism for writing the equations of motion is based on a precisely defined method of indexing vector quantities and the properties of rotation matrices, particularly the representation of the derivative of the rotation matrix using a skew-symmetric matrix. The final form of the equations of motion is written in a local, moving reference frame associated with the launch cart $O_s x_s y_s z_s$. The matrix form of the equations of motion is given by Formula (1). The first equation describes the relationship between the three-dimensional vector of forces acting on the system and the vector of change in the body's momentum. The second equation maps the vector of force moments to the vector of change in angular momentum.

$$\begin{bmatrix} m_{(s)}\mathbb{I}_{3\times 3} & -m_{(s)}\mathfrak{s}\left(r^s_{g_s/s}\right) \\ m_{(s)}\mathfrak{s}\left(r^s_{g_s/s}\right) & \mathfrak{I}^s_{(s)s} \end{bmatrix} \begin{bmatrix} a^{s(ff)}_{s/f} \\ \varepsilon^{s(f)}_{s/f} \end{bmatrix} - \begin{bmatrix} -m_{(s)}\mathfrak{s}\left(\omega^s_{s/f}\right)\mathfrak{s}\left(r^s_{g_s/s}\right)\omega^s_{s/f} \\ \mathfrak{s}\left(\omega^s_{s/f}\right)\mathfrak{I}^s_{(s)s}\omega^s_{s/f} \end{bmatrix} = \begin{bmatrix} F^s_{(s)} \\ M^s_{(s)s} \end{bmatrix} \tag{1}$$

where: $\mathbb{I}_{3\times 3}$ is a 3×3 identity matrix, $m_{(s)}$ is the mass of the launch cart, $\mathfrak{I}^s_{(s)s}$ is the inertia tensor of the cart, $\mathfrak{s}\left(r^s_{g_s/s}\right)$ is the skew-symmetric tensor of the displacement vector of the cart's center of mass, $a^{s(ff)}_{s/f}$ is the linear acceleration of the cart, $\varepsilon^{s(f)}_{s/f}$ is the angular acceleration of the cart, $\omega^s_{s/f}$ is the angular velocity of the cart, $F^s_{(s)}$ and $M^s_{(s)s}$ are the forces and moments acting on the cart.

The launch cart levitating above magnetic tracks is subject to magnetic forces and moments of force $F^s_{(L_s)}$, $M^s_{(L_s)}$, aerodynamic forces and moments of force $F^s_{(As)}$, $M^s_{(As)s}$, gravitational forces and moments of force $F^s_{(G_s)}$, $M^s_{(G_s)}$ as well as the propulsion force from the linear motor $F^s_{(N_s)}$; therefore, the resulting force vector and moment of force have the following form

$$F^s_{(s)} = F^s_{(G_s)} + F^s_{(L_s)} + F^s_{(As)} + F^s_{(N_s)} \tag{2}$$

$$M^s_{(s)s} = M^s_{(G_s)s} + M^s_{(L_s)s} + M^s_{(As)s} \tag{3}$$

The mathematical model of the cart should be supplemented with kinematic relationships describing its angular velocity $\omega^s_{s/f}$ and linear velocity $v^{s(f)}_{s/f}$, as well as a geometric relationship describing the vector between the local coordinate system of the cart and the inertial coordinate system $r^f_{s/f}$.

$$\omega^s_{s/f} = \int \left(R^s_f \varepsilon^{f(f)}_{s/f}\right) dt \tag{4}$$

$$v^{s(f)}_{s/f} = \int \left(a^{s(ff)}_{s/f} - \mathfrak{s}\left(\omega^s_{s/f}\right) v^{s(f)}_{s/f}\right) dt \tag{5}$$

$$r^f_{s/f} = \int \left(R^f_s v^{s(f)}_{s/f}\right) dt \tag{6}$$

where R^s_f is the rotation matrix describing the orientation of the inertial frame relative to the local frame of the cart.

The described mathematical model of the launch cart provides a basis for further research on its dynamic properties, including the numerical analysis of its motion. These studies were preceded by an analysis of the acting forces. The process of the theoretical modeling of forces and moments acting on the launcher system was discussed in detail in [7]. The analyses presented there indicated interactions between the loads and their influence on the UAV launch process. The key role is played by the magnetic levitation force. Its experimental identification process is included in Section 3.

3. Experimental Identification of Magnetic Levitation Force

The levitation force, which lifts the launcher cart above the magnetic tracks, is caused by the Meissner effect [10]. The Meissner effect is a phenomenon in which a superconductor expels magnetic field lines from its interior. It has been studied in various applications, including Maglev trains and energy storage systems [11,12], as well as in the context of superconducting materials used in aerospace engineering, such as in the development of high-performance motors and generators [13].

In the analyzed magnetic launcher, the levitation force is a result of the Meissner effect occurring between the tracks of the magnetic rails and the superconductors enclosed in containers—supports of the launch cart. After filling the supports with liquid nitrogen, the superconductors transition into the superconducting state, and as a result of the Meissner effect, they begin to levitate above the tracks.

To identify the value of magnetic levitation force, a measurement experiment was carried out using the MTS Bionics testing machine. A section of the magnetic track rail was attached to the lower, stationary jaws of the machine, and a support with four YBCO superconductors was mounted on the upper jaws (Figure 3).

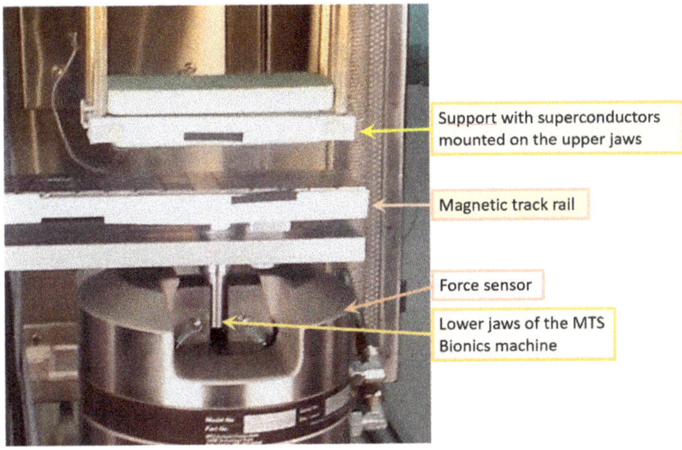

Figure 3. Experimental setup for measuring magnetic levitation force.

Figure 4 presents the results of measurements aiming to determine the effect of the levitation gap height on the value of the magnetic levitation force. In the first series of measurements (red line), the superconductors were cooled with liquid nitrogen at a height of 20 mm above the magnets, where the magnetic field has zero value (ZFC—zero field cooling). Such a superconductor is only affected by the magnetic force resulting from the Meissner effect. In the second series of measurements, the superconductors were cooled with liquid nitrogen in a non-zero magnetic field at a height of 3 mm above the magnets (d_{FC} = 3 mm). The measurements were repeated twice (blue and green lines), which showed the convergence of the obtained results. The upper graph shows the levitation force value registered by the force sensor of the strength machine, while the lower graph shows the size of the levitation gap, i.e., the distance between the levitating support and the launcher rails.

The conducted observations of the superconductor's behavior in a magnetic field showed that the levitation force value is a function of the levitation gap (distance of the superconductor from the magnetic field). Additionally, in the case of filling the superconductor with liquid nitrogen in a non-zero magnetic field, the maximum recorded value of the levitation force is almost half the value obtained when filling the superconductor in the zero magnetic field. This confirms the fact that there are two forces acting on the superconductor: a force repelling the superconductor from the source of the magnetic field, resulting from the Meissner effect F_{LM}, and a force attracting the superconductor to the source of the magnetic field, resulting from the flux pinning effect F_{LP}.

$$F_{Ls} = F_{LM} + F_{LP} \tag{7}$$

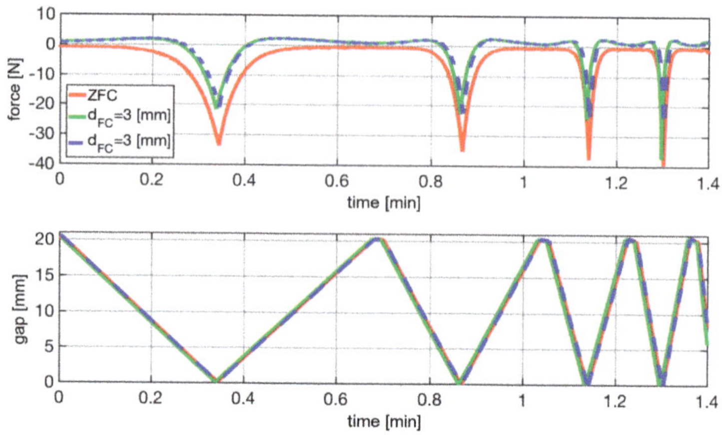

Figure 4. Results of the measurements illustrating the effect of the levitation gap (**lower graph**) on the value of the magnetic levitation force (**upper graph**) for different liquid nitrogen flooding heights (dFC = 3 mm; ZFC—zero field cooling).

As a result of the flux pinning phenomenon, the superconductors placed inside the supports of the cart become magnetized [14]. The widely used advanced mirror-reflection method [15] for modeling magnet–superconductor interactions suggests that the levitation force can be expressed by the following equation [16].

$$\boldsymbol{F_{LP}} = \boldsymbol{m} \cdot \nabla B \tag{8}$$

where B is the external magnetic field and m is the vector of the magnetic moment frozen within the superconductor at the moment of the phase transition.

The approximating functions for the levitation forces used in the numerical model are presented in Figure 5a. The green and red lines indicate the experimentally measured levitation force. The green line corresponds to the superconductor being cooled with liquid nitrogen in a zero magnetic field ($\boldsymbol{F_{LP}} = 0$), while the red line represents the case where the superconductor is cooled with liquid nitrogen at a height of 3 mm. Thus, the red line represents the sum of the force resulting from the Meissner effect (green line) and the force arising from the flux pinning phenomenon.

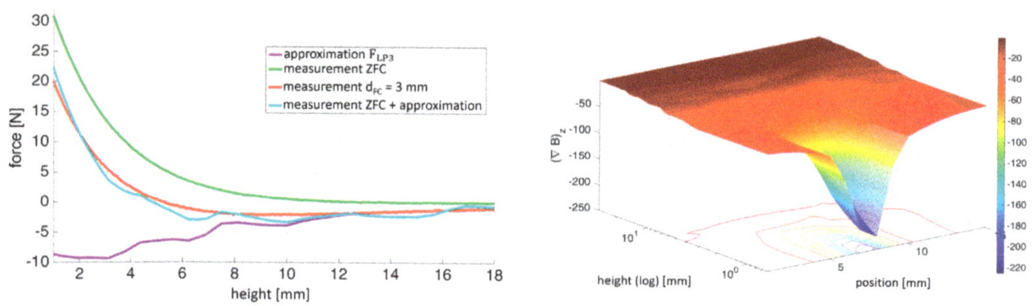

(a) (b)

Figure 5. (**a**) Forces acting on the support of the cart with superconductors; and (**b**) spatial distribution of the vertical component of the magnetic field gradient above the launcher tracks.

At a height of 5.2 mm, the force resulting from the Meissner effect (with a value of 5.4 N) and the force resulting from flux pinning balance each other. By determining the value of the magnetic field gradient along the vertical axis at a height of 5.2 mm (with an

absolute value of $\nabla B = 20.6$ N/T) (Figure 5b) based on finite element analysis [12], the induced magnetic moment in the superconductor was calculated to be $m = 0.26$ T using Equation (8). Thus, knowing the value of m, the force resulting from flux pinning was approximated when the superconductors were immersed in liquid nitrogen at a height of 3 mm (purple line).

The light blue line was obtained by summing the measurements of the levitation force when the superconductor was immersed in liquid nitrogen under zero magnetic field conditions (green line) and the approximated force resulting from flux pinning (purple line). Similarly, the force acting on the superconductor immersed in liquid nitrogen at any height can be approximated using a similar approach.

The identification of the magnetic levitation force provided a basis for further numerical studies of the magnetic launcher system. The results of the numerical simulations of the longitudinal and lateral motion of the cart as well as the configuration of the UAV–cart system during UAV launch under variable atmospheric conditions are presented below.

4. Numerical Studies of the Dynamics of the Launch Cart in a Launcher

The computer program designed for the analysis of dynamics and the visualization of the trajectory of the launch cart was implemented in MATLAB. The simulation model was formulated as an initial value problem of ordinary differential equations based on the mathematical model discussed above.

Two simulation scenarios were demonstrated, encompassing:

1. Longitudinal motion of the launch cart, where the center of mass of the cart only moves in the vertical plane.
2. Transverse motion, considering both longitudinal and the lateral displacement of the cart from the center of the tracks in the transverse direction.

The longitudinal motion of the launch cart is generated by the linear motor. In the GABRIEL technology demonstrator, which was designed as part of the project, the motor stator is mounted in the middle of the magnetic tracks, and its structure restricts the lateral movements of the cart [17]. Therefore, in the simulation analyses of the investigated launcher system, the omission of the lateral movements of the launch cart is fully justified. In the second simulation, the behavior of the cart was additionally analyzed when it is detached from the motor stator, taking into account the effects on its motion in space caused by the trapping of magnetic flux inside the superconductors.

The authors' intention was to conduct numerical analyses based on a nonlinear model of the launch cart's dynamics, described by Equations (1)–(6). Therefore, the analysis of these simulation scenarios is a result of applying initial conditions and physical relationships derived from modeling the forces acting on the system.

In both presented simulation scenarios, it was assumed that the launcher tracks are in a horizontal position. The simulation time was 2 s. During the first second, the launch cart is driven by a constant force of 20 N, while during the second s, the cart is braked and the propulsion force changes direction.

4.1. Longitudinal Motion of the Cart

In the presented simulation scenario, the behavior of the cart was investigated, where its center of mass is consistently located on the axis of symmetry of the tracks, but it does not coincide with the geometric center. It was assumed that the center of mass of the cart was displaced along the x_s longitudinal axis by $\left\{r_{g_s/s}\right\}_{x_s} = 0.1\ \text{m}$. At the initial time, the cart is located at a height of $\{r_{s/m}\}_{z_m} = 6.6\ \text{mm}$. Due to the displacement of the center of mass, the cart is subjected to the gravitational moment of force. The unbalanced moment of force causes the cart to tilt around the y_s lateral axis. The difference in height between the front and rear supports of the cart results in a different value of levitation forces exerted on individual supports. As a result, a moment of force is created that balances the gravitational moment of forces and stabilizes the orientation of the cart. Figure 6 shows the trajectory of

the cart. The blue line represents the motion of the center of the cart, the purple line marks the position of the launch rails, while the green and red lines visible on the left and right reflect the position of the cart supports. It can be seen that the cart has tilted forward and the front supports are lower than the rear supports.

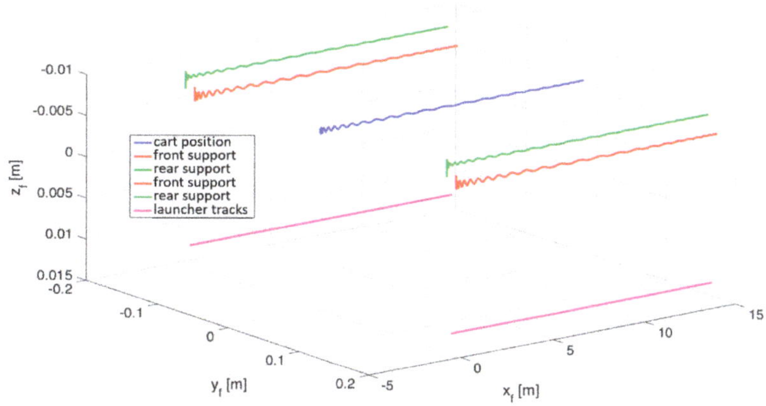

Figure 6. The trajectory of the longitudinal motion of the launch cart.

After 2 s, the cart traveled a distance of 14.2 m (Figure 7a) and reached a maximum speed of 14 m/s, and then came to a stop (Figure 7b). The simulation showed small vibrations of the cart (Figure 8a) around the equilibrium position. The vertical velocity of the cart oscillates within ±0.0035 m/s (Figure 8b). The tilt angle of the cart stabilizes at −0.005 rad (Figure 9a), and then its angular velocity decreases to zero (Figure 9b).

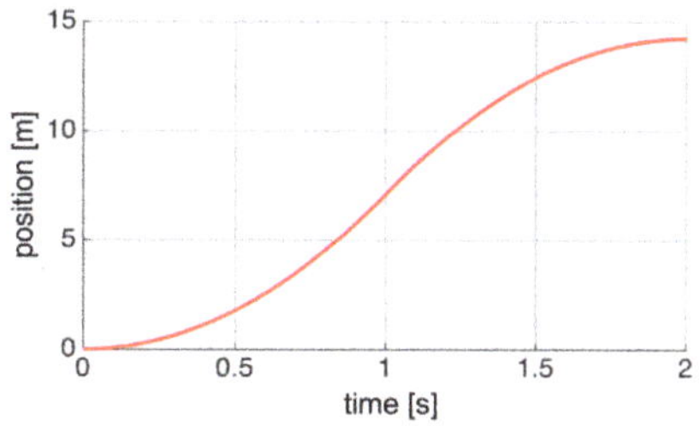

(**a**)

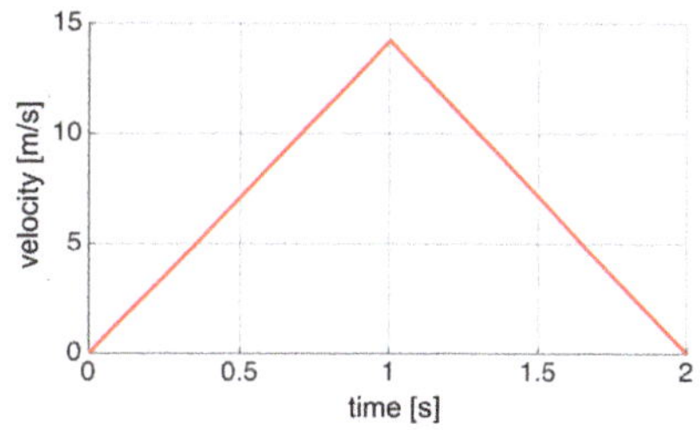

(**b**)

Figure 7. The longitudinal component of position (**a**) and velocity (**b**) of the launch cart.

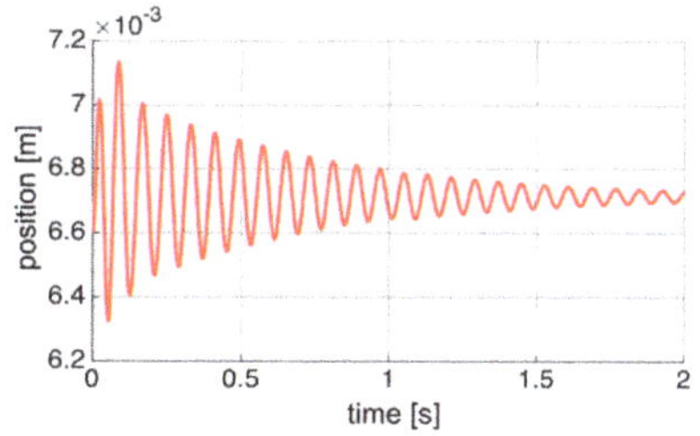

(**a**)

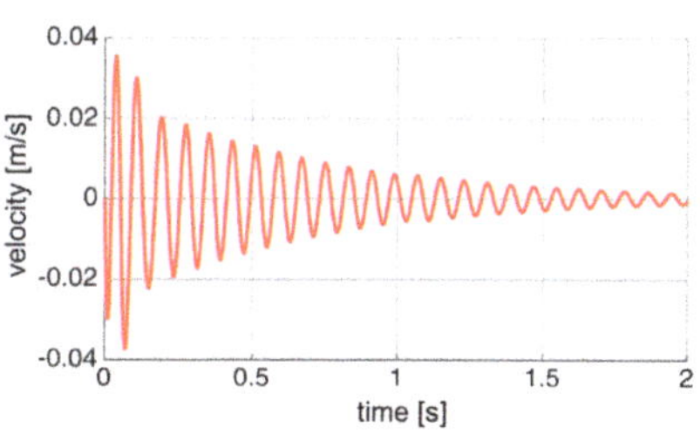

(**b**)

Figure 8. The vertical component of position (**a**) and velocity (**b**) of the launch cart.

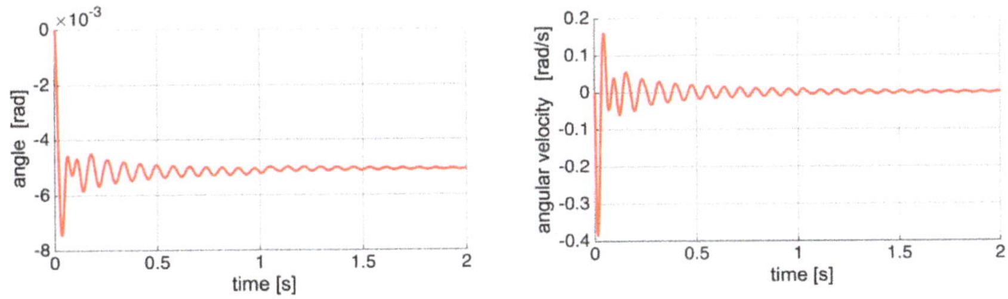

(a) (b)

Figure 9. Pitch angle (**a**) and angular velocity of pitch (**b**) of the launcher cart.

Powered by a linear motor, the cart moves with a constant horizontal acceleration of 14.2 m/s^2 during the first second, and then decelerates with the same magnitude of acceleration (Figure 10a). The vertical acceleration component oscillates around zero (Figure 10b). The vibrations are related to the cart's stabilization around the equilibrium position.

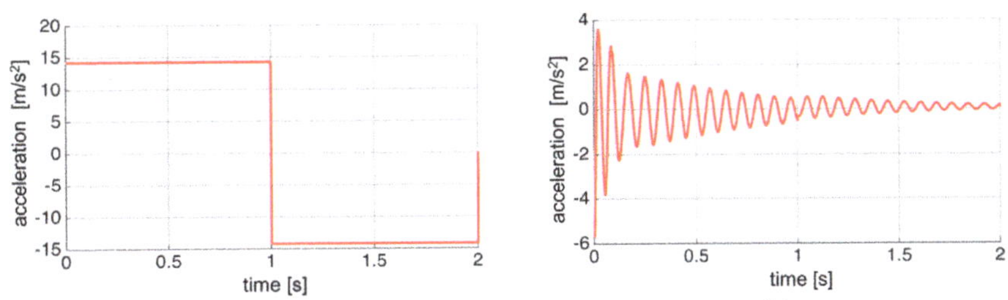

(a) (b)

Figure 10. Acceleration of the cart with respect to (**a**) the horizontal axis x_f and (**b**) the vertical axis z_f.

The launch cart tends towards the equilibrium position (Figure 11). The gravity and levitation forces components along the longitudinal axis having opposite values. The oscillations are slowly damped, which is a characteristic feature of the magnetic suspension of the launcher.

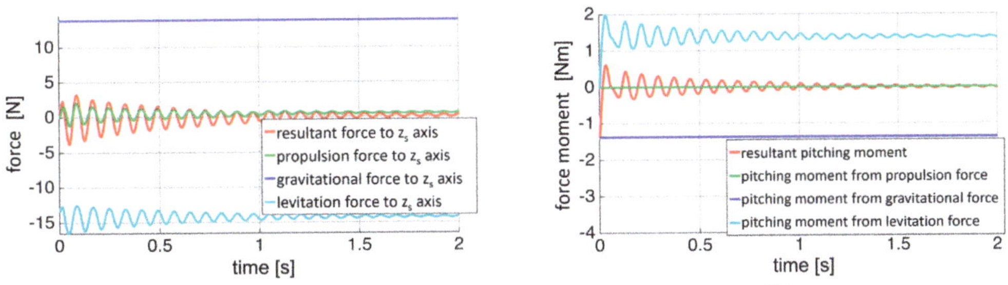

(a) (b)

Figure 11. The components acting on the launch cart: forces relative to the vertical axis of the cart (**a**) and pitching moments (**b**).

The results of the conducted simulations indicate that the equilibrium point at which the weight of the cart is balanced by the levitation force is 6.6 mm; the maximum height to which the cart rises due to the action of the levitation force is 7.16 mm; shifting the center of mass of the cart in the x axis direction results in a non-zero tilting moment, balanced by the moment of forces resulting from the difference in the value of the levitation force acting on the front and rear supports of the cart. The simulation studies confirmed that a characteristic feature of the launcher suspension is the low damping of vibrations, which was demonstrated in experimental studies [7].

An important attribute that distinguishes the levitation launcher from other unmanned aircraft launchers is the maintenance of a constant acceleration value for the launch cart during takeoff (Figure 10), ensuring a consistent acceleration throughout the entire runway. It is worth emphasizing that the value of this acceleration can be easily controlled by increasing or decreasing the thrust force of the linear motor.

4.2. Lateral Motion of the Cart

In the next simulation, the motion of the cart detached from the linear motor was considered, with its center of mass shifted by 2 mm along the lateral axis of the magnetic tracks. The simulation took into account the forces resulting from the phenomenon of flux trapping inside the superconductor.

It was assumed that the superconductor was immersed in liquid nitrogen at a height of d_{FC} = 3 mm, similarly to the experiments described in Chapter 3. The superconductor is then subject to the resultant force $\boldsymbol{F_{Ls}}$ described by Equation (7), as well as a torque aiming to align the magnetic moment vector of the superconductor with the external magnetic field induction vector.

The trajectory of the cart's motion is shown in Figure 12. When detached from the motor stator, the cart moves along the rails of the launcher in a weakly damped periodic motion. The oscillations of the left and right supports are synchronized in antiphase, aiming to maintain the center of mass of the cart aligned with the center of the magnetic rails.

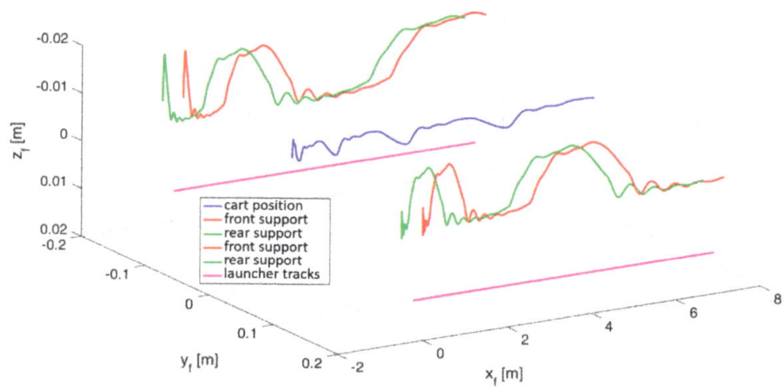

Figure 12. The 3D trajectory of the cart's movement after destabilization with respect to the lateral axis.

Figure 13 shows the course of the position and orientation components of the cart. It can be seen that the cart was set in an undamped vibrational motion relative to the center of the launcher tracks (Figure 13a). In the simulations, the oscillations of the cart deflection angle (around the x_f axis) were observed (Figure 13b), which explains the spatial trajectory behavior (Figure 12)—at the moment when the right supports of the cart are the highest, the left supports reach the lowest position, and then the situation is reversed. The cart oscillations along the longitudinal x_f and lateral y_f axes are synchronized and are in antiphase. The oscillation period is 0.45 s, and the amplitude of lateral motion is 2 mm. The maximum tilt of the cart in periodic motion is 0.033 rad. As the cart moves away from the center of the rails, its height increases (Figure 13c). At zero displacement of the cart along the y_f axis, its height is 5 mm, which is smaller than the equilibrium point determined for a cart solely subjected to the Meissner levitation effect.

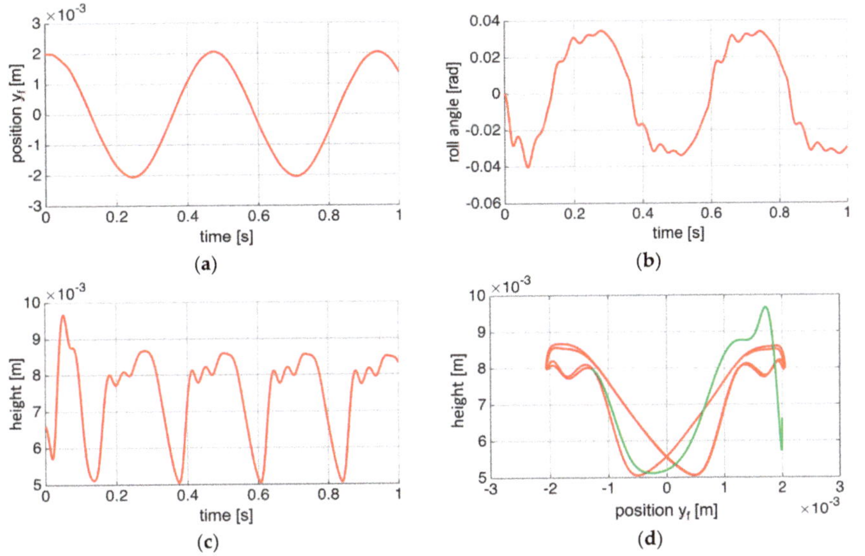

Figure 13. Position and orientation components of the cart with respect to the lateral y_f and vertical z_f axes associated with the magnetic tracks: (**a**) position of the cart relative to the y_f axis, (**b**) roll angle, (**c**) cart height, (**d**) cart height as a function of displacement along the lateral axis y_f. The green line marks the trajectory during the first 0.18 s, when the motion of the cart is aperiodic.

Figure 13d shows the height of the cart as a function of the displacement of the cart along the lateral axis y_f. The green line marks the trajectory during the first 0.18 s, when the motion of the cart is aperiodic. After that, the cart is set in periodic vibrations marked on the graph in red. The graphs show clear nonlinearity and hysteresis of the cart motion.

Based on simulation studies, it was found that the unpowered launch cart, when its center of mass is slightly shifted, does not return to its equilibrium position. The cart vibrations are not damped. This shows that the lateral motion of the cart is very unstable and requires additional position stabilization, e.g., using a linear motor.

5. Numerical Investigations of the UAV Launch Process from a Launcher

The complement to the analyzed dynamics of the magnetic launcher cart is the analysis of the process of the UAV uncontrolled takeoff, which was conducted taking into account the influence of disturbances resulting from air movements in the form of constant wind and gusts. The test UAV considered was a Bullit airplane (Figure 14). The aerodynamic model of the UAV [18] was examined. The main data of the analyzed UAV are gathered in Table 1.

Figure 14. The Micro UAV Bullit by Topmodel [19].

Table 1. Parameters of the Micro UAV Bullit [18,19].

Parameter	Nominal Value	Parameter	Nominal Value
Profile	BELL540	Engine	XPower XC3520/10 (4S)
Total mass	1.27 [kg]	Length	0.68 [m]
MAC	0.3738 [m]	Wing span	0.84 [m]
Inertia moments Ixx/Iyy/Izz/Izx/Ixy/Iyz	0.0184/0.0367/0.0550/−0.00021/0/0 [kg m^2]	Area of the wing	29.5 [dm^2]

According to the takeoff procedure developed within the GABRIEL project, the aircraft is mounted on the launch cart prior to takeoff. A linear motor is used to accelerate the cart. When the system reaches the separation speed, the aircraft is detached from the cart and continues its independent flight.

The numerical simulations were conducted assuming that, at the initial moment, the UAV is located on the launcher cart above the tracks, and the aircraft's propulsion is turned off. The motion of the system is initiated by the launcher cart, which is accelerated by a force of 50 N. The takeoff is performed with the UAV's propulsion turned off. In the developed numerical model, the couplings between the cart and the UAV starting from it were analyzed, taking into account the influence of constant headwind and tailwind, as well as gusts on the process of the UAV takeoff from the magnetic launcher.

5.1. Takeoff of UAV in Calm Atmosphere and with Headwind or Tailwind

In the conducted numerical simulation analyses, the results of modeling the UAV takeoff process were compared under calm atmospheric conditions, and then with the consideration of a constant headwind or tailwind blowing at a speed of 5 m/s.

The motion parameters of the launcher cart and the UAV taking off from it in calm air are presented in Figure 15 by a solid line. In this case, the UAV separation occurs at t = 2.2 s with a speed of 24 m/s after covering a distance of 28 m from the cart. The headwind shortens the UAV takeoff distance, while the tailwind lengthens it, as shown in Figure 15a. This is caused by a faster increase in the lift force in the case of headwind (Figure 15d). The separation velocity measured in the inertial system will then be lower (Figure 15c). During the climbing phase, the UAV without propulsion loses its speed.

The mutual interaction between the motions of the starting cart and UAV is shown in Figure 16. After the UAV leaves the launcher, the oscillations of the velocities of both elements of the system are visible, with the high-frequency oscillations of the velocity vector of the starting cart being damped much more slowly. This is a characteristic feature of magnetic suspension, which has been confirmed in experimental studies [6].

5.2. Gust Impact on UAV Takeoff

In the analyzed cases of uncontrolled UAV takeoff in a gust, it was assumed that the gust of wind occurs along the horizontal axis and is described by the function

$$V_w = \begin{cases} 0 & if\, t < t_s \, or \, t > t_s + d \\ A\sin\left(\frac{\pi}{d}(t - t_s)\right) & if \, t_s \le t \le t_s + d \end{cases} \tag{9}$$

where it was assumed that the wind gust occurs along the horizontal axis and is described by the function, where the gust amplitude A = 5 m/s, duration d = 1 s, and the gust onset time t_s = 2.3 s coincides with the moment of UAV detachment from the launcher.

On Figure 17, the results of the analysis of the start without wind are compared with the results of the start with variable head and tailwinds. The simulations conducted show the effect of the gust on the change in UAV trajectory and speed. The headwind causes the UAV speed relative to the ground to decrease. Conversely, in the case of tailwind, its speed relative to the ground increases (and at the same time, the aerodynamic speed decreases), which in special cases can lead to exceeding the critical angle of attack.

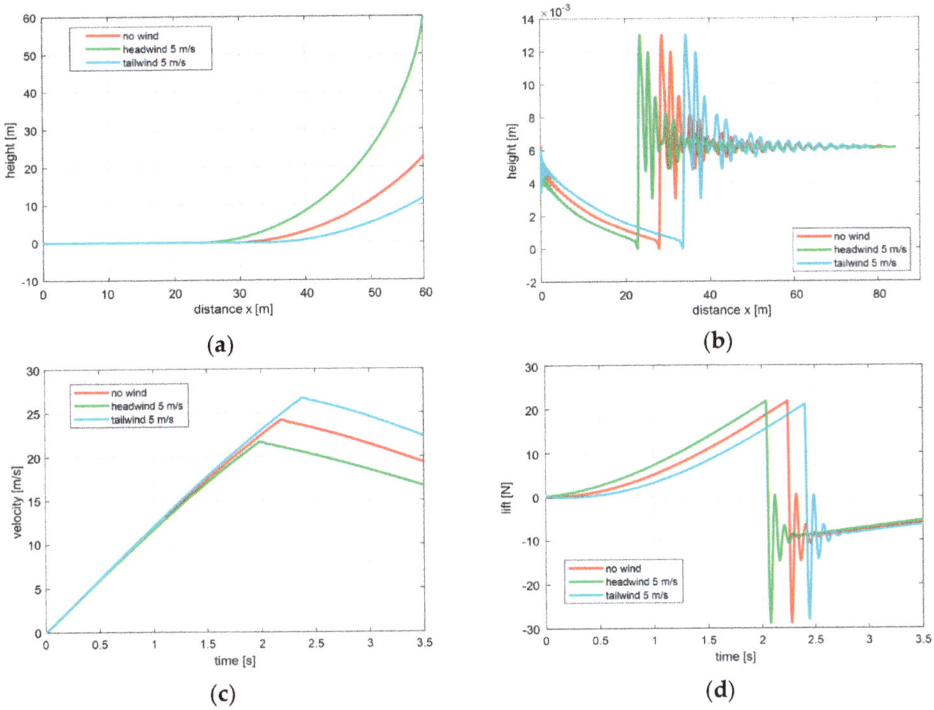

Figure 15. Influence of constant wind on (**a**) UAV trajectory;(**b**) launch cart trajectory; (**c**) UAV velocity; and (**d**) its lift force.

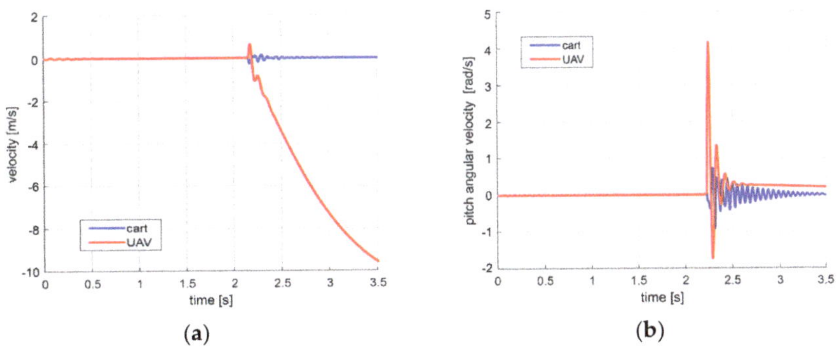

Figure 16. (**a**) Climb rate and (**b**) pitch angle of the cart and UAV during takeoff in calm atmosphere.

During the analyzed gust, the kinematic parameters of the launching cart do not change because the gust starts at the moment when the UAV departs from the launcher. However, it does affect the dynamics of the cart, as shown by the levitation force graphs (Figure 17b). During the gust, the levitation force increases or decreases its value in a manner analogous to the function of the wind speed change.

In the conducted numerical analyses, the relationship between the wind and the motion parameters of the UAV launch system was shown. The impact of wind on the parameters of the launch system, such as the trajectory, speed, tilt angle, and angular velocity, as well as the lift force of the launch cart, was examined. Compared to the trajectory without wind, headwind shortens the takeoff distance while the tailwind lengthens it. The results of the conducted numerical studies confirm the correctness of the developed mathematical model, demonstrating that the design of the analyzed launcher enables a safe takeoff. Based on these results, the optimal parameters for the magnetic launcher system can be selected. Primarily, the dimensions of the magnetic rails must correspond to the size of the launching UAV, ensuring that the generated levitation force is capable of lifting it. Therefore, in the analyzed simulation cases, the Micro UAV Bullit was taken into consideration.

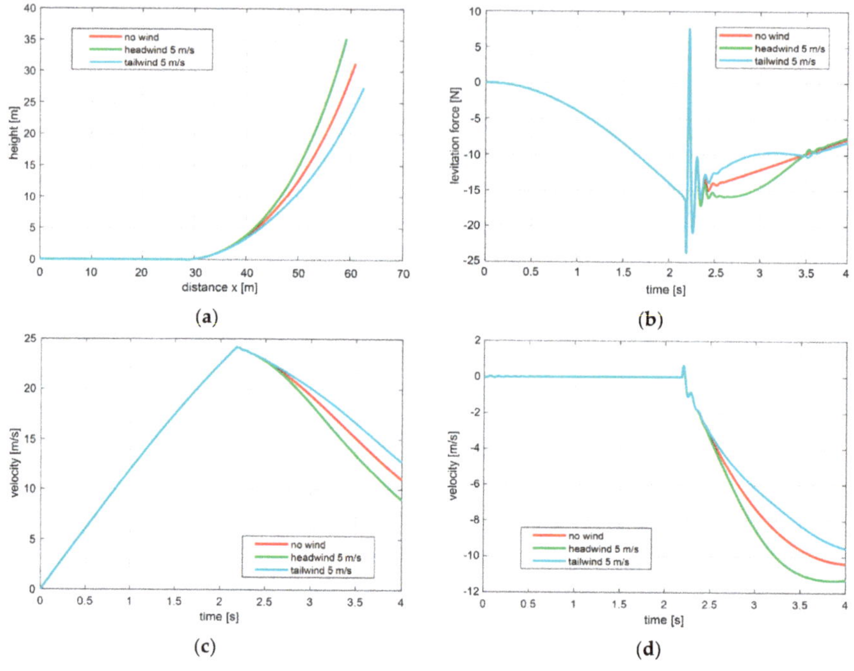

Figure 17. The influence of the gust on (**a**) UAV trajectory; (**b**) the levitation force of the launching cart; (**c**) the longitudinal velocity of the UAV; and (**d**) its vertical velocity.

6. Conclusions

The article describes the process of modeling and analyzing the dynamic properties of the launch cart of an innovative prototype of an unmanned aircraft launcher utilizing passive magnetic suspension with high-temperature superconductors. The aim of modeling the dynamics of the launch cart of the analyzed launcher, constructed as part of the GABRIEL project, was to specify the operating conditions of the system. Two types of cart motion were analyzed. In the first case, the motion of a cart driven by a linear motor, which constrains lateral movements, was considered. Then, the behavior of the driven cart detached from the motor stator was analyzed, allowing for the examination of the spatial behavior of the cart on the launcher rails, taking into account its lateral movements and the influence of forces resulting from magnetic flux trapping inside the superconductors. Complementing the dynamics analysis of the magnetic launcher's launch cart were numerical simulations, in which variable atmospheric parameters and couplings resulting from the mutual interaction of individual elements of the launch system were considered. The analysis included the study of the process of uncontrolled UAV launch, taking into account the influence of constant headwind, tailwind, and gusts.

On the basis of the results presented in the paper, the following conclusions can be drawn:

- The developed simulation model enables the specification of the operating conditions of the modeled catapult system and analysis of the possibilities of counteracting disturbances.
- During the launch, a constant acceleration of the launch cart is maintained, resulting in the UAV's takeoff process occurring with constant acceleration.

- The acceleration value can be easily controlled by adjusting the thrust force of the linear motor, thereby minimizing the applied overload on the UAV.
- The suspension system of the launcher exhibits the low damping of vibrations, as reflected in the simulated motion of the launch cart and confirmed by the vibrations of the cart support recorded during the field tests of the launcher [6,7].
- The simulations indicate that, for the analyzed system parameters, the equilibrium point where the cart mass is balanced by the levitation force is located at 6.6 mm, and the maximum height to which the unloaded cart rises due to the levitation force is 17.3 mm.
- Displacement of the carts' center of mass along the longitudinal axis results in nonzero tilting moments, which are balanced by the differential levitation force acting on the front and rear cart supports. However, this equilibrium is lost when the displacement of the carts' center of mass along the longitudinal axis exceeds 16 cm, causing the destabilization of the carts' orientation and leading to an impact upon the launcher rails.
- Lateral motion of the cart is unstable and requires additional position stabilization, such as through the use of a linear motor, as demonstrated in the GABRIEL project's launch tests and confirmed by numerical simulations.
- Displacement of the carts' center of mass relative to the transverse axis of the rails by less than 5.4 mm induces weakly damped harmonic vibrations with respect to the center of the rails, while larger displacements cause the cart to derail from the launcher rails.
- Compared to the windless trajectory, headwind shortens the takeoff run (as the profile moving into the wind can generate greater lift force), while tailwind lengthens it, simultaneously increasing ground speed.
- Tailwind reduces the UAV's aerodynamic speed, which can, in certain cases, lead to exceeding the critical angle of attack.
- The levitation force adjusts its value proportionally to the wind speed changes.
- The results of the conducted numerical studies confirm the correctness of the developed mathematical model and demonstrate that the design of the analyzed launcher enables a safe takeoff.

The analyzed launcher is an innovative alternative to the available market options for unmanned aircraft launchers. The developed mathematical and numerical model, enabling the simulation of the launch cart's motion on the launcher rail during the takeoff and landing of the UAV, indicates that the design of the analyzed launcher allows for a safe takeoff under variable weather conditions. The longer takeoff path provided by the launcher reduces the overload on the UAV during takeoff, resulting in increased safety for the launcher, aircraft, and onboard equipment. Any vibrations of the UAV system, caused by uneven mass distribution or external factors, are dampened by the magnetic forces and moments arising from the Meissner effect and the trapping of magnetic flux within the superconductors.

The most significant advantage of electromagnetic launchers, especially magnetic levitation launchers, is the reduction in acceleration and its constant value during takeoff. Conventional catapults are characterized by very high accelerations during takeoff, which can damage the aircraft and equipment. Due to the fact that, in levitation launchers, the aircraft is accelerated at a constant acceleration throughout the entire length of the runway, the takeoff process is very smooth. It should be emphasized that the value of this acceleration can be easily controlled.

Author Contributions: Conceptualization, E.Ł.-K., K.S. and A.S.-M.; methodology, E.Ł.-K., A.S.-M. and K.S.; software, A.S.-M. and D.P. validation, E.Ł.-K., D.P. and A.S.-M.; formal analysis, E.Ł.-K. and A.S.-M.; investigation, A.S.-M., D.P. and E.Ł.-K.; resources, K.S. and A.Ż.; data curation, K.S. and A.Ż.; writing—original draft preparation, E.Ł.-K.; writing—review and editing, E.Ł.-K., K.S. and A.S.-M.; visualization, A.S.-M.; supervision, E.Ł.-K. and K.S.; project administration, E.Ł.-K.; funding acquisition, E.Ł.-K. and A.Ż. All authors have read and agreed to the published version of the manuscript.

Funding: Research was funded by the Warsaw University of Technology within the Excellence Initiative: Research University (IDUB) program. The APC was funded by Air Force Institute of Technology.

Data Availability Statement: Data supporting reported results are contained within the article.

Conflicts of Interest: The authors declare no conflict of interest.

References

1. Westinghouse Eng. A Wound Rotor Motor 1400 Feet Long. 1946, Volume 6, pp. 160–161. Available online: https://www.theengineer.co.uk/october-1946-westinghouse-unveils-the-electropult/ (accessed on 11 May 2023).
2. Patel, A.; Hopkins, S.C.; Baskys, A.; Kalitka, V.; Molodyk, A.; Glowacki, B.A. Magnetic levitation using high temperature superconducting pancake coils as composite bulk cylinders. *Supercond. Sci. Technol.* **2015**, *28*, 115007. [CrossRef]
3. Schultz, L. Interaction of Ferromagnetic and Superconducting Permanent Magnets: Basics and Applications of Passively Stable Quantum Levitation. In *Sustainable Industrial Processing Summit SIPS2019*; Kongoli, F., Calin, M., Dubois, J.M., Zuzek-Rozman, K., Eds.; Flogen Star OUTREACH: Mont Royal, QC, Canada, 2019; Volume 3, pp. 38–39.
4. Magnetic Take-Off and Landing for Fuel-Efficient Aircraft. Available online: https://cordis.europa.eu/article/id/92581-magnetic-takeoff-and-landing-for-fuelefficient-aircraft (accessed on 11 May 2023).
5. Sibilski, K.; Falkowski, K.; Kaleta, R.; Ładyżyńska-Kozdraś, E.; Anna Maria Sibilska-Mroziewicz, A.; Wróblewski, W. Development of the Small-Scale Model of Maglev System Assisted Aircraft Safety Take-Off and Landing. In Proceedings of the 30th ICAS Congres (IICAS 2016-0369), Daeion, Republic of Korea, 25–30 September 2016; Available online: https://www.icas.org/ICAS_ARCHIVE/ICAS2016/data/papers/2016_0359_paper.pdf (accessed on 11 May 2023).
6. Ładyżyńska-Kozdraś, E.; Sibilska-Mroziewicz, A.; Czubaj, S.; Falkowski, K.; Sibilski, K.; Wróblewski, W. Take-off and landing magnetic system for UAV carriers. *J. Mar. Eng. Technol.* **2017**, *16*, 298–304. [CrossRef]
7. Sibilska-Mroziewicz, A.; Ładyżyńska-Kozdraś, E.; Falkowski, K. Modelling of Forces Acting on a System of the UAV Launcher, Based on Passive Magnetic Suspensions with Superconductors. *Nase More* **2020**, *67*, 60–68. [CrossRef]
8. Sibilska-Mroziewicz, A. Opracowanie Modelu Dynamiki Wózka Startowego Wyrzutni Samolotu Bezzałogowego, z Wykorzystaniem Efektu Meissnera. Ph.D. Thesis, Warsaw University of Technology, Warsaw, Poland, 2018.
9. Sibilska-Mroziewicz, A.; Ładyżyńska-Kozdraś, E. Mathematical Model of Levitating Cart of Magnetic UAV Catapult. *J. Theor. Appl. Mech.* **2018**, *56*, 793–802. [CrossRef]
10. Yaghoubi, H. The most important maglev applications. *J. Eng.* **2013**, *2013*, 537986. [CrossRef]
11. Maitreya, S.; Soni, S.; Paliwal, P. Analysis of Electromagnetic Aircraft Launching System for Naval Aircraft. In *Technology Innovation in Mechanical Engineering*; Chaurasiya, P.K., Singh, A., Verma, T.N., Rajak, U., Eds.; Lecture Notes in Mechanical Engineering; Springer: Singapore, 2022; pp. 773–781. [CrossRef]
12. Bhatia, D.K.; Aljadiri, R.T. Electromagnetic UAV launch system. In Proceedings of the 2017 2nd IEEE International Conference on Intelligent Transportation Engineering (ICITE), Singapore, 1–3 September 2017; pp. 280–283.
13. Nusran, N.M.; Joshi, K.R.; Cho, K.; Tanatar, M.A.; Meier, W.R.; Bud'ko, S.L.; Prozorov, R. Spatially-resolved study of the Meissner effect in superconductors using NV-centers-in-diamond optical magnetometry. *New J. Phys.* **2018**, *20*, 043010. [CrossRef]
14. McHenry, M.; Sutton, R. Flux pinning and dissipation in high temperature oxide superconductors. *Prog. Mater. Sci.* **1994**, *38*, 159–310. [CrossRef]
15. Kordyuk, A. Magnetic levitation for hard superconductors. *J. Appl. Phys.* **1998**, *83*, 610–612. [CrossRef]
16. Sibilska-Mroziewicz, A.; Ładyżyńska-Kozdraś, E.; Falkowski, K.; Sibilski, K. Estimation of the force causing the levitation of the starting trolley of the unmanned aerial vehicle. *Bull. Pol. Acad. Sci. Tech. Sci.* **2020**, *68*, 1177–1185. [CrossRef]
17. Sibilska-Mroziewicz, A.; Ładyżyńska-Kozdraś, E.; Sibilski, K. VR-Supported Analysis of UAV—Magnetic Launcher's Cart System. *Energies* **2023**, *16*, 4095. [CrossRef]
18. Mystkowski, A.; Ostapkowicz, P. Weryfikacja modelu dynamicznego mikro-samolotu z wibrującymi generatorami wirów do sterowania przepływem. *Prace. Inst. Lotnictwa* **2011**, *216*, 103–125.
19. BULLIT EVO 0.84m ARF. Available online: https://www.topmodel.fr/en/product-detail-22834-bullit-evo-0-84m-arf (accessed on 14 June 2023).

MDPI
St. Alban-Anlage 66
4052 Basel
Switzerland
www.mdpi.com

Electronics Editorial Office
E-mail: electronics@mdpi.com
www.mdpi.com/journal/electronics

www.ingramcontent.com/pod-product-compliance
Lightning Source LLC
Chambersburg PA
CBHW061843210926
51850CB00006B/960
9783036584249